What they are saying

"When I met Sherry years ago, we met when she e-mailed the VCFS Educational Foundation, looking for information. She was too shy, then, even to give me her phone number via email, saying that she was a "scaredy cat" and would call me. One day the phone rang and this quiet, timid voice said, "Hello, Kelvin. . . this is Scaredy Cat", and that was then. What a contrast to today's vibrant, dynamic Sherry Gomez—"scaredy cat" no more —who has overcome her timidity and blazed this information trail with her book, to help brighten the path for other parents and loved ones of children—and adults—with VCFS. She has done a good thing, and without regard for personal inconvenience, to bring the VCFS community this work. When you read this book, know that it represents not only a valued source of information and perspective, but also a tale of personal mastery for woman who used to be just a "scaredy cat." Sherry . . . congratulations. You followed through."
**With great respect,
Kelvin P. Ringold
Administrative Assistant,
VCFS Educational Foundation, Inc. (Retired)**

"I want to applaud the people who have worked diligently to put together the information related to VCFS for families and healthcare providers. This should help the families cope with complex issues and systems. It should also enhance their ability to advocate for their children."
Martha Frisby, M.H.A., Children's Rehabilitation Services, St. Joseph's hospital, Arizona

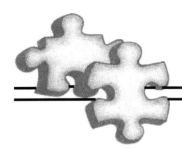

MISSING GENETIC PIECES

STRATEGIES for LIVING with VCFS
Velo-Cardio-Facial Syndrome
The CHROMOSOME 22q11 DELETION

By Sherry Baker-Gomez

Foreword by Dr. Robert J. Shprintzen
Center for the Diagnosis, Treatment,
and Study of Velo-Cardio-Facial Syndrome

Cover Design and Illustrations by Kas Winters

A comprehensive handbook for parents,
professionals and anyone
wanting to know about
CHROMOSOME 22q DELETION (22q11.2)
and its dynamics

Important information with REAL personal stories

SECOND EDITION 2011

Desert Pearl Publishing
Phoenix, Arizona

Second Edition
Copyright ©2011 by Sherry Baker-Gomez
Desert Pearl Publishing
Glendale, Arizona

Illustrations ©2004 by Kas Winters
Winmark Communications
Glendale, Arizona

Copyright ©2004 by Sherry Baker-Gomez
Desert Pearl Publishing
Glendale, Arizona

All rights reserved. No part of this book may be reproduced or transmitted in any form or by any means, electronic or mechanical, including photocopying, recording, or by any information storage and retrieval system, without written permission from the author.

Reviewers may quote passages for use in periodicals, newspapers or broadcasts provided credit is given to **Missing Genetic Pieces: Strategies for Living with VCFS** published by Desert Pearl Publishing, and written by Sherry Baker-Gomez. Contact: Desert Pearl Publishing, 4918 West Phelps Road., Glendale, AZ 85306, Phone 602-789-6416

http://www.22qCentral.com

ISBN #: 0-9745358-0-X

Printed in the United States of America.

DEDICATION

This book is dedicated to my son T.G. who has the Chromosome 22q deletion and still manages to wake up in the morning with a smile on his face and joy in his heart . . . AND to all the people who have made this book a reality AND to ALL children and adults living with Chromosome 22q11.2 deletion and their families.

INTRODUCTION

As the light becomes brighter in the field of genetics, it is said that Velo-Cardio-Facial-Syndrome (Chromosome 22q Deletion) or VCFS, may prove someday to be the _most common_ genetic disorder affecting mankind. Researchers today believe that it can affect 1:2000 births. Some marvelous advancements will come with the increasing knowledge and interest about VCFS!

Some other names that this disorder has been referred to are: DiGeorge syndrome/sequencing/anomaly, CTAF, Catch 22, Robin sequencing, and Potters sequencing. (Refer to sites on the Web such as the CHARGE association.) For the purpose of this book we will use the terms VCFS or 22q. I prefer to use 22q.

As parents, we want and need to know what this condition entails. This book's purpose is to Educate, Advocate, and Reform.

• **Educate**: Teach prospective parents, recent parents, and parents whose child has already been diagnosed with this condition, as well as the general population.

• **Advocate**: Strengthen the voice of those with VCFS/22q and their families by learning to take a stand on their own behalf, and by giving power and validity to family concerns.

• **Reform**: Improve the quality of life for those with VCFS/22q and their families, by offering them a comprehensive learning experience free from blame, shame, and fear.

Since babies are not born with VCFS/22q engraved on their foreheads it takes an awareness of the syndrome to recognize the symptoms and to make a correct diagnosis. We hope that this book serves as an introduction to health care providers, speech pathologists, teachers, social service workers, and other professionals who may not be well acquainted with VCFS/22q. This may not surprise you, but research is showing that there are many more commonly recognized disorders that are now known to be genetic in nature. Doctors are only human, and with so much information to consider and remember, it is not uncommon for many health care providers to not fully recognize what they are observing. As a result many patients get misdiagnosed. VCFS continues to elude the majority, so this book is written with contributions from knowledgeable people in the field of VCFS/22q, as well as many parents *"living with VCFS"* who will bring you, the reader, some important information.

Before being defined in 1978 by Robert J. Shprintzen, the condition we now refer to as VCFS/22q, was a relatively undefined _syndrome_ with many unanswered questions. Since its discovery research related to VCFS has led to many advances in our understanding of abnormal childhood illnesses, conditions and behavior. There are many research professionals to whom we can give credit for this advancement.

Still, to date, the condition is not completely understood. As a nurse and parent of an older child with VCFS/22q, I was compelled to write this book for two main reasons. First, I discovered there are many concerned parents in many countries who need a guide. Second, I was astonished by the general lack of recognition and knowledge of this disorder; not only from the public, but the educational system, as well as the medical professionals, on whom we rely for answers and help.

The book is divided into a number of sections. There are sections about how to identify the symptoms associated with the disorder, and what parts of the body can be affected. There is a chart that shows what is considered *normal* child development and what to look for when one suspects that a child is not developing normally. This book offers guidance about how to help your doctor and how to work with your school system. In addition, there is support group information: 1) How to set up your own support group; 2) How to contact existing groups, and; 3) Tips you can use to become your child's advocate. In one section, there is a frank discussion about mental health and how VCFS/22q can affect the family. It discusses how stress can affect adults and marriages and how we can support others, etc.

The book introduces stories from real families living with VCFS. It includes illustrations, as well as a resourceful mini-medical dictionary in the back.

Genetic research is enlightening the public now as never before. We, as a community, are all touched by 22q. So parents, school officials, organizations for the disabled, social workers and medical professionals owe it to the current and future generations to be pro-active now; and educate the public to make 22q a household name. Remember, doctors can only diagnose 22q correctly if they are knowledgeable about the disorder and its symptoms. I hope that this book will serve not only as a guide for general information about this disorder, but will pique interest in those who forge ahead to support more research. It is this continuing research that is needed to cure, maintain, or stop the increase of this disorder as well as all genetic disorders. To do this takes time, effort and money. Please support our sincere efforts in making a difference.

For more resources go to **http://www.22qCentral.org** or any number of sites that are dedicated to educating the world of mankind. Genetic disorders are not discriminating—they can, and do, affect any nationality in any country or family.

Please feel free to visit my website at:
http://www.22qCentral.org

> "The official published data from our Center is 1:2000 population prevalence in the U.S. Population prevalence means the number of people alive in the population who have it. The incidence, meaning the number of pregnancies that come to term resulting in a baby with VCFS/22q, some of whom do not survive, is pegged at 1:1600."
> *(Dr. Robert Shprintzen 2010)*

FOREWORD

The study of velo-cardio-facial syndrome (VCFS/22q) is a relatively recent field of investigation. As a means of brief introduction, it is important to understand the history of how this condition came to be "discovered" and why it bears so many different names. In the world of clinical genetics, the term "discovered" in relation to a syndrome is not really appropriate. The term "delineate" is really the acceptable and more descriptive one. To delineate a syndrome implies that researchers have associated a number of clinical findings as a package or grouping that occurs in predictable fashion among individuals with the same disorder. Delineating a syndrome is easiest when the pattern is accompanied by striking or singular findings, such as the facial appearance of children with Down's syndrome. In 22q/VCFS, although there are many clinical findings, most of them can be found in many other conditions and are not unique nor distinctive. Facially, although children with VCFS resemble each other, they are not at all unusual looking and do not stand out from the rest of the population except in some rare, very severe cases. In addition, delineating VCFS was more difficult in years past because many children with severe congenital heart disease did not survive the neonatal period.

Today, cardiothoracic surgeons are having great success in repairing even the most severe heart malformations and therefore many more children with VCFS are surviving. As a result, today, the pattern is being seen with increasing frequency and is one of the most common genetic syndromes in humans, perhaps second only to Down's syndrome.

Cases of VCFS were described in the medical literature as far back as the 1950s by Sedla-ková in Czechoslovakia, and subsequently by Strong, Kretschmer, Kinouchi, and others in the 1960s and 1970s. The label velo-cardio-facial syndrome was derived from my attempt to delineate VCFS as a new syndrome in 1978 when I described a dozen cases in the American literature. Although I would love to take credit for having been the *"discoverer"* of a new syndrome, the reality is, like Christopher Columbus, I was actually a relative latecomer to the process, but was fortunate to have published my paper in the American literature at a time when syndrome delineation was an active and interesting new science. My predecessors deserve at least equal recognition for their observations. Because all of us who had a hand in recognizing this syndrome came at it from somewhat different angles (heart, immune deficiency, speech, craniofacial, etc.), the disorder became recognized under many different labels. They include, but are not limited to VCFS, DiGeorge syndrome, conotruncal anomalies face syndrome, Sedla-ková syndrome, and Shprintzen syndrome. Let there be no mistake that these are all the same disorder as long as they are accompanied by a deletion of 22q11.2.

I am contributing only these few paragraphs to this book, and it is written independently of the many pages written by the author. She has gone into great detail in describing her personal experiences from a parental point of view while adding in information about the syndrome written from her own point of view. She is to be applauded for her work and diligence in bringing to the reader a volume that will allow a more informed approach to dealing with VCFS. Of course, the field of study of VCFS is not without controversy and differing opinions about the syndrome and its treatment. This is true for all medical conditions. It is, in fact, these controversies that continue to fuel our strong interest in studying this genetic disorder. Although we do not have all of the answers today, we hope to have them some day. For you, the reader, be assured that we, the scientists (if I may be so bold as to speak for my colleagues) will continue to work tirelessly to unravel the mysteries of this complex disorder. At my institution alone, Upstate Medical University in Syracuse, NY, I have assembled a team of over 30 superb clinicians and scientists who are devoted full time or part time to the study of VCFS, and we are not alone. Scientists all over the world are doing the same, and this should increase your optimism in looking towards the future. The author of this book has now joined our growing number of dedicated pioneers, coming at it from yet another point of view, indeed a much more personal one. All of us thank her for her diligence.

Dr. Robert J. Shprintzen
Director, Center for the Diagnosis, Treatment,
 and Study of Velo-Cardio-Facial Syndrome
Director, Center for Genetic Communicative Disorders
Director, Communication Disorder Unit
Professor of Otolaryngology
Professor of Pediatrics
Upstate Medical University
750 East Adams Street
Syracuse, NY 13210

DISCLAIMER

This book is designed to provide information about the subject matter covered. It is provided with the understanding that the publisher and authors are not engaged in rendering medical, legal, accounting, or other professional service. If medical or any other expert assistance is required; the services of a competent professional should be sought.

It is not the purpose of this book to reprint or quote from all information known on this subject; but to complement, amplify and supplement other texts that are available.

Every effort has been made to make this book as complete and as accurate as possible. However, there may be mistakes both typographical and in content. Therefore, this text should be used only as a general guide for information on the subject contained and not as the ultimate source or authority. Furthermore, this handbook contains information only up to the printing date. It has been revised in 2011.

Since this book's sole purpose is to educate, support and inspire; the authors and the publisher shall have neither liability nor responsibility to any person or entity with respect to any loss, damage caused by or alleged to be caused directly or indirectly by the information contained in this book. Since much of the content of this book is contributed there is no guarantee of any one source used in compiling information or any implied compensation.

As of the publication date, all resources and links have been verified. Our experience shows that Web sites and other contact information are frequently subject to change.

CONTENTS

Introduction v
Foreword ix
Chapter 1: History of VCFS 1
   Anomaly fact sheet of symptoms 5
Chapter 2: Description of VCFS 13
   Most common systems involved might be 21
   List of medical specialists and their fields 22
Chapter 3: Systems Affected 25
   Common heart defects 25
   Pulmonary and respiratory 33
   Kidneys, hernias and anal anomalies 37
   Ears 41
   Eyes 47
   Facial features 49
   Hormone difficulties 53
   Immunization 57
   Pain 61
   Dental 65
   Palate and feeding 67
Chapter 4: Development 81
   Developmental issues 81
   Developmental stages 91
   List of age-appropriate toys 109
   Newborn conditions 110
   Potty training 117
   Psychophysiology 121
   Biofeedback 121
   Emotional development 127
   Social skills 133
Chapter 5: Education 143
   Accessing services 143
   Individual education plan (IEP) 145
   Non-verbal learning disorder (NVLD) 152
   IQ test scores 153
   Grade school ages 157
   Neuropsychology 164

Chapter 6: Speech 175
   Teach VCFS babies sign language 182
   Central auditory processing 189
   Fisher's auditory problem checklist 191
Chapter 7: Behavioral problems 199
   Behavior list 202
   Puberty 205
Chapter 8: ADHD/ADD/Autism 213
Chapter 9: Psychological issues 233
   Bipolar disorder 235
   Schizophrenia 249
Chapter 10: COMT gene 253
   Obsessive compulsive disorder and the COMT gene . . 258
Chapter 11: Negotiation 261
Chapter 12: Advocacy 269
   Alliance coordinating offices 273
   FAPE Solutions 309
   Housing and group homes 315
Chapter 13: Finances and trusts 323
Chapter 14: Support groups 337
Chapter 15: Diaries 359
Chapter 16: Marriage, relationships and VCFS 435
Chapter 17: Family support 445
Chapter 18: Grieving 449
Chapter 19: Parent to parent talk 457
Chapter 20: Parents speak out 469
Chapter 21: Medications 523
Chapter 22: Medical dictionary and glossary 527
Acknowledgements 543
Bibliography 544
Index 549

"Somewhere there is something incredible waiting to be known." Carl Sagan

CHAPTER ONE: HISTORY OF VCFS

HISTORY OF VCFS/22q

This name and that name . . . Why so many names?

You may be asking yourself, *"Why are there so many names for the same disorder?" (See Introduction, page v.)* Confusing isn't it? Well I hope that after reading this simple explanation, you will understand why there appear to be so many names for the same syndrome.

A syndrome is a situation where there are multiple abnormalities that occur together within the same individual. When this happens, medical clinicians usually suspect that there is a related cause, with a common factor or source.

Why so many names? I have learned that there are no guidelines, best way, or system used in naming medical discoveries. In the genetic community there are many different ways this is done. In many cases the syndrome is named after the first person who discovered and described the findings, honoring that person for his discovery. Often, the syndrome is described by the most common symptoms that appear in the majority of the patients affected. Acronyms are used as well. These are the first letters of the most characteristic/common clinical features of the syndrome; as in the case of VCF (Velo-Cardio-Facial). There are literally hundreds of known syndromes and new ones are being described each year. One fact is certain; they are all unique in some way, yet they may have many similar commonalities.

As an example, there are syndromes that have a limited number of abnormalities associated with them and then there are ones like VCFS that have many (almost 200). The difficult thing is for doctors to distinguish which syndrome is affecting their patient; especially when individuals may have different symptoms for the same disorder or when many syndromes have the same or similar symptoms. Then doctors try to determine if there is a syndrome at all or if the individual is just a sickly person. (That was my son's personal experience; the doctors did not know what they were seeing.) It seems that the combinations on the symptom list for VCFS are as unique and as varied as the people who are affected. That is why several people can be diagnosed with the same disorder although they seem to have different identifying problems. This is why we advocate for newborn genetic screening.

The main confusion between VCFS and DiGeorge labels is that some patients with a couple of symptoms similar to VCFS are diagnosed with *only* the DiGeorge sequencing/syndrome label or some other related syndrome. They may in fact have these in the mix of symptoms; but this limits physicians in the diagnosis, both with what symptoms to look for in other areas of the body and with checking for problems. However, 22q11 covers a wider symptom list which may or may not include the symptoms found in DiGeorge and those other labels. This is why I use 22q. It is all encompassing.

You may ask at this point, if some of these diagnoses are not actually syndromes, then what are they? They are a sequence! Oh boy—more confusion. What is a sequence and how does that differ from a syndrome?

Below, you will find the shortest and easiest definitions of the two as they were explained to me by a research geneticist in Los Angeles: Monica Alvarado, MS, CG, Assistant Professor USC.

Sequence: Describes a condition when _one_ _single_ _defect_ or mechanical factor in a fetus' development leads to a cascading sequence or a pattern of multiple but related defects. For example: in the DiGeorge sequence, experts suspect that a failure in the early development of certain structures in the fetus causes underdevelopment of the thymus and parathyroid glands. This results in a decrease in immune response and a decrease in calcium levels; which in turn can lead to a diminished ability to fight infections and an increased risk of seizures. Therefore, this results in a domino effect.

Syndrome: Is a collection of multiple and different congenital defects, often affecting separate organ systems that seem unrelated but share one common cause. For example, all the features of Down's syndrome (developmental delay, craniofacial features, cardiac defects, etc.) are not part of a sequence but are caused by a common chromosome defect, an extra chromosome 21. This is why, in 22q11, the different areas and organs of the body are affected.

Here's a little DiGeorge history

Originally, in the 1960's, Dr. Angelo DiGeorge, a pediatric endocrinologist in Philadelphia, did not set out to describe a syndrome per se. He was searching for clues to some particular symptoms and believed he had found the mechanism that caused that particular group of clinical anomalies to show up together on a consistent basis. These include the following:

- **Hypoparathyroidism:** This is an under active parathyroid gland, which results in **hypocalcemia** (low calcium levels).

- **Hypoplastic immune system:** This means under development of the thymus or absent thymus, which can lead to problems in the immune system.

- **Heart defects:** These include Tetralogy of Fallot, an interrupted aortic arch, ventricular septal defect and vascular rings.

- **Cleft lip and/or palate:** These include several abnormalities.

The name DiGeorge Syndrome was associated with this group of symptoms, only.

Both DiGeoge and Shprintzen discoveries were later identified to be in the same chromosome 22q11 region, but VCFS is manifested in many more areas of the body.

For a more in-depth explanation and history of how disorders are named, please refer to the article by Dr. R. J. Shprintzen titled **The Name Game** *which can be found on the Web site at* www.vcfsef.org

Velo-Cardio-Facial Syndrome Educational Foundation

> *It is important to remember that VCFS is **NOT** a disease but a condition with many facets to it.*

The Foundation's Executive Director:

Dianne M. Altuna, M.S./CCC
SLP
VCFS Educational Foundation, Inc.
PO Box 12591
Dallas, TX 75225

855-800-VCFSEF
execdirector@vcfsef.org

List of anomalies: Almost 200

The following specialist fact sheet was drawn up by the *Velo-Cardio-Facial Syndrome Educational Foundation.* The list of anomalies can be found at **http://www.vcfsef.org**. The following list shows various anomalies which have been found to be associated with VCFS; in association with the 22q11.2 deletion. (As an example: My son has 30 of the possibilities listed.)

None of the findings have 100% frequency. Thus, some patients will exhibit more findings than others. Most people are only going to be affected by a small number of the findings at one time, compared to the almost 200 possibilities.

Below is the list of anomalies that can affect your child. Put an **x** and **date** on the line of those that apply to your child. As time goes by this can prove to be an important record for your doctor(s). There will be other areas in this book for documentation.

The symptoms are listed in alphabetical order, not in order of importance or appearance.

Anomaly fact sheet of symptoms

Abdominal/kidney/gut

_____ Hypoplastic/aplastic kidney
_____ Cystic kidneys
_____ Anal anomalies (displaced, imperforate)
_____ Inguinal hernias
_____ Umbilical hernias
_____ Malrotation of the bowel
_____ Hepatoblastoma and diaphragmatic hernia
_____ Hirschprang megacolm (rare)
_____ Diastasis recti abdominis

Cardiac findings

_____ VSD (ventricular septal defect)
_____ ASD (atrial septal defect)
_____ Pulmonary atresia or stenosis
_____ Tetralogy of Fallot
_____ Right-sided aorta
_____ Truncus arteriosus
_____ PDA (patent ductus arteriosus)
_____ Interrupted aorta
_____ Coarctation of the aorta
_____ Aortic valve anomalies
_____ Aberrant subclavian arteries
_____ Vascular ring
_____ Anomalous origin of carotid artery
_____ Transposition of the great vessels
_____ Tricuspid atresia

Please notice there is a mini medical dictionary in the back of this book for your convenience. (See page 523.)

Cognitive/learning

_____ Learning disabilities (math concept, reading comprehension)
_____ Concrete thinking, difficulty with abstract thinking
_____ Drop in IQ scores in school years (test artifact)
_____ Borderline normal intellect (based on 100% as "normal")
_____ Mild mental retardation
_____ Attention deficit hyperactivity disorder (ADD/ADHD)
_____ Autism/Asperger

Craniofacial/oral findings

_____ Overt, submucous or occult submucous cleft palate
_____ Retrognathia (retruded lower jaw)
_____ Platybasia (flat skull base)
_____ Asymmetric crying facies in infancy
_____ Structurally and/or functionally asymmetric face
_____ Straight facial profile
_____ Cleft lip (uncommon)
_____ Enamel hypoplasia on teeth (primary dentition)
_____ Small teeth
_____ Congenitally missing teeth
_____ Hypotonic, flaccid facies
_____ Downturned oral commissures
_____ Microcephaly (small head)
_____ Small posterior cranial fossa
_____ Vertical maxillary excess (long face)
_____ Tortuous retinal vessel
_____ Suborbital congestion "allergic shiners"
_____ Strabismus
_____ Narrow palpebral fissures
_____ Posterior embryotoxin
_____ Small optic disk
_____ Prominent corneal nerves
_____ Cataract
_____ Iris nodules
_____ Iris coloboma (uncommon)
_____ Retinal coloboma (uncommon)

_____ Small eyes
_____ Mild orbital hypertelorism
_____ Mild orbital dystopia
_____ Puffy eyelids

Ear/hearing findings

_____ Over-folded helix
_____ Attached lobules
_____ Protuberant, cup-shaped ears
_____ Small ears
_____ Mild asymmetric ears
_____ Frequent otitis media
_____ Mild conductive hearing loss
_____ Sensorineural hearing loss
_____ Ear tags or pits (uncommon)
_____ Narrow external ear canals

Endocrine

_____ Hypocalcaemia
_____ Hypoparathyroidism
_____ Hypothyroidism
_____ Mild growth deficiency, relative small stature
_____ Absent, hypoplastic thymus
_____ Small pituitary gland
_____ Poor body temperature regulation

Genito-urinary

_____ Hypospadias
_____ Vesico-ureteral reflex
_____ Cryptorchidism

Immunologic

_____ Reduced T cell populations
_____ Frequent lower airway disease (pneumonia, bronchitis)

_____ Frequent upper respiratory infections
_____ Reduced thymic hormone

Limb findings

_____ Small hands and feet
_____ Long tapered digits
_____ Short nails
_____ Contractures
_____ Triphalangeal thumbs
_____ Soft tissue syndactyly
_____ Rough, red, scaly skin in patches
_____ Morphea
_____ Polydactyly (both preaxial and postaxial)

Nasal findings

_____ Prominent nasal bridge
_____ Bulbous nasal tip
_____ Mildly separated nasal domes
_____ Pinched alar base, narrow nostrils
_____ Narrow nasal passages

Neurological/brain MRI findings

_____ Periventricular cysts (mostly anterior horns)
_____ Small cerebellar vermis
_____ Cerebellar hypoplasia/dysgenesis
_____ White matter UBO's (unidentified bright objects)
_____ Cerebellar ataxia
_____ Seizures
_____ Strokes
_____ Spina bifida/meningomyelocele
_____ Mild developmental delay
_____ Enlarged Sylvian fissures
_____ Generalized hypotonia

Pharyngeal/laryngeal airway

_____ Upper airway obstruction in infancy
_____ Absent or small adenoids
_____ Laryngeal web (anterior)

_____ Large pharyngeal airway
_____ Laryngomalacia
_____ Arytenoid hypoplasia
_____ Pharyngeal hypotonia
_____ Asymmetrical pharyngeal movement
_____ Thin pharyngeal muscle
_____ Unilateral vocal cord paresis
_____ Reactive airway disease
_____ Asthma

Problems in infancy

_____ Difficulty in feeding, failure to thrive
_____ Nasal vomiting
_____ Gastro-esophageal reflux
_____ Nasal regurgitation
_____ Irritability
_____ Chronic constipation (not Hirshprung megacolon)

Psychiatric/psychological

_____ Bipolar affective disorder
_____ Manic depressive illness and psychosis
_____ Rapid or ultra-rapid cycling of mood disorder
_____ Mood disorder
_____ Depression
_____ Hypomania
_____ Generalized anxiety disorder
_____ Schizoaffective disorder
_____ Schizophrenia
_____ Impulsiveness
_____ Flat affect
_____ Dysthymia
_____ Cyclomania
_____ Social immaturity
_____ Obsessive compulsive disorder
_____ Phobias
_____ Exaggerated startle response

Skeletal/muscle/orthopedic/spine

_____ Scoliosis
_____ Hemivertebrae
_____ Spina bifida oculta
_____ Butterfly vertebrae
_____ Fused vertebrae (mostly cervical)
_____ Tethered spinal cord
_____ Syrinx
_____ Sprengel's anomaly/scapular deformation
_____ Small skeletal muscles
_____ Joint dislocations
_____ Chronic leg pains
_____ Flat foot arches
_____ Hyperextensible/lax joints
_____ Extra ribs
_____ Rib fusion
_____ Talipes equinovarus (club feet)
_____ Osteopaenia
_____ Juvenile rheumatoid arthritis (JRA)

Skin/integument

_____ Abundant scalp hair
_____ Thin appearing skin (venous patterns easily visible)
_____ Skin rashes/rosacea
_____ Psoriasis
_____ Rough, Red, Scaly, Dry

Speech/language

_____ Severe hypernasality
_____ Severe articulation impairment
_____ Language impairment (usually mild delay)
_____ Dyspraxia
_____ Velopharyngeal insufficiency (VPI) (usually severe)
_____ High-pitched voice
_____ Hoarseness

Vascular anomalies

_____ Medially displaced internal carotid artery
_____ Tortuous, kinked, absent or accessory internal carotids
_____ Jugular vein anomalies
_____ Small veins
_____ Circle of Willis anomalies
_____ Absence of vertebral artery (unilateral)
_____ Low bifurcation of common carotid
_____ Tortuous or kinked vertebral arteries
_____ Raynaud's phenomenon
_____ Thrombocytopenia, Bernard-Soulier disease

Associations/sequences

_____ 22q11 Deletion Syndrome
_____ DiGeorge Sequence
_____ Pierre Robin Sequence
_____ Potter Sequence
_____ CHARGE Association
_____ Holoprosencephaly (single case)

Medical Trivia

Each square inch of human skin contains twenty feet of blood vessels.

Blank page for personal data:

"People are like stained-glass windows. They sparkle and shine when the sun is out, but when the darkness sets in, their beauty is revealed only if there is a light from within."
Elizabeth Kubler-Ross

CHAPTER TWO: DESCRIPTION OF VCFS

DESCRIPTION OF VCFS/22q

What is VCFS/22q11 in a nutshell?

VCFS is a *collection* of symptoms or findings that have been seen on a recurring basis in patients, usually in conjunction with missing fragments of chromosome 22q11. It is congenital; you are born with it. With this deletion, the medical problems can be quite variable, ranging from mild to severe. Palate and heart abnormalities are the most common indicators and yet there are many who show no symptoms of heart or palate problems at all. VCFS/22q11 is much more difficult and complex than a disease as there are so many facets of the *"chromosome 22 deletion."*

Most patients with 22q are missing a piece of genetic material (DNA). This is information found on the long arm of chromosome 22. There are 2 halves called arms of a chromosome; one short and one long. Chromosomes are divided into the *top* half which is called the "p" arm and the *bottom* half which is called the "q" arm. The *longer* of the two halves is the "q" arm. VCFS/22q and many other chromosome disorders are found by taking a blood test and staining this blood that contains the DNA. By using the **FISH** (Fluorescence In Situ Hybridization) test, which became available in the mid-1990s. A special light makes the stain show up in the lab. Technicians are able to see if the material is intact in all the chromosomes; because it glows like a firefly if the material is present. In VCFS the FISH test shows that at *band* "11.2" of the "q" arm of Chromosome 22, there is a missing fragment of DNA material. A chromosome reminds me of an earth worm with the different segments/bands. *(See picture on page 14.)* In some cases of VCFS, the symptoms can be so subtle, that doctors do not see the connection of the symptoms to suspect VCFS/ or any other syndromes. Thus, the patient is never sent to have any genetic testing done. This has lead to improper diagnosis, which can lead to a delay in treatment.

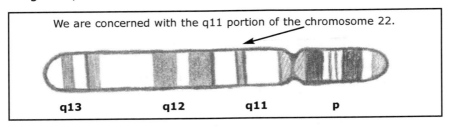

What are genes and a genetic disorder?

Genes are made up of a collection of protein/amino acid chemical strands called DNA that are housed in larger structures called chromosomes. Genetic disorders are caused by one or more missing, duplicated, fragmented (broken) or faulty genes. Down's syndrome, for example, has a whole extra chromosome 21. Changes in DNA, the body's *blue print*, can either be from heredity or mutations (meaning, something just happened to cause it to change while the baby was being formed). In the case of a mutation, once it is created, then it can be **inheritable** for future generations.

Every human cell carries 23 sets/pairs of chromosomes, which makes 46 in all. These are numbered 1-22; with the 23rd pair being the sex chromosomes or teo sex chromosomes which normally determine whether you are male or female. These are inherited from our parents—half (23) from the female egg and half (23) from the male sperm. As we mentioned before, chromosomes are long strands of DNA (the body's blue print) which contain the genes. These genes give our bodies instructions about how our features are to look and function. We depend on these genes to be able to function correctly; and be intact. So when a gene is missing some genetic informational material or is just not functioning correctly it results in what is called a genetic disorder. This usually leads to birth defects or learning difficulties. The handbook **Faces of Sunshine*** described it this way, *"A good way to think about chromosomes and genes is to compare them*

**Faces of Sunshine,* (Philadelphia: Children's Hospital of Philadelphia), 85. Edited by Donna M. McDonald-McGuinn, MS, CGC; Brenda Finucane, MS, CGC; and Elaine H. Zackai, MD., *Faces of Sunshine: The 22q11.2 Deletion, A Handbook for Parents & Professionals,* (West Berlin, New Jersey, Published by Cardinal Business Forms & Systems, © 2000), page 2.

to a train. A train has a number of box cars just as a chromosome has a number of stripes (or bands). *We can see the box cars when we look at a train, just as we can see chromosomes and their striped pattern as we look under the microscope. We cannot, however, see the packages inside the box car without first opening up the door."* The same is true of the chromosome. The genes are the packages inside the train. These packages are all the information the body needs to make a complete, whole, healthy *normal* person. *(See picture below.)*

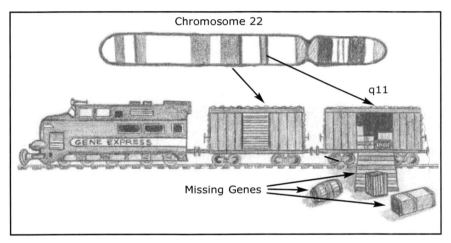

There is an unlimited range of difficulty for any genetic disorder, but the severity will depend on whether it is a dominant mutation or a recessive mutation. Some people's disorders have little impact on their lives, while others are profoundly affected by their disorder. 22q is highly complex and affects a great many tissues and organ systems. That is why the list on the <u>Anomaly Fact Sheet of Symptoms</u> is so long and growing. Severe disability and death in childhood are possible. Many people with these complaints have nowhere to turn for help in certain areas of the country or regions. Awareness of the set of symptoms associated with VCFS is limited and genetic research is just now bursting forth. We are in a rather exciting time in history, yet a frustrating one as well. **We are making an effort to get 22q11/VCFS recognized worldwide.** Parents with VCFS children don't just deal with *ONE* illness or disability but face a literal list of them; which can arrive in any order and at any time during the child's lifetime. Some of the symptoms occur at birth, <u>while others show up later in life; perhaps triggered at a certain age such as puberty, by a health crisis, or chemical imbalance in the body.</u>

Doctors have observed individuals who have been *clinically* diagnosed with VCFS based on their set of symptoms, although they did not show any deletion on the FISH test. Some believe it is because these deletions are so minute that it was undetectable, and others believe that there is yet another unknown combination of factors.

DNA: A little more in-depth discussion that scratches the surface

The birth of a child with medical problems is always a traumatic event in any parent's life. Usually two issues arise first and must be resolved for the parent to have any peace of mind. The first deals with the immediate and future medical care of their child and the second is wanting to know what the future holds for their child. Should they or can they get married, hold a job, and have children of their own? How do other families successfully handle the diversity in symptoms?

First let me say that today we have genetic counselors who can assist with many of our questions. Counselors may be provided after the birth of an affected child; or may be offered to those who are at risk of having a child who is affected. Either a team of trained professionals or someone of your choice, usually a geneticist, can help a family make a decision as to what is best for them or can direct them in any decision they make.

If a couple already has one child with VCFS or one partner of a couple has VCFS, they may wish to know if a pregnancy will be affected. If the affected individual knows they have a deletion, the embryo can be tested for the deletion. This can be done through chorionic villi sampling (CVS) or amniocentesis (amnio).

<u>CVS</u> is performed at 10-12 weeks gestation and involves having a piece of the placenta taken either through the abdomen or cervix (similar to a biopsy). <u>Amniocentesis</u> is the more commonly performed procedure. It is performed after 15 weeks gestation. Amniotic fluid (the fluid that surrounds the growing fetus) is taken through a needle inserted through the abdomen. *Ask your doctor if you are concerned with the chance of miscarriage with these procedures.*

Karyotyping (DNA testing) and FISH testing can be carried out on the fetal cells obtained by amnio-fluid or CVS. Prenatal diagnosis can only tell whether or not there is a deletion. The degree to which an individual will be affected cannot be predicted.

Now, let's get to the meat of where it all happens

Genes, are the basic units of inheritance, they are composed of DNA and are located in the chromosomes. To review:

- DNA is composed of phosphate molecules, and 4 types of repeating nitrogenous bases. These bases are (A)denine, (C)ytosine, (G)uanine and (T)hymine. It is these bases you see only as initials when you look at an illustration of the physical structure called the double strand helix of DNA. Each person has a different combination in these bases; one strand will gravitate to its correct partner or complement base that has been designated by the RNA (the transcriber for DNA) for the second strand to attach itself to create the person you uniquely are. Think of it as a united arranged marriage.

- Transcription and translation are the two basic processes in which proteins are specified by DNA, both involve RNA. RNA is chemically similar to DNA, but it is single stranded with some slight differences. It has a ribose sugar molecule and it has a base called (U)racil rather than the (T)hymine base that you would see in DNA. It is by this transcription of RNA, or messenger (m)RNA, that the DNA can specify the sequences of the information it needs to put all the bases in order, to form the right molecule combination for each of us; to meet its assigned mate, so to speak. It is this translation process by which RNA directs the information to interact or transfer (t)RNA a molecule to the correct attachment site.

- A mutation is an inherited alteration of genetic material (DNA), or it can be the first misreading or damaged translation by the RNA in your family line. Substances that cause mutations are called mutagens. The mutation rate in humans varies from locus (place) to locus (place), from one generation to the next.

It is said that 1 in 150 live births has a major diagnosable chromosome abnormality. Chromosome abnormalities are the leading known cause of mental retardation and miscarriage. Abnormalities of chromosome structures include deletions, duplications, inversions (change of position or out of correct order—a divorce from its assigned mate, if you will) and translocations (a change in location—walks out on its assigned mate) or when homologous (same) genes are interchanged.

Alleles: This is one of the two or more alternative forms of genes that occupy corresponding loci (locations) on a homologous (same-like) chromosome. At any given loci (place) in a somatic cell (body cell), an individual has two genes, one from each of the parents. Each allele encodes a phenotype feature (his or her genetic make-up) or certain inherited traits. The two alleles can be the same or different. Just as both parents could be blue eyed or one blue and one brown eyed. Each individual normally has two alleles one on each gene, one contributed by the mom and one from the dad. When the same gene occupies the same position on both chromosomes (ex: the two blue eyed genes), the individual will be homologous (the same). If the alleles are different the individual will be heterozygous or having two different genes (ex: blue eye, brown eye) at the corresponding location on the chromosome; one of the genes has to be dominant. If the one gene from a parent's contribution is affected with the 22q11 deletion, and is dominant (like brown eyes are dominant over blue eyes); the child from this pair will more than likely have the disorder and the affected gene will be dominant. In most cases there is a 50/50 chance of having a child with the disorder that the gene carries.

The recurrent risks of a genetic disorder specify that the probability that a future offspring will inherit a genetic disorder or disease is said to be 50% or more. (We need to take into consideration any risks on either side of the gene pool.). Recurrence of risks remain the same for each offspring. Males and females are equally likely to exhibit gene dominance and pass it on to their offspring.

Is VCFS a RARE disorder?

No, it is not *RARE*. It just has not been as well recognized or publicized as other disorders, since its rediscovery in the 1960's to 1970's. As a matter of fact, this deletion (22q11) is much more common than once thought. What is ironic is that a syndrome this common can be so widely undiagnosed. It is currently estimated that 1 out of every 2000 babies will be born with it. These statistics vary widely, depending on to whom you speak or in which part of the world you live. This is partly due to the new tests available and the willingness of doctors to test newborn babies if there is any suspicion. Some feel the FISH test should be given at birth as routinely as the PKU test (which is positive in 1:16,000 births).

The high ratio for 22q11/VCFS is one reason why it is vital to broadcast this information. VCFS is said, by some, to be the most common cause of genetic palate and heart defects; second only to Down's syndrome. Potential parents should be screened for *Mosaicism*, which means that an individual developed from a single zygote (pronucleus stage of conception) that had two or more cell populations that differ in genetic formation. This is common in humans with the variations in the number of chromosomes. If you are confirmed to have the deletion or confirmed to be a carrier, *you will have crucial decisions to make regarding whether or not to have children or to have more children.* As we stated before, only you, when educated with genetic counseling, can make the right decision for yourself.

Now that we know it is VCFS, what's next?

Albert Einstein once wrote: *"The main thing is to not stop asking questions."*

Is there the same answer for each person? No. The degree to which a child will be affected is extremely variable. It is *impossible*, actually, for anyone to predict. (The common possibilities are on the list.) The information we have from the list of symptoms is the only clue we have to gear ourselves up for what the future may hold. There are *almost 200 possible abnormalities caused by this deletion*. These abnormalities can affect one from head to toe. Since not all symptoms come at the same time or *ALL* in one person, we could encounter more than one health battle at a time. The best advice is to *"take one day at a time"*. When you are a parent managing your child's health issues; it can sometimes become overwhelming. This is especially true, if another symptom strikes before you have a breather from the last one or the still existing one(s). Sometimes it can feel like you are alone and losing the battle, but have faith. *"This, too, shall pass!"* I have had to live by that motto myself and it has served me well. Try to be optimistic and stay on top of things as best you can. Don't be afraid or have too much pride to ask for help, even if it's for a little time to yourself.

It has to be noted that there are the rare exceptions. There are a few with VCFS who do not have any <u>discernable</u> problems. This does not mean; however, that the flaw or predominance for difficulty is not there, but it may be just slight enough not to be apparent or recognized. Perhaps it is not restrictive enough to have affected their life, to have taken notice as being *"abnormal"* nor to have affected their life to the point of calling it a condition. However, this continues to be in the genes and it can be passed on to one's children, and it may or may not be minor in nature for the next generation.

In the following pages we will touch upon some of the <u>possible</u> medical and psychological issues pertaining to VCFS/22q.

VCFS and your PCP (<u>P</u>rimary <u>C</u>are <u>P</u>hysician): What should he know?

Since VCFS is linked with a variety of clinical symptoms, it is important to find a doctor who is knowledgeable or to educate your own doctor to recognize the symptoms and be aware of any implication of changes within the symptoms of this disorder. Doctors may not realize that VCFS, DiGeorge and 22q11 deletion are actually the same condition. I have learned that the diagnosis name can differ depending on age. When in infancy, it is common for the child to be diagnosed as "DiGeorge;" when older, a diagnosis of "VCFS" or "22q11.2 deletion" is more common. I have found that when your child gets to the magic age of 18 it gets more difficult to find a doctor who is familiar with VCFS or who has the time to learn about VCFS in its entirety. It is very important that we find doctors who are willing to learn and monitor the changes that occur in the bodies of our children as they grow. The doctors need to recognize these changes and their implications, since almost every organ of the body can be affected.

Obviously the earlier 22q/VCFS is acknowledged the better. Then a game plan can be put in place for the child or young adult. This helps tremendously in setting the foundation for their life.

Initially, the most common systems involved might be:

- Cardio-vascular
- Cleft palate
- Feeding difficulties
- Immunization problems
- Growth hormone deficiencies
- Delayed neurological and psychological developments
- Speech problems
- Renal abnormalities

There is also an assortment of other congenital abnormalities and medical conditions of which to be aware. These might include: thrombocypenia (low platelets), juvenile arthritis, hyperlaxity of the joints, weak muscles, inguinal hernias, undescending testicles also known as cryptorchidism (where the testicles do not drop from the abdomen or inguinal canal into the scrotum), hypospadia (abnormal location of the opening of the urethra on the penis), craniostenosis (premature closure of the sutures of the skull, where the head bones meet), vertebral anomalies (abnormally shaped vertebrates), tethered cord (inability of the spinal cord to migrate downward), ophthalmological changes of the eyes, abnormal positioning of blood vessels behind the windpipe causing respiratory problems, etc. *(See the Anomaly fact sheet of symptoms on page 5.)*

In the future, stem cell research may be an option for those who have genetic disorders, using umbilical cord stem cells. Could this be the treatment of the future? We hope so!

When my son reached the age when he no longer qualified to attend the pediatric office, I searched for a doctor to take over his care. I was hoping to find an internist who was informed about VCFS. I found none! I was fortunate to find one doctor, out of all those I contacted in the Phoenix area, who admitted to knowing nothing about VCFS when I first approached him a few years ago; but who was willing to try and learn what he could to help my son stay healthy. He recently admitted to me that he still does not feel he knows enough. I told him that I think of him as the chief or *CEO*, and as symptoms or conditions crop up, he can send my son to the appropriate departments (specialists) for care. Since VCFS is <u>NOT</u> a disease that we can expect to cure, but an accumulation of symptoms and conditions in the different organs and tissues, the objective is to keep these organs as healthy as possible.

Here is a list of specialists in the field of medicine with their specialties

Anesthesiologist	Pain management
Cardiologist	Heart defects
Chiropractor	Physical manipulations to bring health through the nervous system
Cosmetic surgeon	Reconstructive and repair surgery
Dentist	Teeth, gums and underlying bone
Dental/oral surgeon	Performs dental surgery
Dermatologist	Skin & scalp problems
Dysmorphologist	Interprets patterns of growth and structural defects
Endocrinologist	Endrocrine system/glands
Family practitioner	General health practice
Gastroenterologist	Digestive tract
Geneticist	Specialist in all phases of genetics
Gynecologist	Female health issues
Hematologist	Blood
Immunologist & allergist	Immune systems and allergies
Internist	A doctor of diagnosis after the age of 18
Naturopathic physician (Holistic medicine)	Uses natural foods, herbs, light, warmth, massage,

	fresh air, and regular exercise for healing and avoids medications and chemicals
Neonatologist	Premature and underweight babies at birth
Nephrologist	Kidneys
Neurologist	Nerves and brain
Obstetrician	Delivers babies
Occupational therapist	Treats all ages whose daily living is impaired
Oncologist	Cancer
Ophthalmologist	Eyes
Orthopedist	Bones/skeletal
Otolaryngologist	Ears, nose and throat (ENT)
Pathologist	Studies pathogens (diseases)
Pediatrician	Children 0-18
Psysiatrist	Physical health and rehabilitation
Physical Therapist	Treats physical impairments with various methods
Plastic/reconstructive surgeon	Performs surgery to alter, replace, or restore visible parts of the body
Podiatrist	Treats disorders of the feet
Psychiatrist	Provides diagnosis, prevention and treatment of mental disorders
Psychologist	Mental/emotional behaviors
Pulmonologist	Lungs
Radiologist	X-Rays
Rheumatologist	Joint diseases
Speech/language pathologist	Human communication development and disorders
Urologist	Urinary tract and vasectomies

Has our physical environment in this modern world played a role in the increase of genetic disorders?

This is an argument many have had in our generation, because of the ratio of normal births vs. those of children with a defect of some sort; either obvious or not. Yes, there is a good argument on both sides of the coin. One may say it has to be all the artificial "everything" that is eaten, absorbed, breathed, bathed in, and lingering from military testing and foreign tests that are caught up in our air, soil, and water supply, etc. And that is not counting the inheritable odds from generations of old.

The other side will argue that defects have always been at this proportion, we just have more people now and have never had the technology before to know the numbers. It is a fact that such things like mental illnesses, retardation, or deformities, were taboo, and that these were hidden and therefore not open and recognizable to the public.

There are also the learning disabilities. Have they always been at this proportion? Not long ago there was more of a need for practical daily living skills than a proper academic education. Then, many families couldn't afford to go to school beyond a grade school education, so there was no way to gauge.

The fact of the matter is, it probably is a little of both! Since the problem is here NOW and we can see the trend, we have to educate ourselves and others about what is happening and what can continue to happen, so that our children and our future generations do not continue to suffer the health problems and the stigmas that go along with these disorders.

Medical Trivia

The largest cell in the human body is the female reproductive cell, the ovum. The smallest is the male sperm.

"Words are windows to the heart...and if I could reach up and hold a star for every time you made me smile, the entire evening sky would be in the palm of my hand."
Author unknown

CHAPTER THREE: SYSTEMS AFFECTED

SYSTEMS AFFECTED

In this section we will address certain areas of the body that are most commonly affected in VCFS children. However, there are many variations in each of these areas, so if you have a specific question or concern that is not included; please contact your physician or the VCFS Education Foundation or any other organization that specializes in this area.

Common heart defects

Research shows that the majority (80%) of children with VCFS have some kind of congenital heart defect and this is often of a serious nature. Even if your child is one of the 20% who does not have an identified heart defect it is wise to have an evaluation carried out by a cardiologist to rule out any concealed problems. This would possibly involve tests such as a ECG/EKG (electrocardiogram) and/or an ultrasound. Both of these tests are noninvasive procedures (meaning, non-surgical, won't break the skin). They are completely painless.

Some heart problems associated with VCFS are: VSD (hole in the heart); aortic arch defect (flow obstruction); Tetralogy of Fallot (anomaly consisting of 4 defects, ask your doctor to explain this to you if it pertains to your child); PDA (patent ductus arterosis) is an abnormal opening (only abnormal after birth) between the pulmonary artery and the aorta caused by failure of the fetal valve to close after birth; coarctation of the aorta (narrowing of the main artery); truncus arteriosis; right sided aorta; and reverse placement; with many other anomalies that are not as predominant. While most of the more serious defects will require open heart surgery at a young age, some imperfections will not require any surgery at all. *(See Cardiac/thoracic findings listed on the Anomaly fact sheet of symptoms on page 5.)*

Because of heart anomalies, most patients will require a dose of antibiotics before *any* kind of dental work is done. This is to prevent endocarditis (infection that involves the heart muscle) from possibly occurring, so *always* inform your dentist.

Common heart anomalies associated with VCFS

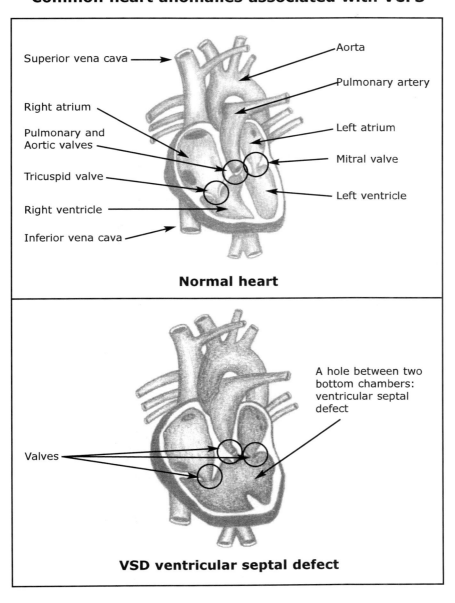

Normal heart

VSD ventricular septal defect

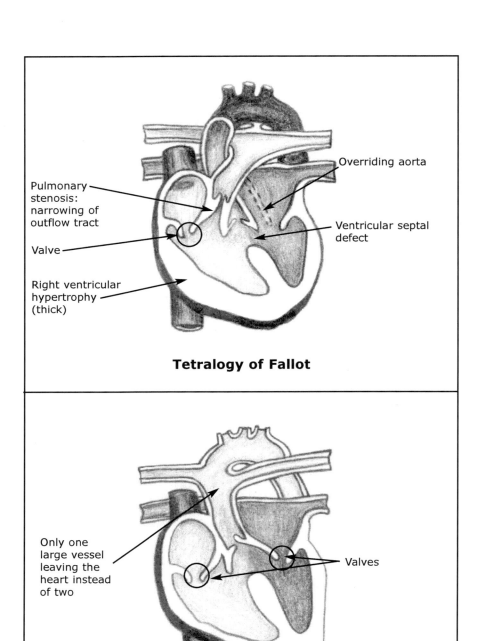

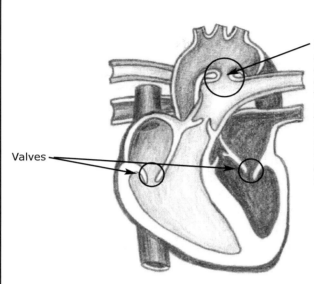

PDA Patent ductus arteriosus

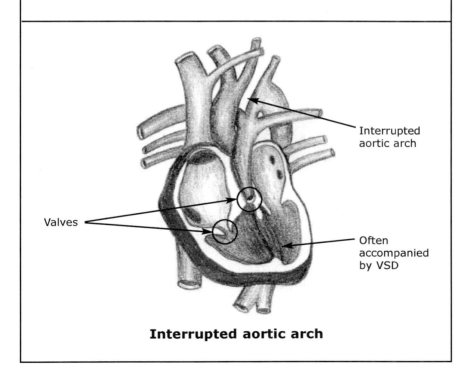

Interrupted aortic arch

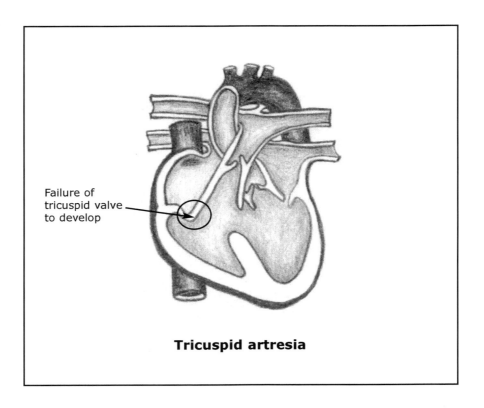

Tricuspid artresia

Let's study the heart a little, to obtain a better understanding of how it works

The heart is a strong muscle about the size of a fist; it is enclosed in a fluid-filled membrane. It has 4 chambers. The _upper_ 2 chambers are called the right and left _atria_. They serve as the receiving chambers for deoxygenated and oxygenated blood. The 2 _lower_ chambers are called the right and left _ventricles_; these are the pumping/squeezing chambers. There is a common wall that separates the left side of the heart from the right side. This is called the interventricle septum. It is comparable to the septum that separates the 2 nostrils of the nose. The atria are separated from the ventricles by valves that permit blood to flow through in one direction (normally). Here is how the flow goes: Deoxygenated blood from the entire body is brought into the heart through the superior and inferior vena cava to the upper right atrium and this blood goes through the tricuspid valve to the lower right ventricle; then the blood flows through the pulmonary valve to flow to the lungs to become oxygenated; then it circles around after it has picked up oxygen again to flow into the upper left atrium, which forces the blood through the mitral valve into the lower left ventricle; finally the heart beats and squeezes the blood through the aortic valve into the aorta to be carried off throughout the body to do its job. And the *BEAT* goes on . . .

Blood vessels carry blood throughout the body. They consist of: Arteries, arterioles, veins, venules and capillaries. <u>Arteries</u> are the large thick walled vessels that carry the <u>oxygenated</u> blood to the body. The arterioles carry blood to capillaries. Capillaries are the small thin walled vessels that allow exchange of material between the blood and tissue fluid surrounding the body cells. This exchange is of oxygen, nutrients and carbon dioxide (waste). Venules are the small transports that branch out into the larger vessels called veins. <u>Veins</u> transport the oxygen <u>depleted</u> blood in one direction, back to the inferior and superior vena cava and then to the heart to make the circuit and to pick up oxygen from the lungs and so it goes.

Does this cycle determine blood pressure?

It is this cycle that is used to measure for blood pressure. This cycle is equivalent to one heart beat. Just imagine all the work the heart is doing even as you read this, and we take it for granted. There are 2 alternating phases.

<u>Systolic</u> pressure is the contraction that forces the blood out of the heart, and diastolic is the relaxed phase allowing the heart to refill with the blood for the next squeeze. How is it possible to measure blood pressure? It is the pressure of exertion of the blood on the walls of the arteries. The <u>systolic</u> pressure is the top number, it measures the force it takes to push or squeeze the blood out of the ventricles. As the arteries dilate a pulse can be felt at certain points of the body.

<u>Diastolic</u> pressure measurement is the pressure in the ventricles when they are at rest, allowing blood to fill them up again. This

is the lower number. A normal range for an adult is a pressure from 100/60 to 140/90 depending on age and weight. Elevated blood pressure is called hypertension, and low blood pressure is hypotension.

Blood Pressure information for children:

Unlike in adults, the diastolic reading in children is the point at which the sounds first become muffled rather than the point at which they disappear completely.

Normal blood pressures vary with age. The table below should serve as a rough guide.

AGE	PULSE	RESP.	BP systolic/diastolic	
Newborn	100-160	30-60	46-92	38-71
6 mo-year	90-130	24-40	72-106	40-66
3 years	80-125	20-30	72-110	40-73
6 years	70-110	16-22	46-92	38-71
10-14 years	60-110	16-20	83-121	45-79
14-19 years	55-90	15-90	93-131	49-85

Medical Trivia

By the time you turn 70, your heart will have beat some two-and-a-half billion times. (Figured on an average of 70 beats per minute.)

Blank page for personal data:

"Appreciation is a wonderful thing; it makes what is excellent in others belong to us as well." Voltaire

PULMONARY AND RESPIRATORY

Can you help me understand the airway's path and what is affected?

There are two phases in the respiratory system, external and internal. It is the air distributor (oxygen supplied) and gas exchanger (carbon dioxide removed).

There are 6 parts of the body that are affected in the respiratory system:

Nose: Within the nose there are hair-like projectiles on the mucus membranes call *cilia.* They sweep away foreign material toward the throat for elimination before it can be drawn into the lungs. Now you know why people clear their throat so much.

Pharynx: The throat is the passage way for air and all that is involved in that passage. (Digestion is also involved in this area in the esophagus.)

Palatine tonsils: These are located on either side of the soft palate on either side of the throat near the uvula (the punching bag). If they are swollen or infected they can cause a restricted airway. This is why it is recommended by some that the tonsils be removed before any surgery on the palate is done. They can also become persistently infected and become rotten. *"Ick,"* that was my experience and as soon as they were removed I had many more healthy days.

Larynx: The large upper end of the trachea is also called the voice box and the vocal cords are contained within the larynx. In someone with VCFS the vocal cords can be weak or partially paralyzed.

Trachea: The windpipe is a cylinder-shaped tube with rings of cartilage around it. The trachea branches into 2 tubes, these are bronchi. Each tube then divides again and each of these goes into a separate lung. Then these sub-divide into smaller tubes, bronchioles, which dead-end at the air sacs. These air sacs have very thin walls and this is where the exchange of gases between the lungs and blood takes place.

Lungs: Each of the 2 lungs consists of smaller sections called *lobes*. The left has 2 lobes and the right has 3 lobes.

To understand the breathing process:

The diaphragm is a thoracic (in the chest) muscle that horizontally separates the abdomen and chest cavity. When stimulated by the phrenic nerve the diaphragm draws the air up into the lungs which is the inhale. Then pressure forces the air out and it is exhaled.

Be aware of some respiratory problems:

Chronic bloody noses: These can be serious, especially if the platelet level is low or the child is anemic. There may also be an underlying condition causing them, such as liver disease, which has afflicted my son. The liver disease is indirectly related to VCFS.

> Bloody noses are not something to take lightly. They are not serious if they are now and again and not for long durations. Many people, however, have bloody noses that are heavy and can become dangerous if not treated quickly. These may even require hospitalization.

Self-help for bloody noses as follows:
- Sit up with your head bent forward.
- Clamp your nose closed with your fingers for 5 uninterrupted

minutes. During this time, breathe through your mouth. You may need to insert a pressure plug made of tissues or a piece of a tampon, before closing the nose tightly.
• If bleeding stops and reoccurs, repeat procedure, but pinch more firmly on both sides for 8-10 minutes. Holding your nose tightly closed. This pressure allows the blood to clot and seal the damaged blood vessels.
• A cold compress can also be applied.
• Don't blow your nose. Don't swallow blood because it might upset your stomach.
• *If this does not stop the bleeding or if it keeps reoccurring, go to the emergency room and set up an appointment with your doctor. It may need to be cauterized or have veins tied off.*

Wheezing: This is a whistling or a sighing sound, resulting from the narrowing of the air pathway.

Asthma: This is difficult breathing, accompanied by a wheezing due to a spasm of the bronchial tubes or swelling of the mucus membranes within the lungs. Asthma can develop from countless factors, such as, allergies, pneumonia, bronchitis, etc. It can be temporary or chronic. If persistent, the doctor typically will prescribe a home-based breathing treatment machine called a nebulizer. Insurance usually pays for this machine.

Croup: This is a cough that sounds like a barking seal. It is a disorder caused by an acute obstruction of the larynx as a result of an infection, aspiration from inhaling a foreign body, or an allergy.

Aspirations: These can be foreign bodies that lodge in the airway. They might cause a cough, hoarseness, or trouble breathing; and can result in an infection, asthma and even a lung abscess. Pneumonia and respiratory distress can result from inhaling contaminated fluids or gastric reflux. This is a major important point to know with children who have VCFS.

Respiratory distress syndrome: RDS is common in newborns and small children. Also common is IRDS which happens at birth when there is not enough *surfactant* secreted to lubricate the air sacs. If this happens, they stick together like cello wrap and collapse the lungs when the newborn is trying to breathe.

Viral and bacterial pneumonia: Pneumonia can cause varying degrees of illness. In a VCFS child it seems to last longer, can be quite serious and often develops as a secondary infection. Commonly this develops from an ear infection or an upper respiratory infection such as a bad cold.

This was what happened in my son's case. He had pneumonia so many times it caused scar tissue to mount, which worried the doctor; causing him to wonder if my son might have to have a lobe removed. He was labeled with *lung disease* and it was at this time that we were instructed to think about moving to a dryer climate. That is how we ended up in dry, warm and sunny Arizona.

Parents Speak Out

A question that I have also been asked about is bloody noses. I must say that my son has had a heck of a time with bloody noses. He has had his nose cauterized many times and at this writing is going through it again. It is discouraging when a referral person in a doctor's office doesn't see the importance of these bloody noses. The frequency and volume involved can be dangerous. In my son's case it is due to the vessels being too close to the surface and the least little bump, crack or blow can start it. He is now, in fact, anemic from the loss of blood. There is a more serious condition called thrombocytopenia that is common with VCFS; this is a decrease in the number in platelets and is the leading cause of bleeding.

Medical Trivia

The left lung is smaller than the right lung to make room for the heart.

"We are what we repeatedly do. Excellence, therefore, is not an act but a habit." Aristotle

KIDNEYS, HERNIAS AND ANAL ANOMALIES

Where are the kidneys? What do they do?

Kidneys come in pairs. These are bean-shaped organs that are involved in the urinary system. They are in the dorsal (toward the back) part of the abdomen. There is one on either side of the vertebrate column or spine and they are positioned just above the waist.

In most people the right kidney is a little lower than the left. In newborns the kidneys are, on average, 3 times the size (in proportion to body size/weight) of those in an adult. Kidneys filter blood and eliminate wastes in the urine. It is technically a complex filtration network and a resorption system. The filtration system filters the blood under high pressure, removing any salts, urea, and other soluble wastes from the plasma of the blood.

What kinds of kidney problems are there in VCFS?

Some children may have abnormalities such as missing, multiple or diseased kidneys. *(See the Anomaly fact sheet of symptoms on page 5.)* All children diagnosed with VCFS should have a kidney ultrasound done, whether or not the child seems to be having any problems in that area. There may not be any pain or obvious problems at present, but an ultrasound will help to determine if there are any abnormalities about which to be concerned. Keep records of frequent bladder infections, blood in urine, pain in the kidney area, or any peculiarity you may notice. Take notes and give them to your physician so that he can check and make sure that there are no hidden underlying problems.

What about hernias? What are they?

Inguinal hernias seem to be a problem for boys with VCFS. This is a protrusion of a portion of the intestine into the inguinal canal or scrotum sac. Consult an urologist for repair.

There are many sites on the lower body that are common for hernias/tears. Hernias are found as unusual bulges in places were there should not be bulges. A common site is at the belly button. It is wise to learn where the common sites are and also what a hernia looks like so that you will recognize one if it develops on your child. However, not all hernias are typical of VCFS and not all are repairable. *(See illustration on page 39.)*

Are there ever any anal abnormalities? What kind?

Yes there can be anal abnormalities, although they are not real common. Anal-rectal anomalies are a hereditary trait themselves. So, it is important to consider VCFS in the initial diagnosis of children with anal anomalies and to look for other features of the syndrome. (See the Anomaly fact sheet of symptoms on page 5.) Although there is a very large spectrum of defects in this area, not all are associated with VCFS.

When a baby is born with an anal-rectal malformation, it is immediately detected as it is very obvious with visual examination. Doctors will determine the type of defect the child has and whether or not there are any associated irregularities. As most parents (who have dealt with this problem) know; children with an anal-rectal abnormality are predisposed to suffer from associated defects (mainly urinary tract, but also spine and sacrum problems). It is important to determine the presence of any associated defects during the newborn period in order to treat them early. Many times there will be another abnormal condition that will follow as a result of the anal-rectal malformation.

One abnormality that is seen in VCFS is anal atresia/stenosis.

This is the absence of an opening, closure, or constriction of the rectum or anus. It includes the diagnosis of *imperforate* anus.

Some common hernia sites

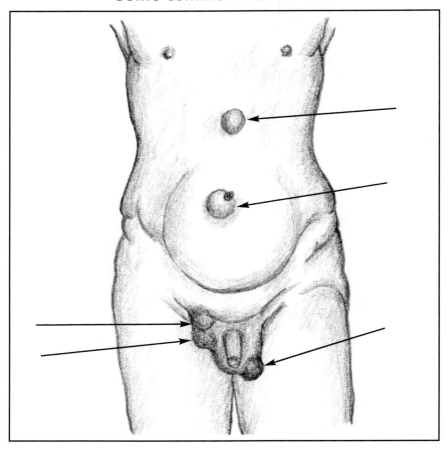

Medical Trivia

The average human produces 25,000 quarts of spit in a lifetime, enough to fill two swimming pools.

Blank page for personal data:

"Be a good listener. Your ears will never get you in trouble" Frank Tyger

EARS

What kinds of hearing and ear problems might my child have?

Many VCFS children suffer from frequent ear infections and packed wax. *Never* use a cotton swab on the inside of the ear canals; this will only pack the wax. A wax plug, made in this way, can lead to hearing loss or a perforated ear drum. The wax can get so solid that it may require an ENT (ear, nose and throat doctor) to remove it. A doctor can instruct you on how to keep the canals clean in a safe way.

If the wax is packed firmly for a considerable period of time, it may lead to fluid retention behind the eardrum. Fluid that normally drains out can become trapped, putting pressure on the ear-drum, and causing an infection. If wax is not the culprit, but the drainage is not sufficient, there may be a need for tubes to be placed in the ears. This should drain any trapped liquid. The tubes will relieve the pressure and thus cut down on the frequency of ear infections. If your child has immune deficiency with the ear infections, the child may appear to be sickly all the time and have a nasty cough. This can also lead to speech delays for a child who is not feeling well enough to communicate and who may hear distorted sounds; words are not clear enough to articulate. As with other disorders of VCFS, the child should be monitored on a regular basis. If you suspect any of these ear problems, have the child's hearing checked periodically. Frequent ear infections can lead to hearing loss. As an example: My son fell into this category. He was 3 and not speaking when I recommended strongly that he have a thorough exam at the local deaf school near where I lived. He did in fact have hearing loss. It was discovered that because of frequent ear infections, the 3 tiny bones (called the malleus, incus and stapes) had stabilized (a form of calcification) and were not able to move as they should for sound. He had surgery to correct this problem and had tubes inserted. He has done well with extensive speech therapy, and is considered within normal range to date with a 30% hearing loss. If the truth be known, those of us who grew up listening to loud music have a hearing loss about the same as that. So I am happy.

Let's now discuss middle ear infections. What are they?

An ear infection (otitis media of the middle ear) is the most common infection in children. This develops when bacteria invades the area of the middle ear, which is located behind the ear drum. Some _symptoms_ can be:

• Acute onset of pain in the ear. An older child can verbalize that their ear hurts, but infants show signs that need to be explored, like crying (a lot!) while rubbing, tugging or pulling on their ear(s).

• You might see some blood or puss running from the infected ear. This could happen if the child's ear drum has ruptured or popped because of the pressure of fluid behind the drum.

• Fever and general fussiness with chills and dizziness in a child are indications that a parent should investigate the ears and adenoids. This is especially true if the child has, or has recently had, a cold, flu or sore throat.

• If young children are nauseous or vomiting it may be caused by an ear infection or it could be a signal that one is approaching. Ear infections can be caused by vomit, which has backed up into the ears; especially when the child has palate problems.

I don't mean to scare you since ear infections themselves are not dangerous, but an ear infection can be quite serious. If left untreated, it can lead into a much more serious secondary infection. An untreated middle ear infection can sometimes lead to complications such as: mastoiditis, brain abscesses, and meningitis. Upper respiratory infections such as pneumonia and hearing loss are common.

How are middle ear infections diagnosed and treated?

A middle ear infection can be diagnosed by your doctor by a direct inspection of the interior ear canal. The treatment usually prescribed is 10 days or more of antibiotics. If there is no real infection yet, only backed up fluid; the doctor will usually prescribe decongestants and antihistamines to dry up the fluid in the eustachian tubes. Your child may have narrow tubes, which could make it harder for the doctor to insert drainage tubes if needed.

Is there anything I can do myself?

Your child most likely will need some professional attention, but as a parent you can make your child as comfortable as possible by laying the child's affected ear on something warm such as a wrapped heating pad on low setting, warm soft towels or a wrapped hot water bottle. Carefully test the temperature of the object to make sure it is not *too* warm. Perhaps test it on your face or inner arm as you would a baby's bottle of milk. A doctor can tell you what OTC (over the counter) decongestant or antihistamine to try. Remember not to give aspirin, because the infection could be caused by a virus. We don't want to add <u>Reye's Syndrome</u> to the list of things about which we have to worry.

What can I do to avoid middle ear infections or reoccurrences?

Make sure you give the child the full course of any antibiotics; even if you believe the infection is over and you don't like the thought of too many antibiotics. Antibiotics can be a friend to a person with VCFS. If you don't give the full course of medication, you run the risk of reoccurrence or maybe of some other secondary infection. This can cause the infection to persist a lot longer. Remember the VCFS child usually has "bugs" that persist longer than they do for most children.

If your child has ear infections several times in one year; you may choose to request that tubes be placed. I wish I had been given that advice earlier. Since many doctors do not know about or understand VCFS, it would be beneficial to explain to them that tubes might be a good prophylactic measure if a child is starting the chronic ear infection route. This may help to ward off more serious long term problems in the future.

This might seem rude to some, but don't be afraid to ask that people not visit or come around your child when they are ill in *ANY* way. People don't understand the immune problems with your VCFS child and will not think past *"I don't think I'm contagious anymore."* This reminds me of a quote by Ed Cunningham, where he says, *"Friends are those rare people who ask how we are and then wait to hear the answer."* I personally remember requesting this consideration to no avail. It was not heeded. We would get visitors; and a simple cold to a visitor became a hospital stay for my son.

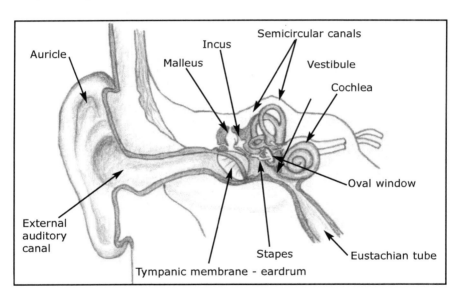

What other ear problems can VCFS children have?

Many children with VCFS have anatomical differences in their ears. Some common differences found are: narrow eustachian tubes, smaller ears than the rest of the family and ears which are positioned lower on the head.

How do you know if they are lower? Here is the way they are measured. *(See Below.)* Take a ruler and place it at the corner of the child's eye and hold the ruler level. The top of the ear should be in line with the corner of the eye.

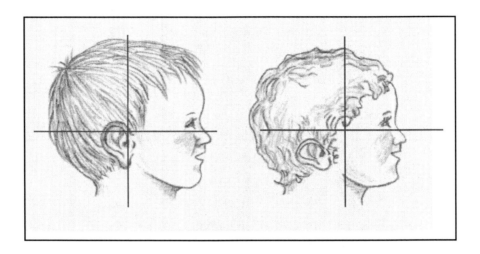

PARENTS SPEAK OUT

One parent writes: Hyperacusis is an abnormal level of sensitivity to loud sounds. It can be seen by the hearing sensation level at which the hearing reflex is triggered. This is the difference in decibel levels between the measured threshold (which is where a person first starts hearing sounds at a specific frequency) and the decibel level at which the reflex is triggered. The reflex is the body's way of protecting itself from sounds that are too loud. The measurement of the reflex is not behavioral. A probe is placed in the subject's ear and a series of increasingly loud beeps are sent into the ear canal until the reflex is triggered. However, the hearing sensation level typically comes from behavioral testing; that is, the person signaling when they hear a sound. If that result is not very accurate (due to inattention, for example), it could lead to an underestimate of the size of the threshold. This could look like hyperacusis. On the other hand, hyperacusis tends to be common with some types of sensorineural hearing loss. This is different from the conductive hearing loss which is most typical of our kids with VCFS.

My daughter is now 12. When she was younger (we did not realize it at the time) she could not handle the stimulation of large groups of people. The noise, movement, and atmosphere completely overstimulated her. She fell apart at church, at restaurants, at her sibling's sports games, etc. After a few years, we learned about _sensory integration dysfunction_ and realized that she wasn't throwing tantrums for no reason. She really could not tolerate certain settings. We also found out that she has hyperacusis, which is simply being overly sensitive to sounds. I wish we could have gotten help for her a lot sooner. It might help you to find an occupational therapist who will provide sensory integration if you see these symptoms.

Medical Trivia

The average human body contains enough: iron to make a 3 inch nail, sulfur to kill all the fleas on an average dog, carbon to make 900 pencils, potassium to fire a toy cannon, fat to make 7 bars of soap, phosphorous to make 2,200 match heads, and water to fill a ten gallon tank.

"Those who bring sunshine to the lives of others cannot keep it from themselves." Sir James Barrie

EYES

What is involved with vision therapy?

I was sent this information from a parent and thought it was good information to include in this section. If your child needs *vision therapy* and you can't afford the out-of-pocket expenses, there are things that you can do. At most learning skill center stores you can find visual perceptual tools. Color cubes, mazes, find the picture in a picture (like those in *Highlights*® magazines) and other items can be suggested by a teacher or special education school.

Some parents have requested that their schools purchase a visual processing remediation program. These are available in various levels and are age appropriate. One parent said she used **Visual-Perceptual Development Remedial Activities** by Karen Gardner Codding and Morrison F. Gardner. This program is distributed by Psychological and Educational Publications, Inc., PO Box 520, Hydesville, CA 95547-0520. **1-800-523-5775**, fax 1-800-447-0907. The program covers visual discrimination, visual memory, visual-spatial relationships, visual form constancy, visual-sequential memory, visual figure-ground, and visual closure. This teaches the children to pay effective visual attention and interpret what they see. The activities go along with the *Test of Visual-Perceptual Skills* (non-motor). There is also a set of activities for remediation of visual-motor deficits as well. Visual-spatial deficits are one area identified as commonly experienced by our kids with VCFS.

One thing that might be wise to do is to have your child evaluated for scotopic vision sensitivity syndrome (similar to night blindness). If the test results are positive, check into Irlen lenses.

They are designed to filter out the frequencies of light that give the problems. Here are a couple of Web sites that are very informative on this subject for visual impairment and testing:
http://www.oepf.org/
http://www.covd.org/

PARENTS SPEAK OUT

My 15 year old son has VCFS and has been wearing glasses since he was 4. We have had many experiences with frames and lenses, because he has very thick lenses. We finally found a great optometrist who was able to give him the lens strength with a tremendous reduction in the weight (and thickness) of the lens. This makes a *BIG* difference on the bridge of his nose, with the soft pads. We have no more marks or sliding. I think they were called Feather Weight® frames and lenses. I was told they can be purchased at Costco® or other discount stores in some areas.

Many of our children with VCFS have problems with ocular-motor, visual-spatial and visual processing problems. We achieved dramatic results for our child within 6 weeks with vision therapy. The tip-off was that he was seeing double. A regular eye exam from an ophthalmologist (MD) will not pin-point these kinds of problems. Try finding a behavioral or developmental optometrist in your area. Ophthalmologists are not trained to do this and they do not know much about it. They may even discourage you from trying it. There is a growing body of research on this subject that shows that up to 93% of children with learning problems could benefit from vision therapy because they are visually confused or cannot maintain binocular vision. A search on the Web will give you some Web sites and providers in your area. I hope this helps.

Medical Trivia

Leonardo Da Vinci suggested using contact lenses to see back in 1508.

*"**B**eware, as long as you live,
of judging people by appearance."* Jean de la Fontaine
BECAUSE
*"**O**ne is taught by experience to put a premium on
those few people who can appreciate you
for what you are."* Gail Goodwin

FACIAL FEATURES

What are some of the "typical" facial features of VCFS?

Children and adults with VCFS usually have some *typical* features which are often quite subtle, although in my experience the changes can manifest themselves more with age. Many have said that they never would have known a person had VCFS had they not been told.

The *typical* features listed below are more visible as a group, when many pictures are viewed as a classification within the disorder, rather than of individuals alone.

The most common characteristics

• As babies, VCFS children may display small features including small ears, small mouths; some have the "O" expression or a pouty appearance of the lips.

• The nose is often wider, especially when they get older. Frequently when these children get older and the nose grows, the face becomes longer and the nose broadens at the tip. This is called "bulbous" (bulb-like). It can be corrected by plastic surgery if desired; but this surgery is not usually covered by insurance. They consider it cosmetic or elective surgery.

- Many have a deviated septum. This is crooked or broken cartilage that separates the two nostrils. It is this cartilage that gives the nose shape. This correction is usually covered by insurance because it is a defect that can restrict breathing.

- Many VCFS children display an open-mouth "O" expression, with lack of much facial expression. It is as if they are in deep thought or distracted. This doesn't mean that they cannot show you expressive reactions when appropriate. For the most part these children are very happy children with an incredible sense of humor. I sometimes think when T.G. (my son) was growing up people must have thought we were related to the actor Jim Carrey, because we were always making faces at each other. Now, I realize that it was good therapy. Unknowingly, it helped to develop a good sense of humor with an expressive façade. So don't be afraid to scrunch up your face at each other once in a while. (No, your face won't really freeze that way!)

- Sometimes there are dark red circles under children's eyes that concern parents. These are called allergic shiners and usually are no cause for concern. See a doctor however, if the circles are dark blue to purple and they have a hollow look about them. A sunken distant look of the eyes can mean a serious case of dehydration and/or lack of oxygen. VCFS children dehydrate easily, so it is imperative that you see a doctor as soon as possible in this situation.

- The child with VCFS may have puffy looking eye-lids. It looks as if they just woke up and are still tired—*ALL THE TIME*. This puffiness will sometimes vanish in time.

- There are some children who are born with Microcephaly (small head).

What about other abnormalities?

Many children with VCFS have loads of hair and l-o-n-g tapered fingers. Friends might comment on the piano fingers your child has. My son put those long digits to work by learning to type and doing data entry. (I knew those long fingers would come in handy someday).

PARENTS SPEAK OUT

My son, Glen, is presently experiencing a terrible problem with psoriasis. It is bad around his, scalp, ears and face. I have not found any over-the-counter ointments to help. But there is a very expensive (prescription only) foam mousse that works wonders. It is $180 a can. A dermatologist can order this and might supply a sample to try.

Medical Trivia

Every person has a unique tongue print.

Blank page for personal data:

"There comes that mysterious meeting in life when someone acknowledges who we are and what we can become, igniting the circuits of our highest potential" Rusty Berkus

HORMONE DIFFICULTIES

Hormonal difficulties in VCFS, what are hormones?

Many children with VCFS are born with or develop some sort of hormonal deficiency. We all are aware of the word *hormones* but do we really understand how important they are? Hormones are chemicals that are produced by specialized glands in the endocrine system (such as thyroid, parathyroid, thymus, pancreas and gonads—sex organs). These chemicals are secreted into the bloodstream, sending signals to the brain to act on or control other organs and tissues of the body. They regulate all of our important functions such as growth, metabolism, temperature, etc.

The most common hormonal problem in VCFS children is *hypoparathyroidism*. This causes hypocalcaemia which is (hypo) low calcium levels. Calcium is very important to the human body. It affects more than just our bones and teeth. It is the most abundant mineral in the body and is responsible for relaying the messages within and between the cells. Calcium effects nerve conduction, muscle contraction, energy metabolism, cell growth, and many other hormonal regulations.

My child had hypocalcaemia at birth. Does that mean he has VCFS?

Many babies have hypocalcaemia at birth (low calcium levels) due to problems with the parathyroid gland (4 tiny structures that sit on the backside of the thyroid gland). This may correct itself in the first few weeks of life. Annual calcium checks are recommended by some doctors because the problem has been known to recur. Some children with hypocalcaemia have also been known to suffer from seizures. Any seizure should be followed up with a visit to a doctor and a calcium check.

Just because you were told your child has hypocalcaemia, it does not mean, in itself, that your child has VCFS. Hypocalcaemia can be caused by many other factors; like a diabetic mother, long labor, etc. There would need to be other symptoms from the Anomaly fact sheet of symptoms list *(See page 5.)* to warrant a genetics test.

My child is small in stature, what about growth hormones?

Many VCFS children have a growth hormone deficiency (GHD). This hormone is produced by the pituitary gland located in the brain. This gland regulates sections like growth, sexual development, energy metabolism, and water balance. The component that affects most VCFS children is in the growth area, particularly in height. Some kids are smaller in stature when compared to their family as a whole. Children with growth hormone deficiency are usually normal size at birth. They start falling behind after their first year. Because some VCFS children have this deficiency, their growth and development need to be monitored. See an endocrinologist for advice. Parents should keep good growth and development records. (There are pages included in this book for this purpose.)

Often times the children will catch up in size. My son did by puberty.

Parents Speak Out

I have learned recently that too much thyroid hormone (hyperthyroidism) can actually cause symptoms that look very much like bipolar disorder, even with psychosis symptoms; just as too little thyroid hormone (hypothyroidism) can cause depression. You'd have to know your child well and the illnesses/symptoms associated with VCFS to ask your doctor to check certain things like the thyroid. If the doctors are not familiar with VCFS (and many are not) it is wise to educate them.

We were very skeptical about trying hormones, but my six year old daughter, Debbie, has been on growth hormones for six months and already has grown three inches, and gained five pounds. We are very glad that we decided to pursue this avenue. It has really helped her in other areas, including gross motor skills, social issues, and self esteem. She is excited to be able to reach the ground when she is trying to swing; so that she can push herself off; or she can stretch out and reach the monkey bars. Her features have matured as well. These facts combined with her new found height, have greatly helped her with peers, and how they see her as a peer instead of a younger friend. She is very much a leader now where she was completely a follower before. We have had no problems at all. She receives a shot every evening six days a week. It is the smallest needle made, a 31 gauge insulin needle. The hardest thing, has been keeping her in clothes and shoes that fit. She used to only change shoe sizes yearly, and in six months, she has grown two shoe sizes, and the same is true for clothes.

Here is a topic I will address that is rarely talked about. Sometimes, children in puberty years will start to masturbate in inappropriate places, such as the play ground, classroom, shopping, etc. Many times the child doesn't react well to "NO". Sometimes redirection of thought will work and sometimes it will not. When this kind of behavior starts, it is usually a sign of a manic episode. This happens with not only boys but girls also. Medication usually helps this symptom.

Medical Trivia

A fetus acquires fingerprints in the first trimester of development, at the age of 3 months gestation.

Blank page for personal data:

*"The greater the obstacle,
the more glory in overcoming it"* Moliere

IMMUNIZATION

My child is often sick.
Is the immune system impaired?

Many children born with VCFS have problems fighting infections, especially in the early years. Breast feeding is the best line of defense, but many mothers cannot nurse due to palate problems or wish not to do so, for personal reasons.

Our immune system protects us from developing an illness or keeps us from remaining sick, helping us to mend after an infection. Some of our antibodies have memories and can actually prevent us from getting re-infected again with the same "villain". Antibodies provide the long-term protection from specific infections. They act like a military force that is there to guard us against an invasion of enemy "infections". There are assorted antibodies much like different branches of the military; each specializing in one area. Combined, they protect us against the germs, viruses, and bacteria that make us sick.

Some children are born with an absent, partially developed, or defective thymus gland (located near the thyroid at the neck). This small gland is mighty in immune importance. The thymus gland helps control the immune system. Providing a child doesn't suffer from any major infections that may compromise her life, the body will generally find ways to build up an immunity to combat any illness that might arise. Many parents think this implies that their children will suffer life-threatening infections for the rest of their lives. This *MIGHT* occasionally happen, but the majority of children seem to improve in their ability to fight illness and infections as they grow older. VCFS children seem to suffer from frequent common infections; such as upper respiratory, pneumonia, colds, flus, and asthma. They take longer to recover from illnesses. The body has two primary defense mechanisms: the non-specific immune defense, which protects the host from all microorganisms regardless of previous exposure; and the specific immune defense, which is a reaction to a specific antigen that the body has previously experienced.

The major reason VCFS children take longer to recover from illnesses, is that the thymus produces an antibody, the "*killer*" *T-cell* that directly kills unwanted intruders like viruses. If the thymus is small or absent, too few T-cells can be produced to help the body. T-cells and its brothers, the *B-cells*, are the bosses or commanding officers. The B-cells set the villain up for destruction by debilitating it and the T-cells finish it off. (Wow what a team!) T-cells also instruct the rest of the immune system (enlisted men, so to speak) about what to do; so the T-cells and B-cells are vital. T-cells & B-cells are antibodies which contain a memory and will attack a known foreign virus if it tries to invade again. However, if there are not enough T-cells the memory is limited. This explains why children with Chromosome 22 deletion can become infected with the same illnesses over and over again. This is noticeable in diseases that are commonly one time occurrences like chicken pox or measles. This reduction in antibodies explains why a common cold in an average person may turn into a more serious secondary infection like an ear infection, pneumonia, or bronchitis, etc. in a VCFS child.

VCFS children are so susceptible to *catching* viruses and bacterial infections, it is wise to request that friends and family not come around the child if they are sick or feel as if they are getting sick. (Yes, I know I repeated myself!)

Immunizations

In order to avoid complications, it is advisable, in some cases, to request *killed* versions of *live* vaccines such as polio. The clinic may have to order it, so ask ahead of time. We recommend that you consult with an immunologist before giving any vaccines to your child. Your family doctor or pediatrician ought to be able to help in this respect.

In some geographical areas the child can be exempt from being immunized. This can occur when the risk for having the immunization is deemed higher, than the purpose for which the immunization is intended. Given this situation the school requirement for immunization can usually be waived.

2011 Immunizations for Children

This is a routine schedule of childhood immunizations:

Birth	Hep B	Hepatitis B
2 Months	IPV	Polio
	DTaP	Diptheria,Tetnus, Pertussis (Whooping cough)
	Hib type b	Haemophilus Influenza - Bacterial Meningitis
	Hep B	Hepatitis B
	PCV7(new)	Pneumoccoccal Conjugate – Bacterial Meningitis
4 Months	IPV, DTaP, Hib, (Hep B), PCV	
6 Months	IPV (3rd dose given 6-15 months of age)	
	DTaP, or DTP, Hib, PCV, Hep B	
12-15 Months	IPV (if not given previously)	
	DTaP	
	Hib	
	PCV	
	MMR- (Measles, Mumps and Rubella)	
	TB- (Skin test for tuberculosis)	
	Varicella Zoster Vaccine (chicken pox vaccine)	

Is fever always bad?

Actually no, fever helps the body fight infection and is the principle sign of illness.

Temperature is regulated by the hypothalamus which is located in the endocrine system, in the brain. Fever aids response to the infectious processes. Higher temperatures kill many microorganisms. Fever involves *resetting of the hypothalmatic thermostat* to higher levels. When fever breaks, it resets back to *normal*.

Fever is produced through chemical reactions which are triggered by the release of pyrogens from leukocytes (antibodies), bacteria, virus, fungi and other cells in the immune response.

Temperature regulation is achieved through a precise balance of heat production, heat conservation, and heat loss. Body temperature is maintained at about 37°C or 98.6°F. This is a basic measurement of temperature; and not an exact temperature for everyone.

There are disorders that affect temperature regulation. Such as:
- **Hyperthermia**: This is an increase in the amount of heat, when the hypothalamus does not respond.
- **Hypothermia**: This is a decreased amount of heat.
- **Trauma**: This can alter temperature regulation. Trauma can also include surgery.

Infants and the elderly require special consideration for safeguarding body temperature. Infants do not conserve heat well, thus the reason for using a cap on their heads immediately after birth. Heat is mostly lost through the head.

Children develop higher temperatures than adults for relatively minor infections. This is your immune system at work.

Temperature can be measured in several different ways:

- **Oral** (Mouth): Glass, paper, or electronic thermometer; Normal 98.6° F/37° C
- **Auxillary** (under arm pit): Glass or electronic thermometer; Normal 97.6° F/36.3° C
- **Rectal** (anus or "core"): Glass or electronic thermometer; Normal 99.6° F/37.7° C
- **Aural** (in the ear): Electronic thermometer; Normal 99.6° F/37.7° C

Fever is defined as an oral temperature (or equivalent) of 100.5° F/38.5° C or above.

Medical Trivia

Every square inch of the human body has an average of 32 million bacteria on it.

> *"Our life is what our thoughts make it"*
> Marcus Aurelius Antonius

PAIN

What is pain? Are there different kinds?

Yes, there are different kinds of pain. Pain is complex and there are categories into which it falls:

- **Somatogenic pain:** Pain from an unknown physical cause
- **Psychogenic pain:** Pain without a physical cause
- **Acute pain:** A signal given to the person indicating that there is something harmful that makes the body hurt
- **Chronic pain:** A persistent pain of either an unknown cause, or a response to a therapy for an undetermined time

The Central Nervous System (CNS) is involved with pain, response, sensation and the perception of pain.

When there is more than one area that hurts, but you only identify one spot because it is the worst, this is called <u>*perceptual dominance*</u>.

One of the quandaries that doctors have is with <u>*referred pain*</u>. This happens when pain is removed from the point of origin and felt someplace unrelated. A good example of this is organ disease or infection; the pain may be coming from the gallbladder, but it feels like it could be from the heart.

Here are some physical signs that may accompany severe acute pain. These are helpful to know when your child has a limited vocabulary, so that you can watch for the symptoms.

- Increased heart rate
- Increased respiratory or breathing rate
- Increased blood pressure
- Flushing or pale appearance

- Dilated pupils
- Sweating
- Nausea or vomiting
- Elevated blood sugar
- Decreased mobility

Some physical and psychological responses to chronic on-going pain can include:

- Depression
- Difficulty sleeping
- Changes in behavior
- Physical adaptation to the pain such as limping
- The child may seem to become obsessive and preoccupied with the painful area

Pain threshold is the point at which pain is perceived. Many of us have a low degree of pain that we live with consistently and don't notice it much of the time unless the pain increases. This leads to pain intolerance where you just can't stand it any longer and pain begins to affect your daily living and activities. Some medical journals state that babies and young children have a low threshold and tolerance for pain, but I know of many parents who would question this statement regarding their VCFS children. *(See COMT gene on page 253.)*

My child complains of leg pains. Is it just growing pains?

Leg pain has now been accepted as a manifestation of VCFS. (Visit **www.vcfsef.org** for an updated article.) Many children have been found to be suffering from leg pain, either when awake or while sleeping. Often the child will wake up several times at night, seemingly for no apparent reason. They might have been awakened by the thrashing of their legs as they kick to try and relieve the discomfort. Many have stated that the leg pain found in VCFS children may be caused by a calcium deficiency. Some have claimed that giving calcium supplements helped since calcium is coupled with muscle contraction. Some children outgrow this condition.

There appear to be a small number of leg and foot abnormalities that are connected with VCFS patients. One is flat footedness. A podiatrist may want to insert insoles that raise the heel and support the arch. A cushioned insole may be needed or recommended to soften the impact of their walk as well.

Parents Speak Out

I am concerned with the fact my son (5) has the narrowest and flattest feet I have ever seen. They look odd like a cartoon. I try to buy good shoes but they all seem to make blisters on his feet. He doesn't seem to be able to stand wearing them with any arch support.

Medical Trivia

A fingernail or toenail takes about 6 months to grow from base to tip.

Blank page for personal data:

> *"What we do today, right now, will have an accumulated effect on all our tomorrows"*
> Alexandra Stoddard

DENTAL

On general dental issues
Contributed by Patricia Schiffman D.M.D.
Used with permission.

"I am the full time orthodontist on the cleft palate and craniofacial teams at The Children's Hospital of Philadelphia. Here is some of what I know about the dental issues confronting patients with a 22q11.2 deletion. About half of patients will have moderate to severe decay. We know that the enamel can be hypoplastic (poorly formed) and we suspect that the thickness may be reduced. We are not sure why this is the case but I believe it is due (at least in part) to calcium metabolism problems early in life. These are things you can do. Ask the pediatrician to determine your child's fluoride intake and then prescribe an appropriate fluoride vitamin if indicated. Brush your child's teeth as soon as they are erupted. Use a cloth if necessary to wipe them clean. If they have oral aversion issues continue to try to get them to accept brushing and try to brush very well one time while they are asleep. Confine juice drinking to meal times. If children are allowed to sip all day, they will alter the pH of their mouth with each swallow. This causes a constant demineralization/remineralization cycle that is harmful to the enamel. Pediasure® is high in sugar. Do not allow the child to take a bottle with anything other than water to bed. Start pediatric dental visits at age 2. Ask if a fluoride rinse is important. Discuss sealants on any surfaces that are spotted or demineralized. Ask about composite restorations as a preventative measure for questionable areas. Don't despair; problems with the baby teeth do not indicate that there will be problems with the permanent teeth."

Best wishes, Patricia Schiffman, D.M.D.

There are topics like teeth that don't get enough credit in my estimation. Teeth are pretty important, we get one permanent set and they were made to last a lifetime. Children with VCFS can have some peculiar and expensive dental problems. My son had 2 sets of baby teeth and a few extra permanent teeth. I have been told this is not common, but I know that many VCFS children do have comparable dental problems, and many have stated that they had no problems with their primary dentition. A number of people said that the enamel was bad from the beginning. From what parents have told me and from what I personally experienced, the ideal is to have a dentist from the beginning of the baby teeth eruption; one who is willing to read about VCFS children and their potential dental problems. This dentist can save you a lot of financial grief in the long run. In many cases the type of procedures necessary may not be covered by insurance; but it will behoove you to get the cleanings done, the extractions if needed, the sealant applied, etc. You might be out-of-pocket a little now, but when the child gets to the age of braces, crowns, etc., it may save you from needing them; which would include saving their smiles!

Parents Speak Out

Some children have multiple sets of teeth or extra teeth (My son had 2 sets of baby teeth and some extra permanent ones). A number of VCFS children have permanent teeth missing. Yes, an orthodontist is more than likely in your future.

Many VCFS children have sensitive gums and teeth. There is a toothpaste called Prevident®, which is a prescription, high fluoride toothpaste. It comes in gel and also in several different flavors. There are tooth brushes available with an ergonomic grip. There are right and left hand models that have a very large oval head with soft bristles. It sounds counter productive, but they work well because no matter where the child brushes, they are making a lot of contact with teeth and gums.

Medical Trivia

It takes 17 muscles to smile, but 43 to frown

> *"Where there is great love there are always miracles."*
> Willa Cather

PALATE AND FEEDING

What palate abnormalities are normally seen in VCFS?

In VCFS the palate is one of the most common areas where an anomaly (problem) is noted, the other is cardiac. It can be associated with a cleft lip and/or palate, but is not always. The problems are variable. Cleft palates (an abnormal separation of two parts) can sometimes be seen or felt. However, oftentimes an examination of the mouth and palate reveals no evidence of an abnormality, even when the muscles of the soft palate are divided. So, tests, called velopharyngealoscopy (velo--meaning palate, pharyngeal-meaning pharynx/throat and scope for looking) and nasopharyngealoscopy (naso--meaning nose, pharyngeal--meaning throat) may be in order. This uses a fiber-optic telescope, which is introduced into the nose, to gaze into the nasal cavity and down into the palate. Then it is video recorded. Request to observe this procedure. Personal observation is a better learning tool then reading about it. I was fortunate to have a doctor (a professor at ASU) who was patient and walked me though the procedure. After that experience, I felt confident we were on the right track.

Video fluoroscopy is a moving x-ray of the child's throat when she is saying words and making sounds. This is quick and painless, yet some children may be distressed by it. The nasoendoscopy involves a small amount of local anesthetic being placed in the nostrils with a cotton bud or long Q-tip®. Then, a thin wire with a camera on the end (the fiber-optic telescope) is positioned in the nasal cavity toward the back of the nose and throat to look at the movement of the palate during speech and sound exercises. These tests give the surgeon and/or speech language pathologist a very clear view of how well the palate and the muscles in the pharyngeal walls move, if at all. It is not painful but might be uncomfortable for the child. This test is video taped so everyone on the medical team, who needs the information, can view it. If surgery is performed, the video is used as a reference for comparison.

There is another crucial benefit to nasoendoscopy. During the nasoendoscopy the surgeon or therapist might occasionally see a pulsating in the cavity behind the nose. This can be an abnormal misplaced carotid artery. To make sure that the carotid arteries (which go to the brain) do not pass too close or through the area of any proposed surgery; an MRI scan of the neck and throat are customarily done to see where the artery is positioned.

In my son, T.G.'s case, he did not exhibit any visible signs of a cleft palate but he had signs from birth of the possibility of palate problems (nose regurgitation). Although never established until he was 18 years old, he had what we now know as a condition called *Velo-Pharyngeal Incompetence* (VPI). This is a situation where the soft palate is short or too weak to close off the nasal cavity from the oral cavity during speech. He had the classic hyper-nasal voice (sounded like he always had a cold) and a higher pitched voice. Articulating clear words was very hard for him and many people could not understand what he was saying. Often they would talk to me instead of him, asking questions through me, with T.G. still standing there. This only caused T.G. to lose confidence, become embarrassed and not be eager to try to socialize. He didn't want to have to talk, even after he was able to do so. This is a classic example how children's social interactions can be altered or inhibited. This abnormality causes a number of children to form different speech habits/patterns to compensate for the sounds of words they are unable to pronounce with clarity due to this misdirected/escaping air.

When there is a cleft problem, a *Cleft Lip and Palate Team* is usually consulted. The advantage to this TEAM approach is that the team works well together. The team usually consists of a constructive plastic surgeon, ENT, pediatrician, geneticist, audiologist, speech and language pathologist, psychologist, and social worker among others. When these different professionals work together they can coordinate all the medical procedures required, resources needed for referrals, support groups, therapies, etc. To see if there is a *Cleft Lip and Palate Team* in your area you may want to contact the *Cleft Palate Foundation* at: **1-800-24-CLEFT. www.cleftline.org**

If certain symptoms exist, such as milk/food coming out the nose (repeatedly) or choking on feedings, consult your doctor right away. Some babies with VCFS have difficulty gaining weight (failure to thrive) because it is difficult for them to consume enough food. Try smaller feedings at frequent intervals throughout the day. Changing the position you feed usually helps, for instance, try sitting the baby more on an upright slant. There are also alternative food choices and methods if the baby can't breastfeed. (Many nursing devices are available now).

Although overt (full) cleft palates are sometimes seen in VCFS, submucous clefts (means behind a mucus membrane) are much more common. VPI (Velo-Pharyngeal-Incompetency) is also extremely common in VCFS as we mentioned before. To recap on VPI, it occurs when the congenital defect is in the structure of the velo-pharyngeal sphincter (located behind the uvula, the little punching bag like structure at the back of the throat). This means that the closure of the oral cavity beneath the nasal passages does not completely close. They do not make a good seal.

Surgery to correct VPI is frequently used. There are different surgical techniques such as Z-plasty and pharyngeal flap, that are used by different surgeons in different parts of the country. A sphincter-plasty is not the same as flap surgery. In the sphincter procedure, muscles on the sides of the pharynx (throat) are repositioned to make it easier for the velum (soft palate) to close off the connection between the throat and the nose. In the flap procedure, a *flap* of tissue is separated from the back of the throat and attached to the velum, making a constant bridge, against which the sides of the pharynx (throat) can close off. Healing takes a while and then there is a need for therapy to retrain the muscles.

Ask your doctor about the options if surgery is recommended. Surgery is *NOT* always necessary to correct these problems. Working with a speech therapist should be done first. It is note worthy to tell you that surgery *DOES* *NOT* improve language deficiency, only sound/word clarity.

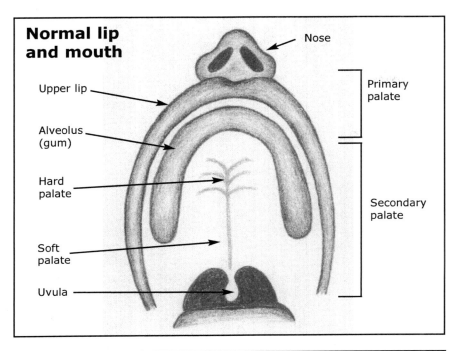

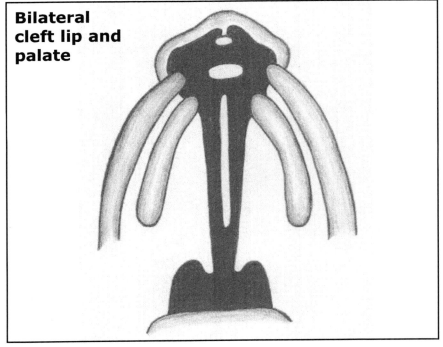

CHAPTER 3 SYSTEMS AFFECTED

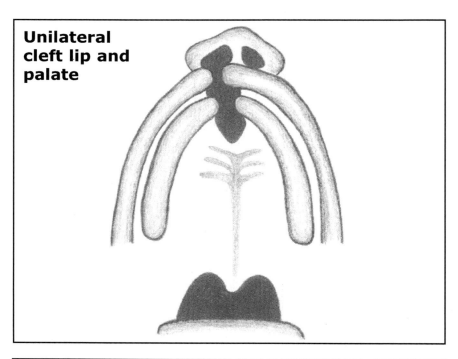

Unilateral cleft lip and palate

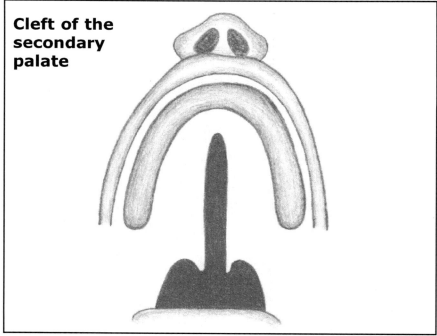

Cleft of the secondary palate

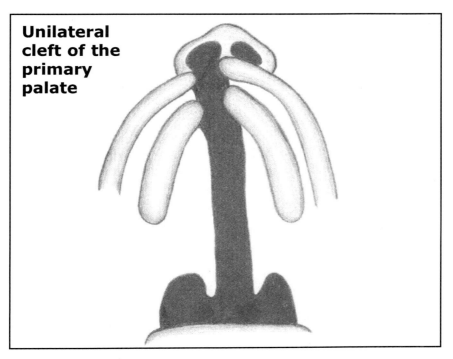

Unilateral cleft of the primary palate

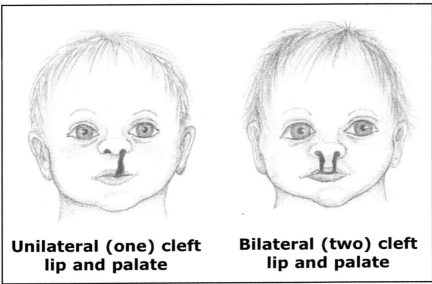

Unilateral (one) cleft lip and palate

Bilateral (two) cleft lip and palate

Before we talk about common feeding problems in infants, let's get familiar with the digestive system, OK?

This system is also known as the gastrointestinal tract. It starts at the mouth and ends at the anus. Organs along the journey work together to prepare the food for absorption into the blood stream so it can be used by the body's cells.

There are three food processes that go on in the body:
- **Digestion**
- **Absorption**
- **Metabolism**

Digestion and absorption are prepared by the system itself but metabolism is done by the cells. The organs involved are: mouth, throat, esophagus, stomach, small intestines, large intestines, rectum and anus. Some other accessory body parts that help in this process are: teeth, tongue, salivary glands, liver, gallbladder, pancreas and appendix. It's quite an event.

First the food is chewed with the teeth and saliva (if there are any teeth). Then it is swallowed and received by the esophagus in the throat, which propels it to the stomach. From the stomach, it is secreted into the three parts of the small intestine:
- **Duodenum**
- **Jejunum**
- **Ileum**

The third part connects to the beginning of the large intestine at the ileocecal sphincter.

The large intestine is also divided:
- **Cecum**
- **Colon**
- **Rectum**

The rectum is attached from the sigmoid or lower part of the large intestine to the anal canal. During _peristalsis_ (a snake like movement that uses this contracting muscle to force food through the digestive tract) waste is moved so that the anus can eliminate it to the outside world. Organs such as the liver and pancreas aid in digestion. The liver aids by secreting enzymes like lipase to help in digesting fat; and the pancreas secretes digestive enzymes, insulin and glucogen (sugar) just to name a few.

The organs that are viewed in an _upper G.I. test_ include the mouth, throat, stomach and duodenum. Those observed in an _lower G.I. test_ include the entire small and large intestines.

My child is constipated a lot. Is that associated with VCFS?

A large number of children with VCFS are found to suffer from constipation, from mild to severe. This could be due to any of the G.I. problems we mentioned earlier or (hypotonia) weak muscles. This produces poor peristalsis. Since the system is from mouth to anus, it can start at the palate and go all the way down to the large intestine and anus. We don't want any colon impaction so encourage your child to go to the bathroom regularly and drink plenty of fluids. If there is ever a black tarry stool, stomach cramps or vomiting, call your doctor.

Some young adults have been recognized to have quite a tough time with constipation. Exhibiting stomach aches, gassy cramps and the whole gamut of distress. The initial problem could be weak muscles but that might possibly be enhanced by anxiety. Since calcium is known to help muscles contract, ask your doctor if it might help to add a calcium supplement. The calcium will not bring relief in the constipation department but may help with the peristalsis and stomach cramps.

If your child is taking several medications they may have a side effect of constipation. So constipation intervention might include stool softeners, adding fiber to the diet or a much more powerful helper (no, not Roto Rooter® the plumber), a doctor may prescribe or suggest a bowel stimulant like Senokot®.

On the other end of the elimination spectrum, if your child has frequent stools it may be a painful inflammation from bacteria, or severe dehydration (or could lead to it). _You must see a doctor ASAP,_ especially if the children are infants or young-ones. VCFS children seem to dehydrate very rapidly. I seem to say this often, but, _"If I only knew then, what I know now,"_ my son would not have gotten so sick at about age 8. I wish I would have been better informed, but there was no one to tell me and I hadn't gone to nursing school yet. Here is a brief description of what happened, so you don't repeat my blunder. My son had gotten the flu just at the tail end of the rest of the family's bout, but instead of bouncing back as we were able to do, he deteriorated quickly, so I took him to the doctor. The doctor was very upset with me for not bringing him in sooner. He acted appalled and put my son on an IV saline drip in his office before transporting him to the hospital. My son was very dehydrated, had bowel movements that were the consistency of lime Jell-O® and was found to have pneumonia. I wondered how that could have happened so quickly, and I felt like the worst mother. I have since learned that VCFS kids can get sick suddenly, often, and sometimes have been feeling ill for longer than we realized. My son, at times, seems to have a high tolerance of pain in that area.

What are the most common feeding problems in infants?

As a nurse I have discovered that feeding infants with clefts can be tricky. They may have little or no sucking power, can choke easily, and tend to have a lot of nasal congestion because of this. I have had to use a nasal bulb and suck or siphon the mucus out for an infant to breath. These babies need to be fed in an upright position (30 degrees or more). If the special bottles are not available or don't work, a paper cup can be used in an emergency by pushing together the sides to make a funnel. (Be careful to not drown the child with the full uneven flow.) In some cases use large syringes which are safer. Perhaps add a little Thick-It® to the liquid if it is flowing too quickly. 30 cc is one ounce, 60 cc is 2 ounces, etc. This is how we are able to keep track of a baby's intake for nutritional evaluation.

You may want to ask your doctor about a good quality liquid vitamin supplement to add to the feeding. I do not sponsor any specific product, but I personally have been pleased with a product that has an aloe vera base; said to be excellent for the stomach. Most products of this nature have a respectable vitamin and mineral balanced formula. Ask your doctor to refer you to a dietician/feeding specialist to address these issues. I mention this because some VCFS children suffer from a reflux, which can be due to the inability of the intestinal tract to move or propel foods or solids throughout its normal course of digestion. There can be so many other feeding difficulties that it is hard to identify them all. Anything that prevents the baby from eating normally may be considered a problem. This ought to be evaluated by a doctor. I personally know of a couple of babies who ended up in the hospital about every month or so. The pediatrician believed it happened because, while eating, a portion of the food went into the lung and caused infections and pneumonia. This situation might require surgery to correct it. In the meantime, it may justify insertion of a _feeding tube_. There are many possibile eating problems and there might be a need for a gastrointestinal evaluation and testing.

What are the most common eating problems in the early years?

Many toddlers prefer a spoon and cup feeding over bottle feeding. This is thought to be because there is less chance of reflux and congestion in the nasal cavity. Cup feeding, for some, can be challenging as it may cause them to gulp because of the inconsistent pour of the liquid. Training cups (a fluid restrictive cup) work well for the majority.

Parents of some of the VCFS children may need to be conscious of the posture of the child while they eat. Visualize how you would feel and how difficult it would be for you to eat if you were slouched down in a high chair or booster; bent over with your arms too high in a high chair; or leaned backward too far to be comfortable. Positions like these may also cause choking.

Some children cannot use their utensils well or maneuver them as needed. The reason for this is a matter of grip, perhaps from weak muscles. There are special utensils that can be used, with styles that are common to those used by the elderly in nursing homes.

Many children with VCFS find it difficult to move on to different forms of food, preferring soft-textured foods. Those who do advance to more mature forms of food may have a tendency to gag more than the average toddler. It might be that it takes them longer to get used to a different feel in the mouth and/or different tastes. This is demonstrated when they refuse or avoid eating things that have altered textures such as lumps, or are sticky, crunchy or chewy. They seem to select one certain category of food over another. (This is excluding differences like preferring ice cream over spinach, of course.) For some reason *junk foods* seem to be a favorite to *ALL* kids, including those with VCFS. Maybe we could try to trick them by saving fast food wrappers. Hmmm . . . Think it would work?

A number of children seem to remain immature in their eating habits, having to be encouraged to try different foods. Some have a problem with not wanting to eat when foods touch each other. If feeding becomes a real problem, a child may eventually be labeled *failure to thrive*, and then a specialist will need to be called in for help. Below are a few suggestions to help you encourage your child to eat.

> **DISCLAIMER:** All of the information below is suggested only and is _NOT_ meant for _EVERY_ child. Many will have food allergies or cannot swallow certain things. _PLEASE USE CAUTION AND CONSULT YOUR DOCTOR BEFORE YOU ATTEMPT THESE SUGGESTIONS._

Finger foods with a messy flare for that Kodak® moment?

While not usually something we would encourage, playing with food may be the only way to get your child interested in different foods. Pudding, yogurt, cottage cheese, and peanut butter, etc., are messy foods, but they are also fun. Playing with these foods is a pleasant oral experience. By licking their fingers, it not only causes the child to get nourishment but it gives *hand and tongue* dexterity. Pudding doubles as fingerpaint and mashed potatoes are the perfect medium for sculptures! Just have some paper towels ready!

Fruits and Veggies: Different colors, smells, textures, tastes and shapes come with this mix. Slice the items in safe-sized pieces (not too small, they may choke). Licking, smelling, rubbing of the gums in teething, squishing the soft ones through the fingers is all part of the experience. Use a variety of different fruits and vegetables to expand the experience; both fresh and cooked.

Handmade foods: Anything simple to prepare, like cookies or sandwich triangles, is nice. Allow your child to help make these things with you. The experience will be good for them and they will want to taste what they made. Just for fun, cut sandwiches or toast in some unusual ways. Use a simple cookie cutter to cut a peanut butter and jelly sandwich to look like a flower or a bunny. Cut two slices of toast into triangles and arrange them on a plate to look like a pinwheel. Slice a sandwich and make it look like an airplane by rearranging the pieces on a plate.

A quick and easy recipe for them to start with might be Jell-O® shaped in molds, or homemade Popsicles®.

Hand-held foods: Foods that can be held come in a variety of assorted choices: bread sticks, crackers, Popsicles®, raw veggie sticks, cheese sticks, etc.

The most essential thing is that mealtime should be a positive and pleasurable experience, and not associated with a feeling of dread. *NEVER* force or push the child to eat more than they comfortably can, even if that means not eating all that is on the plate. It is often challenging to understand why they don't want to eat. It may be that they are not feeling well or that they have had enough to satisfy them. Remember, even as an adult, your stomach is only about the size of your fist. There are going to be periods when children get those growth spurts and seem to eat wonderfully and then there are the dormant cycles when their eating lessens. When they are in the <u>reduced food cycle</u> they may cause a parent to think that they are not eating enough or are sick. Don't worry, you will get to know your child and what is an adequate amount for them. Wait until they become teenagers, then you will change your tune! Just make sure that at those dormant times, they eat healthier. Instead of quantity aim for quality and don't forget to add vitamins and mineral supplements that are recommended by your doctor.

If your child is tube fed he will not be able to experience the pleasures of food if we do not stimulate that desire. These children will not feel hungry while being tube fed, and will not be very interested in table food or familiar with the different tastes of food. So the desire to crave food may not be there yet. You can't just turn off the feeder and let them go hungry, so that they are forced to crave food, so what do you do? <u>Work with a good pediatric dietitian</u>; they may have you slowly cut back on the tube feedings while gradually introducing small amounts of regular food to your child. This should stimulate the appetite and inborn desire for food with its many tastes and smells. Always consult a medical professional when you have a child with feeding problems.

Drinks: Babies and small children are customarily more interested in sweeter drinks (Hmmm . . . so am I.); so it is a challenge to get them to try other liquids, and steer them away from all the sugary beverages. You can figure out a few things to help. Add a drop of flavoring, such as peppermint, to their water; or give them a little taste of what you are drinking, perhaps something like a little gingerale in a cup of fruit juice for a zing!

There is also a variety of flavorful decaffeinated herbal teas which can sooth and calm a child, and they are healthy too. Consult a competent licensed naturopath for suggestions. There are many options if your child is one who has a hard time sleeping. You might want to ask a naturopath about an acidophilus drink, which has the beneficial bacteria or culture found in yogurt. It can be purchased in health food stores. If your child is one that has been on a *LOT* of antibiotics and may have had thrush because of it, it is worth trying. My son had thrush twice and it was not fun. (Another name for thrush is candidiasis, caused by the antibiotics killing off not only the bad bacteria but the beneficial ones as well.) Thrush is frequently seen first in the mouth where it looks like white cottage cheese in spots on the inside of the cheeks and throat. It can also be intestinal and might appear like a severe diaper rash, that looks like scalding.

Parents Speak Out

A happy mother writes: My daughter, who is now soon to be six, was tube-fed until she was four. She also has a full nissen fundo, and did still manage to reflux around it. Actually, she refluxed enough to aspirate, and be hospitalized twice with bronchiolitis. Then we saw a wonderful feeding specialist. We had had to suction our daughter so very often, and treat her nasal drainage with allergy medication, that it was awful. The new doctor informed us that our daughter needed a J-tube rather than the G-tube. Within two weeks, we were able to put the suction machine away for good, and our daughter's runny nose left; she began to accept baby food, and start her feeding therapy. It is a huge pain to have a child hooked up to a pump with a J-tube, but they do get most of the day "free". It was the best thing we ever did for her. She learned how to walk even while hooked to the pump! She was eight months when we got the J-tube. The J-tube goes into the intestines, so it bypasses the stomach, and the child can't reflux. The pump must be used since the intestines cannot hold the larger amounts that the stomach can. Our daughter truly thrived after we did the tube change. Our local doctors didn't believe it would work. They thought she could not be refluxing around an intact fundo, but she was; and it did work, to our elation.

My daughter was diagnosed with VCFS and cleft palate at 8. She had the Furlow Z plasty. Her surgeon is a lovely man and told us exactly what was going on and what our options were. The operation has greatly improved her speech. She is much more easily understood now; although she still seems to talk through her nose (hypernasal). The doctor stated that he would have done the pharyngeal flap but said it was more of a major operation. He will do it; however, if this procedure does not achieve the desired affect. Barbara was in the hospital for 4 days, and back on solids after 2 weeks.

Medical Trivia

The average person releases nearly a pint of intestinal gas by flatulence every day. Most is due to swallowing air. The rest is from fermentation of undigested food.

"Happiness is the sense that one matters.
Sarah Trimmer

CHAPTER 4: DEVELOPMENT

DEVELOPMENTAL ISSUES

My child seems to be behind developmentally and emotionally for his age group. Is this typical for VCFS children?

Every child is unique but it is particularly true of those born with a 22q11.2 deletion. We are steered by our individual personalities and temperament; some are inherited traits while others are born out of our environment, our interactions with family and those with whom we come in contact. This is especially true in the early years when most of the basic learned patterns are set and children are the most impressionable. This will, no doubt, carry over into the education system. Therefore, early diligent intervention is a must if you know there is a problem.

In the early years, the main area of difficulty and concern appears to be with speech and understanding speech. This, in most cases, is the reason for emotional immaturity.

Children with VCFS usually (but not always) suffer from some form of developmental and learning delays. These can be with crawling, walking or talking. This might be, in part, due to spending so much time in hospitals and being isolated at home; but, as a rule, it is due to their genetic condition. Many children have poor motor skills, although some do extremely well with them in things like sports (gross motor skills). The children that show difficulty in this area may lack muscle tone (hypotonia- muscle weakness) and have reduced coordination. Most children are walking by about 18 months of age. Nevertheless, not all VCFS children are delayed to the same degree in *ANYTHING*, if delayed at all. This makes it difficult for parents who want to know specifics to watch for in their own child. There are some adults/parents who never knew they had the 22q11 deletion until they were tested for it; usually after the birth of their own VCFS child. Their symptoms were so minor that they were not noticeable, or, depending on the age of the parents, there may not have been a test for it when they were young.

Achievement in education is a common area of parental concern. We seem to base our child's success or independence on their ability to learn. It is important that this issue be taken into consideration as early as possible in your child's life. An education _strategy_ should be put in place, in the event it becomes necessary; even if it is not yet apparent that the plan will be needed. Parental participation in your child's education is valuable so that both your child and teachers have your support. This, in turn, supports *YOU* and your goals for your child. Added help and accommodations may be needed in school, so that everyone concerned will be on the same page. This will be essential for later years as well, so that your child will have a constructive learning experience. Computer learning programs are said to be an enjoyable teaching tool for home assistance, thus improving the school work experience.

At this point you may be wondering why I don't have any recommended software listed. This is because each of the special education teachers I consulted said about the same thing: that parents need to ask the present teacher with which subjects their child seems to be struggling. Then contact next year's teacher and ask about the curriculum they will be teaching. Combine this information and work on those subjects regularly throughout the summer or on days off. This will keep your child's mind active and will give her an edge for the coming year. If you ask, many teachers have extra work sheets that they will give you for exercises. Special education teachers also mention that it is a good idea for your child to read either to you, a sibling or into a recorder at least once each day. This will improve reading and articulation.

Below are some VCFS learning issues that have been noted, but which will not manifest themselves in all VCFS children.

- When VCFS children are young, learning challenges are not as obvious and they measure up well with their peers; but, as they get older, the gap widens. The VCFS child may still take pleasure in more elementary things whereas their peers have outgrown them. Their peers change in their likes and dislikes while the VCFS child seems to stand still. For this reason some VCFS children seem to prefer the company of younger playmates. The VCFS child may not get involved in fads of the day (such as designer clothes) as seriously as their peers; or they might get obsessed with a fad to the detriment of their parents' wallets and patience.

- Conceptual thinking seems to be the most difficult when it comes to concepts not easily defined like "honesty," "beauty," "appreciation," etc. These are all cognitive issues that usually stay a problem throughout the child's life if a child is deficit in this area. By the time children are in about the third or fourth grade, schools are teaching in a deeper, philosophical, and abstract way. VCFS children appear to learn better by experience and solid facts that don't change, using their senses to process information.

- VCFS children are literal thinkers. So, using terms and idioms (figures of speech) like, *"I'm going to skin you alive if you . . . "*; or *"Cat got your tongue?"*; or *"You're driving me up the wall!"* will not be readily comprehended. You will need to explain the meaning of these expressions at the time you say them. If you don't, your child may be confused, and perhaps frightened (depending on what you have said) because they don't understand the intended meaning.

- VCFS children are wonderful imitators, especially if they have a problem with impulsiveness. For this reason it is wise to monitor their associations and friends. They can imitate so well that their *REAL* condition may perhaps be masked as they copy others. They may not really be *learning* when they are imitating because they just go through the motions. They don't understand *WHY* they are doing a certain thing, or *WHY* they got the end result, whether positive or negative.

- Other areas where VCFS children have some difficulty are in comprehensive speech and expressive language skills (once again cognitive problems). Usually they struggle in English classes. Sentence structure is difficult.

- They may have memory problems and these can either be from the disorder itself *or* from the medications they take, or a combination of both. I have to admit it took me a long while to understand that my son wasn't just being disobedient; but actually had forgotten many things, and very quickly at times. Unfortunately, until I learned that fact, my poor boy was disciplined for not doing what he was told to do. I had to learn to discern whether he was telling me the truth and really had forgotten or he was just being lazy.

- Some suffer from *auditory processing* problems. This is the inability to hear a sound and know *WHAT* it is or from *WHERE* it comes. Think about hearing a buzzing sound and not knowing what kind of buzz it is? Was it made by a bee? A buzz saw? A motor? Auditory difficulties can also limit the amount of verbal information that is understood at one time. For example: children with auditory processing difficulties may be frustrated or uninterested in listening to a story being read because the words of the story may come out too fast to make sense of it; or the words may be understood but in a mixed-up order. The children can't understand what they are hearing, so the story doesn't capture their interest.

- It is very difficult for the VCFS child to remember multi-step directions, instructions or tasks. *One* task, direction or *one* step at a time works best. Simple is better.

- Another area of difficulty for many children is fine motor skills (handwriting, threading a needle, etc.) and/or gross motor skills (jumping, playing hop-scotch, etc.). On the other hand the opposite may be true of your child as it was with mine. Some VCFS children may have exceptional hand to eye coordination. When we took our son to the batting cages he hit 9 out of 10 fast balls, yet he had never actually played baseball much. Recently I bought him a dart board and he came real close to the bull's eye right off the bat several times. As far as I know he had never played darts before. So it is not possible to know where your child's capabilities might be.

• For some VCFS children general math is not a problem, as it is factual and cannot be changed (concrete). These are basics like addition, subtraction, multiplication, and division which may come relatively easily. Since VCFS children can learn by repetition and memorizing, they often do well with things like memorizing multiplication tables. It is when the children move into more complex math with problem solving and *story* *problems* that some of them may need added assistance; particularly when it comes to fractions or figuring out how to tackle new real problems in their lives. For this reason I believe *home economics* is good for both genders. They can use the skills in a practical situation and *SEE* the fractions in a recipe in a hands-on form. Like other areas of VCFS, there will be those who excel in math in its entirety.

• Many VCFS children learn by repetition or memorizing, instead of really discerning. As stated, this can be beneficial for some things, such as memorizing the times tables in multiplication. Where it is a two-edged sword is in *spelling*. A large majority of the VCFS children are proficient spellers, but they have a hard time knowing what the words mean. They need to understand use of words *AND* how to spell them, so they don't digress in that ability.

• Learning to write physically is not usually a problem (unless there is hypotonia weakness); but writing in complete, comprehensive sentences with an organized thought process, is another story. They may find it difficult to get their point across, or explain a story in a way that is understandable to the listener. This is usually because of problems with event sequencing and abstract thinking. They might have a hard time getting organized. Studies indicate that the cognitive area of the brain is also the area that causes problems with event sequencing.

• VCFS children might have difficulty in large groups. (This is especially true if they have a social-anxiety disorder, as my son has. I am happy to say that, with medication, he has overcome this debilitating anxiety.) They find it uncomfortable to talk in front of their class or other audiences. Of course, there are those exceptions who love to "ham it up" and do it quite well.

• Taking notes or researching a school paper is very difficult. VCFS children get confused and overwhelmed by the very thought. If given step-by-step, one-on-one instruction, and an outline, they do better. For the most part these children are visual learners and, therefore, do well with videos. Not only is it visual, but it can be rewound and played again for emphasis. Verbal is the next best way to learn (especially instructions for test papers). Being able to have questions rephrased verbally enables the child to understand what is being asked on the test sheet. I know from experience, that my son would do poorly on a written test; but when he brought the test home and I was able to ask the questions verbally, rephrasing them, he would answer an average of 7 out of 10 questions correctly. In my opinion, the least effective learning is written. This depends on what the lesson is. Not all people learn in this order, but input from families of VCFS children who were surveyed proved this to be accurate. As always, each child is different.

• VCFS children seem to have poor memories on everyday matters (unless they are obsessing about something they want and then dynamite can't blast it out of their minds). For the most part there may be need of cues put in place to help them recall information and to learn with the gift of repetition. Continuous repetition is a must. (I must say _relentless_ _repetition_ because frankly, that's what it is.) It is especially important when learning significant things, such as personal information. This has to be reinforced regularly, until it becomes ingrained. Use conspicuous little personal notes: in the bathroom, on their bedroom door, or on the wall next to an item to which you want to draw their attention, like the vacuum. (I'm dreaming!) This does work until they get so used to the notes being there, that they start to ignoring them. So I try to do a variety of different things. Along with this, it is helpful to teach them memory techniques or word associations. It might also be beneficial for them to carry a small pad and pen with them. (This worked well with my father when he was in the first stage of Alzheimer's.) The pad and pen habit can lead to a better, long term habit of using a day planner when they get older. Let's not forget that the memory problems could also be medication, ADD/ADHD (attention deficient disorder) or a learning disability. So these issues will need to be addressed with your doctor and other professionals who will work with your child.

• What we are about to discuss may borderline on non-verbal learning, but VCFS children often have difficulty with concepts such as maps and understanding that there are continents with countries, in the countries there are states and in the states are cities, etc. and calendars, with days, weeks, months, years, etc. Time, shapes, colors, size differences and money values are foreign concepts. They seem more foreign to them than they do to the average student.

• Because of their inability to understand the value of a dollar, it is wise to monitor their spending. Even in adulthood this can remain a problem. To keep track of money spent; it can be written on a pad or in a daily planner. This is ideal for teaching them *"how much they spend"* and *"where the money is spent"*. To teach the spending concept, I used *"Do you need it, or do you want it."* I started this with my husband and myself when we were first married in 1975 and we were on a very strict budget. We made it a point to always shop together and to keep each other in check. If I were to pick up an item, Gil would ask me, *"Do you need it or do you want it?"* Then I would have to convince him that I needed it and that there was nothing at home I could use in its place to substitute. If I was unable to accomplish that, I would put it back; and I would do the same with him. This worked very well, and I try to do the same with my son. Has it been 100% successful? No, but I can see it has helped and I believe it will get better with continued practice.

Playing games at home that encourage learning these concepts can make a real difference. Use activities such as playing store, where the child is the cashier. This teaches money counting and making change. Read labels with ounces and have them use math to figure out if an item is a good value and thus the best price. This knowledge can be vital for the long term, when they become adults.

Using measuring cups and spoons for fractions and actually using them to cook something helps in two ways. They not only learn math but they learn to cook too. This is what T.G. and I are working on at present, finding things he likes to cook. Then he writes out a recipe and puts it in his recipe box (for when he is out on his own, as well as to give me a break from the kitchen once in a while). When we have 10 main dishes that he likes to cook and has mastered, we are done. Then he can rotate the recipes until he tires of them, and so on.

• VCFS children, for the most part, do not do well with changes in their life. Some examples include: moving to different houses, changing schools, and sleeping in strange bedrooms. This becomes even more obvious if the child has bipolar disorder or some anxiety disturbances. They usually feel comfortable with a familiar structured environment, with people they know and with whom they feel safe. Because of this, places like summer camps may seem out of the question, but many can handle them. The first day or so may not be the most comfortable, but there are those who will adjust, get to know people and begin to relax. Your doctor may even prescribe a short term medication to help with the anxiety. You may find the child will be excited to go back the next year. It just depends on the child and you will know your child and whether or not this is a possibility.

Parents Speak Out

A mother shared this with us: Recently my daughter was being particularly annoying about a lot of things and by mid-afternoon I had had it! I lost my temper and said, *"You need to go to your room I do not want to see your face downstairs anytime soon."* After about 10 minutes she came sneaking down the stairs with a stuffed animal held carefully *in front of her face*! She was completely following what I had said (literally). It cracked me up! But in many situations she doesn't always know what the truth is or what she should do. We talk about things like this all the time and it seems I will tell her the same thing over and over and she never gets it. We say repeatedly, *"Don't leave the yard."* only to look out and find her across the street and a block away—clueless!! Her processing is so slow and scrambled she just doesn't do well in situations where she should automatically know what to do.

I was scared when my husband said that our daughter with VCFS could go to summer camp. So we worked on social skills, and her older sisters were able to go with her to camp. She also did 2 weeks of theatre day camp with her siblings before summer camp. I told the counselors what I thought would be helpful in the social skills area, and that it sometimes took a long time for her. I am happy to say, *"She did great!"* She did well at theatre camp and even made a friend at summer camp! Her older sisters were proud of her and she surprised everyone!

If anyone is interested in sending their child to camp, here is a list of camps that cater to disabled children.
http://www.kidscamps.com or search for "disability camps."

Medical Trivia

Chocolate contains the same chemical, phenylethylamine, that your brain produces when you fall in love.

Blank page for personal data:

*"**B**loom where you are planted."*
Mary Engelbreit

DEVELOPMENTAL STAGES

We know that children differ from adults in their physical, emotional and cognitive responses. The level of maturity dictates their developmental level. Because of this, the children's developmental levels must be assessed.

Birth to kindergarten

Children with VCFS typically learn slower than other children in the general population. These developmental delays can vary from mild to severe. It seems there are no set rules for any area that deals with this deletion. In the early years many of the children will develop within the *normal* range. This is why, as parents, we need to be aware of the stages of development. It is wise to keep a record or journal of your child's progress. Remember that this is a discovery period for your child and he/she should be allowed to finish each stage at their own speed. Allow them to enjoy, practice and master a current stage before coaxing them on to another one.

Now, let us take a look at what is considered *normal* according to the *American Nursing Association* statistics. There are blank pages provided for you after each age group so that you can write down any questions or concerns. Use the headings and "a," "b," and "c" listings to key your questions.

Skills at 1 to 3 months of age

Movement:
    a   When on stomach, raises head and chest
    b   Opens and shuts hands in play
    c   Brings hands to mouth

    d   Swipes at toys (as from a hanging mobile)
    e   Kicks feet or pushes with legs
    f   Rolls from back to side (3-4 months)
    g   Holds head erect (3-4 months)
    h   Twists and squirms freely (3-4 months)

Visual:
    a   Watches faces
    b   Follows objects or people freely with their eyes
    c   Shows interest in environment when awake
    d   Stares at bright objects

Communication:
    a   Smiles when hears a familiar voice
    b   Vocalizes cooing sounds
    c   Turns head toward sounds heard
    d   Makes sounds when happy (coos and gurgles)

Socializing:
    a   Cuddles
    b   Enjoys being sung and talked to
    c   Gazes into faces
    d   Enjoys mobiles

Self-awareness:
    a   Needs to cuddle and have human touch
    b   Lets you know if they are not comfortable
    c   Stays awake longer as they get older
    d   Learns how to calm themselves

Blank page for personal data:

Blank page for personal data:

Skills at 4-7 months

Movement:
- a Raises head and chest while on stomach
- b Rolls (back to stomach) and begins to sit up with some propping
- c Raises hands when being lifted
- d Splashes with hands while being bathed
- e Reaches and grasps things tighter
- f Passes a toy from hand to hand

Play:
- a Plays in the bath tub
- b Explores the toys in more detail (can prefer one over another)
- c Reaches for and can pick up an object, like a block
- d Chews on, bangs and shakes toys (and kittens if you don't watch it!)
- e Bangs with spoon on table

Communication:
- a Responds to own name
- b Laughs out loud
- c Babbles, makes repetitive sounds like *"da-da,"* *"ba-ba"* and tries to imitate sounds
- d Starts to understand speech
- e Responds to *"NO"* (sometimes)
- f Objects verbally when an object they want is taken away

Socializing:
- a Likes to fake a cough and seems to know it is cute
- b Enjoys peek-a-boo
- c Squeals with delight

Self-awareness:
- a Holds bottle
- b Lets you know if they need attention by crying
- c Has some fear of strangers (safety issue)
- d Starts thinking their private parts are pretty interesting

Blank page for personal data:

Skills at 8-12 months of age

Movement:
- a   Sits well without support
- b   Crawls on hands and knees
- c   Creeps and pulls to stand
- d   Walks around furniture
- e   Begins to enjoy throwing things
- f   Walks by themselves but if not, they are *NOT* behind for their age

Play:
- a   Plays with image in the mirror
- b   Picks up larger objects and checks them out more thoroughly
- c   Bangs things together
- d   Puts objects in and out of containers
- e   Rolls a ball
- f   Picks things up with thumb and first finger (such as lint or something else to put in their mouth from the floor)

Communication:
- a   Understands *"bye-bye"* and many other things
- b   Makes more sounds like *"da-da"* and *"ma-ma."* May even attempt to say many words
- c   Responds to requests; such as *"This BIG," "Pat-a-cake,"* and *"ALL gone"*
- d   Gestures to get your attention
- e   Imitates more sounds and words

Socializing:
- a   Enters the shy stage with strangers until usually the end of the first year
- b   Prefers the primary care giver
- c   Repeats sounds and actions to get attention
- d   Likes to play rhyme games

Self-awareness:
- a   Holds bottle or training cup
- b   Takes food from spoon
- c   Prefers one hand over another

d. Feeds self some finger foods like crackers. (Crackers become your friends when waiting in restaurants.)

Blank page for personal data:

Skills at 12-18 months of age

Movements:
    a   Walks well alone or with a little help for balance
    b   Lowers self from a standing to a sitting position
    c   Holds a cup and drinks from it
    d   Enjoys pull toys
    e   Begins to run (in some cases)
    f   Climbs stairs by crawling or with help

Play:
    a   Blows bubbles (if no palate problem like VPI)
    b   Builds a tower with 2 or more blocks
    c   Scribbles (paper, walls, floor, coffee table . . .)
    d   Enjoys playing with different shapes (like the Tupperware® ball)
    e   Begins to play make-believe (with pretend foods and drink, for example)

Communication:
    a   Says a few words, usually 5 or more single words
    b   Knows and points to parts of the body (like tummy)
    c   Knows the meaning of many words and follows directions, sometimes 2 parts (ex: *Pick up your toys and put them in the toy box.*)
    d   Stops when spoken to
    e   Imitates more words

Socializing:
    a   Wants to imitate actions like combing their hair
    b   Likes to offer a toy and then snatch it back
    c   Loves to ham it up with an audience
    d   Laughs at things they find funny (ex: daddy tripping)

Self-awareness:
    a   Begins to be conscious of their dirty britches
    b   Uses a spoon but is still quite messy with it
    c   Shows some cooperation by helping with dressing themselves
    d   Wants to switch from bottle to cup

Blank page for personal data:

Skills at 18-24 months of age

Movement:
- a   Kicks a ball
- b   Walks and climbs (ex: chairs and stairs-always on the go!)
- c   Begins to jump

Play:
- a   Starts to learn to color, wanting to imitate strokes
- b   Finds scribbling more enjoyable
- c   Enjoys make-believe more
- d   Enjoys playing in the drawers and cabinets in the kitchen
- e   Enjoys building higher towers with the blocks
- f   Likes to fill cups up with things, like blocks, rocks, etc.
- g   Enjoys looking at books and turning pages

Communication:
- a   Uses many single words now
- b   Forms short 2-3 word sentences
- c   Points to and knows names of things (ex: ears, nose, eyes, hair, etc.)
- d   Follows 2 step directions
- e   Learns new words all the time (So be careful what you say!)

Socializing:
- a   Increases in independence (ex: brushing teeth and potty training)
- b   Starts to show defiant behavior (ex: saying *"NO"* to everything you ask)
- c   Develops a sense of humor
- d   Starts to show more aggressive behavior (ex: hitting, biting or pushing a playmate)
- e   Becomes a whiner (maybe)

Self-awareness:
- a   Starts to show resistance to naps and bed time
- b   Eats more confidently with a spoon
- c   Drinks almost exclusively from a cup
- d   Weans self from bottle
- e   Pulls up elastic waist pants and tries button-up clothes
- f   Tries to take off own clothes

Blank page for personal data:

Blank page for personal data:

Skills 2 & 3 years of age

Movement:
- a Climbs well
- b Feeds himself totally
- c Walks up and down stairs
- d Kicks a ball
- e Jumps with both feet
- f Picks up things (like bugs) with thumb and index finger
- g Enjoys dancing

Cognitive (thinking) skills:
- a Understands much of what is said to him and is interested in the world around him
- b Draws shapes, like circles and lines (up and down)
- c Matches pictures to shapes
- d Knows where objects belong and can help put them there (ex: laundry articles)
- e Obeys simple directions
- f Starts to develop fundamental habits (without much thought and conflict in doing them)
- g Understands the concept of *"one"*
- h Starts to learn more from educational programs on television

Communication:
- a Uses 4-5 word sentences
- b States his or her name when asked
- c Speaks understandably most of the time
- d Picks up *"naughty"* words easily
- e Starts to boss the parent/care giver
- f Outwardly shows fear (ex: going to a doctors office)
- g Relates some experiences (ex: how they fell and got hurt)
- h Enjoys books more and appreciates music

Socializing:
- a Plays alone well, inside and outdoors
- b Understands the concept of taking turns
- c Develops friendships
- d Imitates behavior of others (for good or bad)

e   Needs a routine to be comfortable and might object to alterations to the routine
   f   Does things on purpose to please another

Self-awareness:
   a   Goes to bed cheerfully most of the time without much protest
   b   Completes toilet training (maybe)
   c   Pulls up his pants
   d   Brushes teeth with help from an adult
   e   Moves from crib to a youth/toddler bed

Blank page for personal data:

Blank page for personal data:

Skills at 3-5 years of age

Movement:
- a   Throws a ball overhead and catches a ball
- b   Rides a tricycle, then a bike with training wheels
- c   Hops on one foot
- d   Enjoys playground equipment
- e   Does somersaults
- f   Handles small tools like a hammer with more coordination
- g   Unties shoe laces and takes off shoes

Cognitive (thinking) skills:
- a   Draws a picture to depict a person
- b   Uses scissors
- c   Knows many color names
- d   Plays simple board games
- e   Starts to memorize ABC's
- f   Begins to count
- g   Copies shapes such as a circle, square and triangle

Communication:
- a   Tells stories of recent experiences, what happened
- b   Uses up to 6 word sentences
- c   Understands some concepts (big and little)
- d   Asks "WHY"
- e   Defines familiar words

Socializing:
- a   Acts mischievous
- b   Helps mother
- c   Has a more vivid imagination
- d   Stops aggressive behavior (usually)
- e   Starts showing preferences
- f   Considers some friends *"special" or "best friends"*
- g   Cooperates more with playmates, takes turns willingly
- h   Starts to want to know *"where babies come from"*
- i   Uses more sophisticated toys

Self-awareness:
- a   Washes and dries hands and face
- b   Uses a spoon and a fork now

- c   Puts toys away
- d   Buttons and unbuttons clothes with big enough holes
- c   Tries to distinguish *"left from right"* (ex: shoes)
- d   Washes self in tub
- e   Begins to learn manners

Blank page for personal data:

Blank page for personal data:

> If you have any concerns, remember to ask your doctor or call your local school district if you have questions about your child's development! You can also call a *Children's Information Center* in your state.

You can enhance your child's learning by providing age appropriate toys.

What they need: What they use:

Birth to 1 year
looking, listening, mouthing batting, kicking, touching, and many sensory experiences

crib mobiles, soft chimes, soft animals with noise makers, rattles, textured balls, measuring spoons, see-through containers with colored objects inside, hard cardboard books

1-2 year olds
climbing, sliding, balancing, crawling, swinging, rocking, rowing, teetering, pushing, pulling, rolling, riding, springing, walking, running, jumping, taking part in activities, stacking and grasping toys

steps, ladders, slides, cases for packing things, tunnels, swings, rocking horse, wagons, riding vehicles, kiddie cars, tractors, trucks, busy activity boxes, telephones scribbling, sorting boxes, stacking blocks and cones, books, spoons, cups, opening mailboxes, water play items

musical toys, *"Mother Goose"* rhymes

toy piano, bells, toys with sounds, hearing rhymes, records, tapes, CDs

identification of environment

name some toys and items in the household; members of the family; identify pictures ex: cat, dog, train

What they need:	What they use:
1-2 year olds, cont'd.	
role playing, begin language	puppets, plastic hammers, cars, riding self propelled vehicles, animals
2-3 years old	
household items and toys	lower cupboards with pots, pans, plastic bowls and utensils
variety of edible foods	sample a variety of foods with less sweetener
familiarize with odors	flowers, fruit, soaps, etc.
musical toys	bells, squeeze toys with sounds, telephones, speak-and-spell toys records, cassette tapes, CDs, music boxes
toys that move	large toys to ride and push, with objects to handle riding toys, trucks, push and pull toys, opportunity to run
cuddly toys, cardboard books	teddy bears, variety of textures of cloth or stuffed animals, and test temperatures
3-4 years old	
climbing, swinging, sliding, balancing, crawling, rocking, running, rowing, teetering, pushing and pulling, carrying, rolling, riding, springing, walking, jumping, balancing on one foot	dollbuggies, playground equipment, steps, wagons, tricycles, toy automobiles, scooters, balls, large boxes, tunnels, sand, water play

What they need:	What they use:
3-4 years old, cont'd. matching, sorting, naming, stacking, using form boards, challenging building blocks, counting things	crayons, paint, puzzles, boxes, form boards, telephones, miniature animals trucks, hammer toys, blocks, stacking toys, color cones, pegboards, clay
rhymes, non-sense syllables, listening activities, stories, activities, stories, understand direction, music	stacking cones, nursery rhymes, singing games, finger games, books, records, string beads, matching cards, puzzles
dressing, housekeeping, roll-playing, relating to others, group play	dolls, playhouse, hats, dress-up costumes, small jobs in the home, play acting
4-5 years old hopping, galloping, jumping riding, climbing, doing somersaults, and using a balancing board	playground equipment, tricycles, balls, scooters, wagons, climbing materials, walking on beams
materials to manipulate	crayons, pencils, paint brushes, color books, puzzles, sand, modeling clay, chalk boards
telling and listening to stories and nursery rymes, role playing, imagination playing, building, reciting, and using problem solving materials	more complex puzzles, construction sets ex: Lincoln Logs®, large building blocks Tinker Toys®, set of cards, trains, farm sets, drawing tablets, chalk board
areas that can excite the imagination, places to work that can stir the interest, place for group play, role-play,	play house, secret spots or forts, costume boxes, their own housekeeping equipment, dolls, dishes

What they need:	What they use:

4-5 years old, cont'd.
dress-up, and doing some household chores.

Birth to five years of age is the most productive learning period in the shortest amount of time that the child will ever experience. They learn a lot about the world around them: sights, sounds, names of objects, concepts such as *hot* and *cold* and interaction with others. They learn how to express joy, sadness, excitement and curiosity. (In a child with VCFS these emotions may or may not come with a lot of facial expression.) Early interaction with other children is important at this stage. It helps to develop language skills, manners, honesty, sharing, and other social skills. One little book many have found to be a good first tool for learning is a booklet titled **Babies' Simple Pleasures: Our Beautiful Bodies** by KellyAnn Bonnell. This book is an excellent *Parent Guidebook* for teaching little ones about their bodies and their five senses while building healthy self-images. This book can be purchased at:
http://www.winmarkcom.com/caroobook.htm.

VCFS children are usually (but not always) the quiet, shy and most well behaved children in a group. They may appear more emotionally immature. There can be several reasons for this, I believe, in part, it is due to the fact they have been ill so often in their short lives and have, more than likely, been held by someone a lot of the time. They usually do not initiate playtime, nor do they *lead* in playtime. They are customarily followers and respond well to an invitation to join in an activity. Since they learn best by imitation, they learn an activity in a short period of time.

A pre-school education can be a beneficial tool to prepare the VCFS child for kindergarten in social acceptance as well as in speech and language, which is often delayed. This can increase chances of success now and in the next 12 years of school. Preschool gently prepares the child to be away from their parents for a certain period of time and to get accustomed to being with a group of children their own age. This is a time to be observant and note what special needs the child may have now and in the future.

Unfortunately a vast number of children do not get the extra help they need. Because they are so good natured, laid back and don't often cause problems; their problems are not always seen. You have heard the expression that *"The squeaky wheel gets the grease."* Unless we, as parents, squeak and see that our little ones are squeakers (when they don't understand something; help them learn to ask) we won't get preferred and needed help for them.

At this age and in a daycare setting, be prepared for all the childhood illness to hit one right after another, since they will be exposed to so many other children. However, the experiences the child will gain being in the daycare environment will be worth it in the long run in a good number of cases. Unfortunately immune systems need to be exposed to illnesses to build and become stronger for the future.

Newborn conditions

Below I have included some terms of conditions that are _normal_ for many newborns; with or without VCFS. They are included because they were some of the most frequently asked questions by new mothers in my work at the hospital in post-partum. I must assume some new mothers will be reading this book.

- **Epstein's pearls:** Small white-yellow epithelial cysts found on the hard palate. Looks like teeth.
- **Foremilk, also called colostrum:** It is the first watery milk, from the breast after birth. It is high in lactose, like skim milk.
- **Hindmilk, also called lactate milk:** It follows colostrum, it is high in fat content, leading to weight gain. It is satisfying to baby while boosting the immune system.
- **Lanugo:** A fine downy hair that covers the fetus/baby's body.
- **Meconium:** The first bowel movement of a newborn (black in color).
- **Milia:** White pin-head size distended sebaceous glands on the cheeks, nose and chin. Looks like whiteheads
- **Mongolian spots:** large patches of bluish skin on the buttocks of dark-skinned or interracial infants.
- **Nevus flammeus:** Large reddish-purple birthmark usually found on the face or neck that does not blanch (turn white) with pressure.

- **Nevus vascularis:** Birthmark of enlarged superficial blood vessels, elevated and red in color.
- **Phimosis:** A condition where the opening in the foreskin is so small that it cannot be pulled back over the glans (head) of the penis.
- **Syndactyly:** Fusion of two or more fingers or toes
- **Talipes or club-foot:** Congenital deformity in which the foot and ankle are twisted inward and cannot be moved to a midline position.
- **Vernix caseosa:** White, creamy substance covering a fetus' body.
- **Witch's milk:** Whitish fluid secreted from a newborn's nipples.

Problems that many children go through

Infancy
- **Nutrition:** Disturbances, feeding problems, food allergies, gastroesopharngeal reflux, colic
- **Failure to thrive:** Inadequate growth

1) Organic: Not obtaining or using nutrients to grow, such as with an illness
2) Inorganic: Physiological factors preventing adequate intake of nutrition, such as parental neglect

Injuries most common
- Aspirations of foreign objects
- Suffocation
- Falls
- Poisoning
- Drowning
- Burns

Problems with adolescents, starting at age 10, can be:
- Conflict with parents, siblings and authority
- Conflicts or problems with school, peers, relationships
- Personal identity is questioned
- Situation anxiety and mild depression

Parents' concerns at any age:
- Rebelliousness
- Wasting time
- Risk taking
- Mood swings
- Drugs
- School accomplishments
- Perceived psychosomatic complaints
- Sexual activities
- Accidents, homicide and suicide
- Conflicts with physical, cognitive and emotional maturity in development

Parents Speak Out

 Karate, tai chi, tae kwon do, ballet, swim lessons or any of the beginning classes that use the entire body can be really beneficial and enjoyed by many VCFS children. Not only does it help their self-esteem but it can improve their coordination, control and body strength too. Video games that are considered by many to be *evil technology*, can be utilized in promoting hand-eye coordination (like a wii) if you purchase software that is educational and age appropriate, like *Wheel of Fortune* etc. It can do two jobs at once by encouraging thinking, spelling, and more.

For those of you who need special toys and utensils, here is one of the neatest sites I've seen with reasonably priced OT-type toys! I wanted to share the site with you: **http://www.pfot.com**

Medical Trivia

Our muscles can't push. They can only pull.

CHAPTER 4 — DEVELOPMENT

Blank page for personal data:

*"**L**earn to get in touch with the silence within yourself and know that everything in life has a purpose."*
Elizabeth Kubler-Ross

POTTY TRAINING

Here is a step by step format that I found on *Dr. Phil's* web site (**www.drphil.com**) that can help parents start sensible potty training. Remember this is just a guide. You can and may need to alter it to fit your child and situation.

What you need to begin

- A doll that wets
- A potty chair
- Toddler underwear for doll and child (instead of diapers)
- Lots of liquids for your child and the doll to drink

<u>Note</u> that the following instructions using liquids also apply to potty training for bowel movements.

Things to consider before you begin

- **Development:** *The American Academy of Pediatrics* suggests waiting until two years of age to potty train, unless the child is showing an interest earlier. Many children end up training themselves and some need to be introduced to it earlier than 2, but should not be badgered into it until they are comfortable.

- **Modeling:** You personally can demonstrate how it is done for the child or have the doll demonstrate the process of *going potty.* Most children are curious and will want to see and hear what is going on.

- **Motivation:** Determine what would motivate your child to want to use the potty.

Potty training step 1:
A doll that wets
Yes, for little boys too! Your child can learn by teaching the doll how to go potty. Have your child name the doll and give it something to drink. Then walk the doll to the potty chair with your child. Pull the doll's *"big kid"* underwear down and watch the doll go potty together. Then . . .

Potty training step 2:
Create a potty party!
When the doll successfully goes potty, throw a potty party! Make it a big blowout with party hats, horns and celebration. Give lots of attention to the doll so that your child understands that going potty is a good thing. If you do not have the elaborate party, make sure you really emphasize in your voice what a good dolly it was and maybe give a hug and treat. You get the picture, make it an anticipated event.

Let your child know that when he goes potty, he will have the same potty party or positive attention too.

Potty training step 3:
No diapers, big kid underwear
At the beginning of the process you placed *"big kid"* underwear on your child's doll. Now it's time to take away the diapers and put *"big kid"* underwear on your child. You can opt for Pull-Ups® or something along that line for this next step, if you prefer; but use them as something special.

Potty training step 4:
Drink lots of fluids
Give your child plenty of fluids to drink. The sooner he has to go potty, the sooner you can begin potty training.

Potty training step 5:
Nine trips to potty when accident happens
Ask your child if he needs to go potty. Your child might say no and that's OK. They may not understand yet what the feeling means, but it would be wise to have them try anyway. Because you've given your child plenty of fluids, he will soon need to go potty.

If your child has an accident in his underwear, don't scold him. You want this to be a positive experience. Instead, take your child to the potty, pull his underwear down, and have your child sit down. Do this nine times. This builds muscle memory and your child will eventually go. If you are busy and need some reminder yourself; set a timer for every 20 minutes or so. You make the judgment call.

Potty training step 6:
Child calls favorite person
When your child successfully goes potty, and you are throwing them a potty party; the most important part is having your child call his favorite person and tell them about what he just did! Enlist the help of the friend or relative to play along and take the phone call, being excited.

When your child has an accident, simply take him to the bathroom nine times in a row as you did before. This will continue to build muscle memory. Remember to keep up the positive reinforcement.

This potty training regiment was meant to help with normal training, but as with all things with VCFS, many children have other issues that can get in the way of potty training. Many children are either scared or have pain when they have a bowel movement. I have known kids to hide behind a door or in a closet trying so hard not to go. This of course only causes other problems like impaction of the bowel and hemorrhoids in these little ones because they have to push so hard and long to have their movement. Or there could be kidney problems, urinary tract problems or some other issues to consider. So consult your doctor. There are many things nowadays that can help in this area.

There is a nice little book called **Dry All Day: Potty Training Skills** by Elaine D'Ippolito that many have found successful. This book is available at:
http://www.winmarkcom.com/pottyworks.htm
This book gives parenting information as well as children's activities that will entice the child to want to learn. Inside this book you get parental instruction, skill building activities, charts, stickers, fun potty songs, rewards, a certificate and more. Make learning fun—not scary.

Blank page for personal data:

Medical Trivia

Beethoven poured ice water over his head when he sat down to create music, believing it stimulated his brain.

"If your fears and dreams existed in the same location, would you fight your fears or run from the path of your dreams?" Author unknown

WHAT IS PSYCHOPHYSIOLOGY?

Psychophysiology is a field which studies the relationships of mental activity and physical functions; in other words, investigates the mind and body connection from a scientific point of view. Remember hearing of *psychosomatic* symptoms? This is when there is a physical complaint with no obvious physical cause for the complaint. Psychophysiologists are interested in how mental characteristics can affect the body, and how our bodies translate this information from the brain. They want to know exactly what changes happen when people experience particular emotions, and what experiences can trigger these emotions as applied to the discipline of psychophysiology. This information can be used for practical purposes. As an example, research is done to find better ways to teach people how to control their internal body functions, which may be causing health problems. This science is studying ways to predict and prevent stress-related health problems of individuals. Biofeedback is one method of self-regulation.

What is biofeedback?

Biofeedback is a method of training an individual to control an involuntary body function such as blood pressure or temperature by means of sight and/or sound signals on a recording device wired to the person. When alerted by the signal of the device to try and change the blood pressure or temperature for example, the person makes an effort somehow to produce a change. Another signal then informs the person when they have accomplished that change. People who are trained in biofeedback can then take this training into their everyday lives.

Biofeedback is a laboratory procedure which can train individuals to alter brain activity, such as blood pressure, muscle tension, heart rate, sweat production, brain wave activity; and perhaps to control things like OCD, anger, hyperactivity, just to name a few.

Biofeedback involves a mechanical device that is commonly used to detect a person's bodily functions with great accuracy. This information can be valuable to the person, but also to the research community. Studies have shown that we have far more control over our bodies than we thought, particularly of the so-called _involuntary_ bodily actions than was once thought possible. As a result, it is said that biofeedback can train an individual to use techniques to live an over-all healthier life. Just think how beneficial it could be to learn this therapy before you actually need it. You could stay in good health longer.

What kinds of problems can biofeedback help?

Biofeedback techniques are expanding. There is an ever increasing list of conditions that they have been documented to help.

Some of these include:
- Migraine headaches
- Tension headaches
- Various types of chronic pain
- Disorders connected with the digestive system
- Urinary incontinence
- High blood pressure
- Cardiac arrhythmias (abnormalities of the rhythm of the heart)
- ADD/ADHD
- Circulatory disorders
- Epilepsy
- Some paralyses (movement disorders)
- PMS (pre-menstrual syndrome)

It is said that more than 700 groups are using EEG biofeedback (neurofeedback) for ADD/ADHD with success. Clinicians have reported that patients experience 60 to 80 percent improvement in their conditions and have had a marked reduction in their medication.This procedure is just lately being recognized and used in medical facilities that are dealing with VCFS with

ADD/ ADHD diagnosis. This procedure has actually been in practice since it was developed in about 1940.

It is said that more than 90 percent of children under the age of 12 who have a sleeping disorder such as bed wetting, have fully recovered from this condition with biofeedback treatment for the disorder.

Numerous studies have shown that people with panic and anxiety disorders can significantly control these states of mind with the help of biofeedback training. I believe this is just one of the new techniques available that will advance to eventually benefit some of the symptoms of VCFS. In time it may reduce or eliminate taking so many medications; thus giving the liver and other organs a much needed rest.

What is your child like now?

Answering these questions, can guide you to decide what needs attention and where you need to turn for help next. VCFS can be overwhelming at times with so many issues to attend to, that we don't know where to start. This exercise will help you prioritize. Do this exercise annually to help you keep track of changes in your child.

- **General health:**

How are eating and sleeping habits, and general fitness? Are there many absences from school? Do they have minor ailments, or frequent coughs or colds? Have there been any serious illnesses, accidents, or periods in the hospital? Are they on any medications or special diets? How is their general alertness? Are they frequently tired? Do they show signs of use of drugs, smoking, drinking, glue or paint sniffing?

- **Physical skills:**

Do they walk, run, climb, ride a bike, play football or other games? Do they like to draw pictures, write, do jigsaw puzzles, use construction kits, household gadgets, tools, or sew? Or do they seem not interested in doing anything?

- **Self-help (depending on age):**
What is their level of personal independence? Do they dress themselves, have good hygiene, make their bed, wash their own clothes, keep their room tidy, and cope with day to day routines? Do they budget their pocket money? How is their general independence when going on errands or out for fun? Can they be trusted, or is there a crisis many of the times when they go out? Do they drive?

- **Communication:**
What is their level of speech? Do they explain things understandably and describe events and people so that they convey the correct information? Are they able to deliver messages (e.g. messages to and from school)? Do they willingly join in conversations, and use the telephone?

- **Playing and learning at home:**
How does your child spends their time? Do they use a computer, watch TV, and read for pleasure or information, or have hobbies? Is their level of concentration normal or limited? Do they share or interact with others?

- **Activities outside:**
Do they belong to any clubs or sporting activities? Are they happy to go alone or do they NEED a companion?

- **Relationships:**
How are their relationships with parents, brothers and sisters, friends, and other adults (friends and relations)? What are their relationships generally like at home? How are their relationships outside the family, generally speaking? Is your child a loner? Are they interested in the opposite sex? If they are older, what are their self expectations? Do they feel they can bond? Do they know how to date and act appropriately?

- **Behavior at home:**
Do they cooperate, share, use manners, listen to and carry out requests, and help in the house? Do they offer help or just help when requested? Do they fit in with the family routine and obey the rules? What are their moods? Are they good and bad, sulking, demonstrative, or affectionate? Are there any temper tantrums? Is there any difference in moods with any of the household members?

- **At school:**
How are their relationships with other children and teachers? Are they making progress with reading, writing, numbers, other subjects and activities at school? How has the school helped or hindered your child? Have you been asked to help with schoolwork such as listening to your child read? Have you perhaps been given advice to have your child read out loud into a recorder in front of a mirror? If so, what were the results?

Preparing your child for dangerous situations

True safety can only work when in an actual situation. As parents, we must make sure that what we're teaching our children about safety will sink in; so that when needed, it is there on instant recall.

- **When it comes to strangers:**
I wrote this section one day while watching a show on TV about safety, because I thought it was so appropriate for those of us with VCFS children who are so sweet and gullible. I also feel it is vital information because my daughter was almost abducted right out of our front yard, when she was 2 years old. She would have been gone had I not been watching out the window and seen the man reach for her. So yes, it can happen to any of us.

- **Teach them that there are NO exceptions!:**
Predators are very adept at looking harmless, and will never look like what the child expects. Be sure to teach your child that there are no exceptions when it comes to dangerous people, and someone who could hurt them won't necessarily look like a *"big bad ugly person"*. This doesn't mean your child should be mistrustful of every person she sees, but it does mean that there is no reason your child should get into a car (or any dangerous situation) with someone you or they do not know. Our VCFS children are so trusting it seems odd to feel I need to put this in here.

- **Give them a definite concrete plan of action:**
Our children do well when they have very clearly defined reactions prepared. Even if it means giving them specific words or

phrases to use, make sure that they have a definite course of action in mind. For example: you may find it useful to teach them to use the phrase *"This man is not my daddy!"* when confronted by a strange or threatening man. Or just plain say *"NO"* really loud and blow a whistle if they have one. I used to give my kids whistles that went around their necks while walking home from school. It may be that no one would have paid any mind, but if close enough to home, I sure would have.

- **Rehearse, rehearse, rehearse:**

Talking about what to do is helpful, but nothing will ingrain it in your child's head like actually doing it. <u>*Practicing how to react is the single most effective way to ensure your child will react safely if confronted with danger.*</u> Let them know it's OK to yell at an adult in this situation and they won't get into trouble if they are wrong. It's OK to make a game out of it, just as long as your child knows that it's very serious business. We also had an escape plan for what each person would do to get out of the house, should there ever be a fire in any part of the house. Once out of the house, we had a place to meet with one another. This is another way of protecting your child and working with the possible confusion of 22q.

- **Have them teach safety to someone else:**

Once your kids have the drill down, try getting them to show it to someone else, like a neighbor, a schoolmate, or even a household pet. The best way to learn is to teach, and you could be saving another child's life at the same time

Medical Trivia

Human bones are not solid, dry, brittle and white. Bones are porous, pulsating, blood soaked living tissue. They are soft, relatively lightweight and have a sponge-like interior.

"When you get to the end of your rope, tie a knot and hang on." Franklin D. Roosevelt

EMOTIONAL DEVELOPMENT

What, if any, are the changes in emotional stages of VCFS children?

Emotional stages

- **Infants:**

This paragraph primarily concerns emotional adjusting at home. But if your child is in the hospital there will be little difference. Your child will still need physical touch, warm nurturing, and parental reassurance to feel they are not abandoned.

The VCFS infant generally has the same physical emotions as any other baby. They become attached to their caregiver. They cry and become fussy like any other baby when uncomfortable or upset. However, take into account that they could have more than their fair share of problems, such as feeding difficulties and the medical complications that should be addressed. These can, of course, increase crying from discomfort.

Like other children they usually don't like having siblings take things away from them. However, there are some infants who react very passively with their siblings, actually handing everything over to the sibling, then they just stare and watch.

It is hard but parents also need to learn that having a baby with a disorder or health issues does not mean that we have to pick them up at every little peep. If you have checked for physical distress and the baby is warm, has been fed, is wearing a dry diaper and appears to be fine, you can *give yourself permission* to complete a meal, go to the bathroom or finish some chore without guilt. This, of course, is a personal decision and only you can determine what your comfort level is. I make these comments because I was given a pep talk about wearing myself out, to the detriment of not only myself but my whole family. That is why I mentioned the term *"Give yourself permission."* Not only was I worn out from jumping at every peep the first few months, but I also felt I was neglecting my other child who was only 19 months old. I didn't know what was wrong with my infant, T.G., but I knew that something was not right and I was so afraid of SIDS, that I felt I had to be vigilant. From lack of sleep, and worry over my baby, I was on automatic pilot half the time and had no energy to play with my toddler or to give her the attention she needed and deserved. She is now 33 years old and assures me she doesn't hold it against me. Smile!

Another mistake I made, that took some convincing to see, was keeping the house too quiet when the baby napped. I had to learn to put music on and do all the things I would normally do if the baby was awake to get him used to it. It was much appreciated counsel as I was always on pins and needles. I don't think I got a good night's sleep for 5 years. I want to try and save you those sleepless moments, if I can.

I believe it is at this stage that we, as parents, go through some emotional stages ourselves.

- **Toddlers:**

This time period in the VCFS child's life is unlike that of infancy. Although, as parents, we still need to care for and look after our child, we start teaching them independence and guide their behavioral patterns, such as impulsivity, anger, manners, etc. This is the age of language and words. This is when the speech delays will begin to show as our children try to communicate a desire to express their feelings. Since VCFS children have difficulty in expressive language, it is at this period where we can see the most anger being shown in the way of *temper tantrums*. This is when intervention for communication difficulties is so beneficial and when we can teach our child certain words, signs or ways he can express himself. Our child may feel less frustrated and therefore act out less if they feel they are being heard and understood. Some experts believe it is not beneficial to teach sign language, where some professionals feel it can be an advantage. In this regard, go with your own feelings about what suits your family. *(See Teach VCFS babies sign language on page 182.)* I personally believe it's an advantage.

- **Preschool:**

Generally by this stage the child has learned a few words or sentences and instead of acting out their feelings in behavior, they can think of ways or strategies to get the point across (even if it is in the form of a grunt) without a tantrum. They begin to learn strategies to get things they want, like, *"How am I going to reach that box of candy on top of the shelf?"* They also learn that they enjoy pleasing others, so they are zealous to please. But this can also cause behavior or emotional problems if they feel they have failed to please. Then self-esteem can become an issue. With this type of personality the child can easily be taken advantage of, which can lead to abuse by others. So keep a *"sharp eye out."*

It is in this age group that VCFS children might show signs of OCD (Obsessive Compulsive Disorder). It starts to manifest itself in several ways; by picking up things and lining things up, being overly clean, showing repetitive behavior, or obsessive thinking with or without compulsive action, etc.

> *Routines are good and important, but they need to change as the child does.*
>
> *Routines include: regular mealtimes, school time, play time, homework time, bed and bath time, chores, etc.*

- **School age:**

School age is a very busy time for the children and parents. This is when most of the symptoms of learning disorders manifest themselves. There may be diagnoses of various kinds, such as attention disorders called ADD and ADHD *(See ADHD on page 213.)* as well as an assortment of other learning disorders like dyslexia, cognitive problems, auditory processing problems, visual problems and a list that can go on-and-on. This is different for each child. Then there is the level of education of the school faculty on VCFS. Now is the time for us to learn the different styles of teaching by the different teachers in each grade, and what kinds of programs they offer. Then there are the IEPs to be customized to match your child. *(See Education-IEPs on page 145 and Advocacy on page 269.)*

It is at this time when the peer groups are beginning to be of great consequence to the child and they will want to fit in. Social skills will need to be addressed as well. *(See Social skills on page 133.)* The child with VCFS has so much to deal with at this age that it is easy to see why many go into depression and/or shut down. This emotional roller coaster can stimulate other symptoms already exhibited such as ADD, etc. It is at this period that the differences in children show through and VCFS children may struggle to keep up physically and mentally with friends; not only in schoolwork but in sports, noticing the opposite sex, speaking the same dialogue and other areas of change that children go through at this age.

Many parents have found that keeping their child involved in activities which interest them, helps them do well emotionally and is beneficial throughout this stage.

- **Teens:**

When managing this age group, routines are still important. It is the adjustment to the age that needs consideration. Give the teen a little more responsibility, gradually; such as: caring for their own laundry, and making sure that all their clothes are clean and ready for the school week before doing something fun. They can pack their own school bag (right after homework), so they don't forget anything in the morning. You might want to help the child update their daily school diaries, which helps to alleviate stress the next day. This approach not only teaches the teen how to organize and look for oversights of lessons not done in the daily school diaries, but it gives you a chance to oversee if they are keeping up with studies or falling behind. Many schools now offer the opportunity for parents to check on-line and see what their children did on any given day, what the assignments are for the next day, and what is coming up in the near future. Our teens can also learn to cook, learn to shop for groceries, make a menu, and learn to drive (if applicable). To keep the child both mentally and physically happy during the summer, check out the *Parks and Recreation Department* in your area for activities or investigate appropriate youth camps. I must add here that this is also the time when life gets busier. So remember <u>one task at a time</u>. A list helps.

Teen years can be a frightening age for us as parents, our babies are growing up and it is now time to allow the child to socialize in groups without us. Set boundaries and a curfew, and demand adherence to them. This can also be a time to talk to your child about not being impulsive, and learning self-control. They should not buy on impulse or act impulsively just because others in the group do. Point out specific things if you know of any that may be a problem for your child.

I personally was nervous the first few times I left my son on his own with no supervision from an adult. I loaned my son the cell phone and had him call me every hour or so. Later, I got him a cell phone for my peace of mind. He eventually learned on his own to call mom periodically. Over protective? Maybe, but I was also learning as a parent. As I always told my children, *"I have been a kid and I know what to expect, but I have never been a parent, so I work on a learn-as-I-go basis."*

Emotional and self-esteem issues can occur at any age but are most prevalent in the teen years. There can be anxiety and moods, feelings of not being *good* at anything. Expressions like, *"Why can't I do anything right like my sister? It's not fair!"*. This is when you have to reassure them that they do many things right, but separate things and in different ways; that we are all unique and none of us do things the same way. There is always more than one way to get a job done. This is a time to encourage and praise every effort. Exhibit and point out all their good qualities as you see them. When VCFS kids are in the low image mode, they find it hard to cope with negative feelings from others. They can become easily upset by unkind words. This can lead to depression, which I will comment on more thoroughly later. *(See Psychological issues on page 233.)*

Medical Trivia

Every time you lick a stamp, you are consuming 1/10th of a calorie

Blank page for personal data:

"Your talent is God's gift to you.
What you do with it is your gift back to God."
Leo Buscaglia

SOCIAL SKILLS

My child has few social skills. Why?

Many VCFS children have endearing senses of humor and can surprise you with their wit and antics. However, some show lack of or little expression, so their humor is not always noticed by other people. This lack of facial expression is sometimes due to poor muscle tone, to shyness of personality, or both. These flat expressions can affect their education as well as their social life. Teachers may misread the child, thinking that the child looks indifferent or uninterested in class, when they are actually puzzled or are, in fact, interested. This lack of expression is also socially difficult for them because they seem *blah* and uninteresting to their peers. It may also be a factor in whether they get a job or not or will have a long term relationship with the opposite sex in the future.

Some VCFS children are loners and need to be encouraged to interact with others. As a parent, you may have to take the initiative to invite children over to your house to spend time with your VCFS child. Find children whose company you believe your child will enjoy and who could be a good influence. This encouraged interaction helps the child to build relationships and learn how to relate to others. Because many VCFS children have noise sensitivity and/or anxiety disorders, it is wise to test the waters. Set time constraints on visits, and limit the number of children invited at one time.

Once the VCFS child feels more at ease, a group setting may be welcomed. You could have a few kids go as a group to a function, like skating, bowling, or a volleyball game at the park, and then go out for pizza or something along that order.

Many VCFS children never get to that point; however, and parents will need to become more involved and creative in coming up with activities that the child can or would want to do with others. In doing this, parents need to take their child's limitations into consideration.

Since the teen years are an especially scary time for parents. (Boy do I know this. I've earned every gray hair.) This is the time when children want to go out by themselves. Perhaps they want to get their driver's license to feel more independent and grown-up. Independence is to be encouraged; but, as we mentioned before, it can be nerve wracking if your child has impulsive tendencies. Parents can feel like they never know what they have gotten themselves into on a daily basis. First, it is wise to teach your teen about all the different things that can happen before they ask to do these things. This helps to keep them from getting mixed up in anything that is questionable. For example: teach them how to deal with getting lost, panhandlers, or a situation where a friend has taken them to a place where they feel uncomfortable. Give them ideas about what to do and how to handle the situation. It is wise to limit the amount of cash they have on hand.

Simple after-school clubs and short-term sport courses (if they are able) can be beneficial in helping a child get better grounded in knowledge and experiences. This also encourages social interaction in casual settings away from home. Most of the time however, their peers are not familiar with VCFS like we are at home, where our child feels safe and comfortable; so their peers might shy away and not make lasting friendships. Their peers may in fact treat them as strange and this can make it difficult to get our children to _want_ to do things outside the home. They often do not want to venture outside their comfort zone; unless they have already been introduced to a variety of different experiences throughout their life, or are on the hyperactive side of the spectrum in their personality. I don't mean to imply that they are anti-social, they just tend to cling to the familiar. There are many parents I have had the privilege of talking with who have some wonderful ideas for interaction, and who have implemented them and found great success.

Keep in mind that some of the following suggestions are for different age groups.

What about music or sports?

If your child has a passion for something such as an instrument, sport or hobbies, it never hurts to introduce them to that subject. You never know when it may be just what is needed to fulfill them or to develop a hidden talent. It may take several tries, but each exposure of a new interest is a learning experience, for both you and your child.

A surprising number of VCFS kids are good at playing instruments or do well in a sport. Many can master more than one instrument and have even played in bands or small orchestras. Some have exceptional hand-to-eye coordination.

I am worried for the future, what can we do?

One of the priorities for early intervention is developing interaction skills with others, and using meaningful communication. Teachers, daycare workers, and teacher's assistants can help encourage many of those skills listed below. They may offer other suggestions or discuss some of the methods they have used that worked for them. It is through this sharing of observations between parents, teachers and others that we find the strengths and weaknesses of our children. In this way we know how to help them.

What we need to faithfully work on with the children are:

• Tolerating closeness to others:
We all know how it feels when people invade our space; it makes us feel awkward and uncomfortable. Well, VCFS kids can be even more sensitive in this area. They have to be helped to overcome that feeling when it is not warranted. They may perceive that a person is too physically close, where we may see it as a natural distance. For example, they can misinterpret the purpose of sitting in a group setting, where people can be shoulder to shoulder, head to head, hand in hand, playing and sharing, etc. At the other end of the spectrum, a child may be too clingy, needing unusual amounts of attention and continuous human touch. This

also needs careful examination. Is there a cause or need, or are they just a cuddly loving person?

• Paying attention to adult instruction:
Teach respect and patience. This means learning to listen and follow rules, instructions and directions from whomever is taking the lead. Children need to pay close attention to learning safety rules, like: *"stranger danger"*; bike street safety; being aware of the area around them for potential danger; looking before walking somewhere; not petting strange animals; not eating something that a strange person gives them; waiting in a place designated without moving from that place; and following other instructions that are beneficial.

• Eye contact:
Eye contact is something we may not think is important, but children need to learn to look people in the eye, even those they don't know. I am not talking about strangers on the street, but those with whom we come in contact. This shows courtesy to others. Teach them to shake hands when greeting others and look directly at the person. They should learn to be courteous and use polite words when asking for things such as a glass of water, directions, etc. Many people believe that people who do not give eye contact are lying or deceptive. This is not true of our kids, but because it is a common belief, it can hinder them in their future endeavors like job interviews. Teach them to use eye contact when they are young. If you have ever watched *"Judge Judy®"*, you know she is a stickler for eye contact. This is a classic example of why our children need to learn this habit. They may be judged wrongly for such a minor oversight.

• Responding to verbal summons:
Children need to be taught that they *MUST* acknowledge that they have heard you, when you call them directly. They need to realize that it is rude not to respond, and, in this day and age, it is also scary for people who care for them if they don't respond. A parent or concerned individual might worry about the safety of a nonresponsive child and wonder if something has happened to them.

- **Spontaneous *(normal one-on-one)* communication:**
Children need to be able to convey a spontaneous thought to an adult, no matter what the subject, give or receive directions, retrieve objects when asked and put things away after use. Good habits should be set in childhood. Teach them that you mean *NOW* and not 5 hours from now after you have made the request. When many of these children get older, they tend to be procrastinators. To them it is always, *"OK, but later . . . "* and they try to rely on memory to do it later. Most of the time, it doesn't get done without repeating the request more than once. If this type of procrastinating behavior becomes habitual and they do the same at school and/or on a job, they may not be successful throughout life in these areas. What comes to my mind is that they might get married, and the procrastination can drive their mate nuts. Hmmm . . . Am I showing my pet peeve?

- **Recognition:**
What does your child really recognize? Can your child name ordinary items? Many people might think it is not necessary to bring this up, because all children need to learn these things, but with VCFS it is more difficult to broaden their knowledge and recognition. It may take several times before they have mastered the word for a particular item and to recognize that the word refers to a specific thing. If we were to itemize as many things as possible with the names we use to describe them to our children, we might be surprised. Our child might not know as many items as we think they do. It is a very important task to teach what things are and not take for granted what a child really knows or recognizes. It can be something simple like knowing what a countertop is or perhaps understanding that countertops can be in kitchens, bathrooms and other areas. They may need to understand terms for similar types of items such as a chest of drawers, a bureau or armoire. Why is an item that looks the same to them called something different in another room? What is the difference between a chest of drawers and a bureau? They will need to know what things are and how to use different things like tools: household tools, kitchen tools, father's woodworking or mechanic's tools, etc. Ultimately, children need to know how tools work and why do they work the way they do. Also, teach them verbal tools such as the way to remember how to tighten a bolt, *"righty-tighty and lefty-loosy"*, and others.

- **Learn early to take turns and share skills:**
Working on *"taking turns and sharing"* and other interaction skills should be done as early as possible, even before speech emerges. This will need to be continued and reinforced after speech begins. Some of this will overlap with behavior, which is reactive to communication. These are learned behaviors and the child doesn't have to talk or be verbal to learn many of the lessons. Some of these exercises will not come easy. You will have to work out a discipline that works with your child's disposition and personality. Reasoning alone will work for some sensitive children, especially if they are older. However, using only reasoning will not work well with the younger child. Then again, reasoning may never work for some. These children may need more persuasion and direction because they can't figure out why you requested this or demanded that. They may think that you are just being mean to them. If that is the case, it is not discipline they need, it is further teaching. Discipline will come when they are willfully disobedient.

Our children may find it difficult to find and keep friends of their own age who are willing and patient enough to get to know them as a person. Much of the problem may stem from lack of communication as a whole and lack of cognitive (understanding) skills. Because the oral expressive language can be limited in the VCFS child, and they may also be hard to understand; <u>*normal*</u> kids tend to shy away from having to work at understanding what is being said and/or meant by the VCFS child and they may give up. Trying to understand them reminds me of the way a person may react to someone with a thick accent. It is embarrassing to keep asking that person to repeat the same words, and to ask if this is what it means. Hence, many find it easier to find other friends who are less trying. For this reason VCFS children tend to enjoy the company of either much younger or much older people who seem to be more tolerant and patient.

One thing I want to express is that we mustn't fool ourselves into thinking our kids don't understand what is going on around them. They usually do, and this hurts them at times; but they may not say much about it. So it is of great consequence that we stay alert to depression. The differences in people need to be kindly addressed to them. About the time the movie *Forrest Gump*® came out, my son asked me, *"Mom, am I your Forrest Gump?"*. I replied, *"Sort of, look at how smart and talented he*

REALLY was in different ways. A person is not smart just because they do well in school. He did marvelous things. People, not just life, are 'like a box of chocolates; you never know what you are going to get.' Each has one's own recipe for excellence that is sweet to the taste."* I believe that if a child feels that he is seen as just another recipe but all candy is sweet, he will not become downhearted.

I like the way Maya Angelou puts it: *"My mother said I must always be intolerant of ignorance but understanding of illiteracy. That some people, unable to go to school, were more educated in life and more intelligent than college professors."* In many ways this fits our VCFS kids and our circumstances. Therefore, experience is the best teacher.

PARENTS SPEAK OUT

I found that engaging our son in family board games was valuable and fun. It opened the family to laughter and bonding. Simple games like *Trouble*® for instance, are fun for the whole family and easy to learn to play.

The VCFS child raised in an environment of laughter, learns to enjoy participating in harmless practical jokes; like putting large plastic spiders in the bath tub when they know that company is coming. Hmmmm . . . how do I know this? Laughing and being silly with lots of lightheartedness around them makes for a happy, less stressed child. Some parents forget to laugh and drown themselves in seriousness. We can fail to remember to enjoy life and our children. It is easy to feel so bombarded with duty that it becomes hard to demonstrate happiness or feel lighthearted. When we take ourselves too seriously, this can happen and could result in a depression problem for one or both parents or caregivers.

My contribution is in the area of music, although personally, I am musically challenged! My son has been interested in music since the age of two. We had an old piano and just left it open for him to play whenever he wanted. At the age of five we arranged for some structured but fun lessons for him with no pressure. He enjoyed them immensely.

As time went on, we realized that our son had difficulty learning to read music. It seemed that the music went from his ears to his hands without stopping in his brain. But we continued with the lessons because reading music is so important, particularly if you want to join school bands, etc. The funny thing about it was that when we started each September, the music instructor would start to teach him fundamentals and then give him his recital piece. He would struggle with the fundamentals, but learn the recital piece within a week. (For most students, it would take a *LOT* longer than that to learn a recital piece.) He is now fourteen and plays piano, drums and guitar. (Electric is easier for his smallish hands.) From what I understand, piano (keyboard) is a great instrument for starting to learn music; but it is also important to realize when a child is ready to participate in lessons. If they are too young, they think they are going to learn music almost immediately and it is a slow process. I have heard that eight-years-old is a good age to start lessons, but it all depends on the child, of course.

With regard to wind instruments, we have not encouraged our son to take up a wind instrument. He had flap surgery when he was about five years old, but this had little to do with our decision to avoid wind instruments. Our son has always had stridor breathing and his stamina is not excellent. And then, there are the braces on his teeth. I am sure there are VCFS kids who would do wonderfully with a wind instrument.

I STRONGLY encourage you all to consider adding music to your children's lives. It has been a source of comfort and a builder of self-esteem for our son. An added benefit this year, has been a resource teacher who realized what music means to our son, and allows him to bring his guitar or keyboard to school for special events.

My son has had a best friend for 10 years. He is a neighbor and is a year younger. They have done everything together for most of their lives. All of a sudden, the friend has decided that he is going on to greener pastures.

He has completely shut my son off and even asked him to leave his house once because he had another friend coming over to visit. I have struggled with this on many levels. One, I am proud that my son is a good friend to everyone and would not be hurtful or be unkind in any way. I have often said that he doesn't understand *meanness*. And, of course, the other boy's parents don't seem to get the concept of friendship, they have taken the "*Oh well*" attitude. It has surely prompted discussions about friendship in our house but it is difficult to explain to our son why his best friend is acting in this way.

On another level, I see that this friend is maturing at a faster rate than my son. I am not excusing his behavior just realizing that the friend is probably outgrowing him. Our VCFS children do not socially mature at the current rapid rate for *typical* teens. This has caused me to once again realize how difficult it must be for our children and how scary the future is for them. My son has always been an easy kid to raise—we have been very fortunate—but I am becoming increasingly concerned for his future. Will we ever be able to look at him as a self-confident, mature young man? Will he ever be able to take responsibility for his own life? Will he ever be able to live on his own?

We recently had a special needs trust made in concert with our wills. This was done to protect our son and ensure that he will be financially monitored when we are gone. Having this done was another one of those "*hit in the face with reality*" moments. But, as I say all the time, even with all the challenges, I would *MUCH* rather have my son than any of the other kids in our neighborhood. They all have their issues—many scarier than ours—even though they are considered *normal*.

Medical Trivia

A cough releases an explosive charge of air that moves at up to 60 miles per hour.

Blank page for personal data:

"All kids are gifted; some just open their packages earlier than others." Michael Carr

CHAPTER 5: EDUCATION

EDUCATION AND THE CHILD WITH VCFS

Accessing services:
Referring a child for special education

Major portions of this chapter contributed by Donna Landsman, M.S.

Donna Landsman M.S. is a teacher in Madison, Wisconsin, and the parent of a child with VCFS. I wish all of our children could have teachers like Donna, with her knowledge of special needs in our schools. She is who Robert Frost described when he said, "I am not a teacher, but an awakener." *Now THAT would be progress!* Author of: **Educating Children with VCFS**. Web site: **www.cutler-landsman.com**

Most children with VCFS will require some type of *special education* service as they progress from preschool through college age. Managing the educational maze can be one of the most frustrating aspects of raising a child with special needs. The U.S., federal and state laws regulate what school districts are obligated to do, even though shrinking school budgets make it difficult to provide students with maximal levels of service. (You may have to strongly request to receive services in some areas.) Many children with this syndrome are served in the USA public schools through the IDEA Act *(Individuals with Disabilities Education Act)* in the categories of *Speech and Language, Learning Disabilities* and (OHI) *Other Health Impairment*. Parents may refer their child for special education testing by contacting their local school district. This initial contact can be made as early as age three.

Once a referral has been made, the school district must convene a special education team to evaluate the suspected area(s) of need. School districts have 90 days in which to do this evaluation, develop an education plan and, if required, offer placement. Parents must give their permission for their child to be tested. Parents are a part of this team and can make suggestions regarding which areas to test. They can also provide information to the team which they have gathered, through outside testing, medical reports, articles, studies, this book, etc.

Once a child has been evaluated, the team will meet to determine if the child meets the criteria for needing special education services. This determination should be made based on <u>*norm based*</u> test scores, classroom performance indicators, medical records, and interviews with teachers and parents. No single test (such as an IQ test) can be used as the sole determining factor as to whether a child should qualify for special education services. The team must consider several assessments to make this determination.

Children with VCFS often have many deficits which should be explored when considering special education placement. Although the type of program needed will vary from individual to individual, there are areas of need that seem to be shared by a great many children with this syndrome. Professionals should take a close look at these target areas when a child is referred for evaluation.

Once a student has qualified for special education services, schools <u>ARE</u> <u>REQUIRED</u> to make modifications, provide therapy, assistive technology, and accommodations to help the student to succeed. In addition, the special education teachers will provide a bridge for the parents to work with the classroom teacher to modify or clarify assignments, provide special materials and help with the transition to a work environment or a higher education setting. Most teachers will try their best to understand VCFS if given information about it; therefore creating a nurturing environment for providing assistance.

Many children with VCFS can be successful in a regular classroom with modifications and assistance. Educating all of your child's teachers, including those in art, music, physical education, etc., about your child's needs, will help achieve a school experience that is positive and productive for all.

The team should consider some or all of the following "target" areas:

- **Speech and language needs:**

Articulation problems, expressive language delays, auditory processing deficits, problem solving difficulties, reasoning difficulties, word finding problems, difficulty understanding idioms or words with multiple meanings, and problems following multiple directions

- **Learning disability/other health impairment issues:**

Memory difficulties, math reasoning impairments, problems with written language elaboration, reading comprehension delays (decoding skills may be at a normal level), difficulty understanding cause and effect relationships, reasoning difficulties in social studies and science, lack of ability to apply learned knowledge to novel everyday situations, attention and organization problems, hypotonia (children are tired and lack stamina), fine motor coordination delays (writing, cutting, keyboarding, coloring), hearing deficits (many children have frequent ear infections or a hearing loss), physical therapy needs, lowered immune system causing frequent illnesses, vision/tracking problems and behavioral difficulties (easily frustrated, low self esteem, poor coping skills, difficulty getting along with peers, teased, etc.)

What is an IEP? What is written in the Individualized Education Plan?

The (IEP) program is one of the most useful tools for the child and teacher. Once a child is determined to have a special educational need, the team will convene and draft an IEP (Individual Education Plan) to address the child's deficits. The IEP is a very important document because it spells out the school district's commitment to provide services for the child. The IEP should consist of easily measurable goals, assessment methods to determine progress, the number of times the child will receive special education services and who will deliver each service.

Parents should familiarize themselves with IEP's *(See IEP also under Advocacy on page 271.)* so that they come to this meeting prepared to offer suggestions regarding accommodations they feel will benefit their child. Parents need to become their child's informed advocates. If you disagree with anything or want to add something *SPEAK UP*. *YOU* know your child better than anyone. If your child needs some attention or privilege in an area that is not normally addressed like *"more potty breaks," "help with learning to button their pants after the lavatory,"* or *"If the child gets too excited/hyper for learning, maybe a chaperone could walk them up and down the hall to settle them down and then return"*. These can be added to the IEP with a doctor's recommendation. On the other hand, remember that the teachers only have so much time, too many students, limited resources and may not know much about this disorder. *(A copy of this book might help.)*

Below is a list of accommodations that might be considered. These will have to be recognized as the child starts to experience difficulties at school and will require added help from home. There should be a partnership developed between the two places of school and home. These are the two areas where a child spends the most time and where they get their learning structured.

What can teachers and schools do to help?

Some accommodations that might be considered are:

- **Specialized instruction in the academic areas**, emphasizing simple direct instruction and multiple practice opportunities. Depending on the teacher and their style, you may have to let them know that your child needs this direct instruction. No, abstract or *"We just went over that, you figure it out."* type of instruction.

- **Modified tests and assignments**, tailored for the child. The teacher must remind the child of the need for study on tests that are coming up soon and make sure the test is on the child's daily sheet (to take home each day). It would be helpful to have a handout with chronological outlines for the test and what areas should be studied again. This can also be a good time for parents to get involved and request that the child give you the notes (preferably a copy of teachers notes) and you can help them to follow the outline to cover all the material being covered. Being prepared is the key.

- **1-on-1 or SMALL group settings** within the main classroom have proven best for learning.
http://www.specialchild.com/archives/lf-014.html

- **Additional time for tests.** It is the *law* to give *disabled* children more time on testing if needed. Also if the child has a hard time with fine motor skills and cannot write well or fast enough, they would qualify for this privilege. A computerized test or all multiple choice question tests might be a good option.

- **Note taking services** by some TA (teachers aid or older student) can be helpful. This way the student does not miss out on the material, since most are not able to take adequate notes themselves. These can be looked over and used for homework.

- **Preferential seating**, which is more inducive to listening and learning. Does your child have visual or hearing difficulties?

- **Specialized behavioral plans** (ex: described above in IEP) This includes training and help with problem solving in social settings.

- **Provisions to allow special items** such as: note cards, calculators, and formulas

- **Enlarged print, when needed.** This is required by USA law, so insist on it if it is not already available.

- **Books on tape,** when available. It is very helpful if the book can be read along with the verbal book on tape. I found it helped to teach appropriate places to accentuate voice tone, pause, grammar, spelling and to provide an overall more enjoyable learning experience.

- **Assignment notebook** to keep track of assignments, not only for tests but also to keep assignments for daily homework. Practice, practice, practice—that is the key in memory retention with a little visual reinforcement. This will do wonders for their test scores and for remembering important points.

- **Extended time for assignments.** There will be times when the assignments are too much and due too fast for the child with disabilities. This is where, like in test taking, the child will need more time to complete the assignment; or, a special assignment in the same subject, that is less complicated, can be given. It should be one requiring less work, thus less time to complete.

• **Test retakes**. There are times when a child should have the opportunity to take a test over again.

Building confidence and self-esteem is an important issue to discuss with the IEP group. Find out what is available that matches your child's abilities, interests and aptitudes that can give him or her an opportunity to excel. This may be hard for a new teacher to know, but for one who knows the child and with the input from the parents it can be a great asset. As they say, *"Two heads are better than one."*

While some VCFS children are phenomenally creative, many do not show much creative interest unless you excite them, and invite them to participate. They need to be encouraged to do things and make choices. Coloring, painting, drawing, assembling puzzles, etc., from young ones on up, will help to develop a creative streak, and encourage them to use their minds and imaginations. I know it sounds funny, but we all have to be taught to *think*. Some children are just taught better or faster than others. *(I used to tell my husband that and he would laugh, but when he became an instructor in the Drywallers' Union and had to instruct young adults, he suddenly realized what I meant.)*

There are a few books that I recommend for activities and entertainment. **Mother Lode: The Ultimate Collection of Ideas for Keeping Kids Busy**, contains over 5,000 ideas for activities for toddlers through teens. Many of these work well for our kids. This book and others like it are available through one of my favorite sites: **http://www.winmarkcom.com/motherlode.htm**. I have come to know the owner, Kas Winters, who is the author of several of the books for children. Many of her books are collections of ideas for children's activities. She is wonderful to work with and represents more than 100 other authors, including many grandparents, on her gallery-like Web site.

A computer has proven to be a VERY useful tool for our children in many ways as we have already mentioned. Hand to eye coordination is developed by using the mouse, and keyboarding, etc. Keyboarding reminds me of searching the web and that reminds me how good that experience is for using logical thinking (or in some cases *NOT* so logical) to find things for which you are looking. So some of the lessons should be computer based for this reason. Special learning software is encouraged. We have found there are many inexpensive learning software programs from which to choose, so go wild!

What can parents do?

While schools can offer help, parents must play the primary role in keeping the child *"on track"*. Some suggestions that might help are:

- **Be positive, but firm** in your expectations. Make sure the work is completed, and that the best effort has been made to do it in a timely manner.
- **Study for tests with your child** in short periods over several days. Be familiar with the material in case of questions. (Yes, welcome back to school. I must have earned at least another diploma myself.)
- **Arrange regular contact with the teachers**, not only on conference nights.
- **Check the daily notes with your child** about assignments, due dates, and class requirements each day before homework time.
- **Help your child gather materials** and re-explain concepts in terms they will understand. Make sure they do understand, so they won't waste time on a school paper or other work that has to be redone.
- **Enroll your child in a supplemental educational program**, if possible, to reinforce learning. You can also search libraries for tools to use.
- **Be realistic about the limitations** of your child with VCFS and reward their best efforts.
- **Explain difficulties** within the context of a *medical* problem, not as a lack of ability.
- **Trust your instincts.** You are the expert about your child and understand his or her needs. Do not be intimidated. Do not feel that you are forced to accept school or class placements which you feel are wrong for your child. Seek alternatives so that you know what choices you have.

We can all learn from each other's experiences. Sharing our successes and failures will help us understand VCFS and assist in developing strategies to help our children's futures. Many inexpensive books can be purchased to help in this area, including the one below.

The book, **Super Student/Happy Kid** by Sally D. Ketchum is recommended. It can be ordered at: **www.winmarkcom.com/superstudent.htm**. This is *"A Practical Student Success Guide for Everyone"*. Included in this book are 40 chapters and over 500 *"Hot Tips"* which are designed to: build self-esteem, build character, build confidence, build real skills and increase potential.

Resolution of disputes:

If parents disagree with the evaluation results, the IEP contents, or the implementation of the IEP there are provisions in the U.S., IDEA Law for mediation and/or a due process hearing. Parents are strongly encouraged, however, to work with their child's school to address their concerns without progressing to a more adversarial arena. An IEP meeting can be called at anytime by anyone on the team, if there is evidence that progress is not being made or if there are concerns that promised services are not being provided. Usually these problems can be resolved in a friendly manner. If this approach fails, consideration should be made for a totally different placement, a new set of teachers, involving a *VCFS expert to better explain the syndrome*, or perhaps a new environment within the district. Mediation and due process hearings should be used as a last resort.

Auditory awareness in therapy:

(See Central auditory processing on page 189.)

Classroom management for children with special needs: *(Strategies for the teacher)*

- **Preferential seating:** Means an optimal place to sit, but not first row, rather second or third row near the teacher for fewer distractions
- **Aides:** Use of visual and hearing aids
- **Monitor comprehension:** When repetition of a statement does not work, rephrasing the idea in another way might work. (There is a need to consciously monitor the child, and not to assume they understand.)
- **Delivery style:**
    1) Clear, animated and audible voice
    2) Limit hand gestures. They tend to be distracting

3) Use clear concise directions
4) Increase predictability of events with low key outcome, thereby reducing stress levels
- **Organized presentations:**
  1) Isolate key words, emphasize those words
  2) Highlight key concepts with voice or in written form, using colors to show important points
  3) Use eye contact to reinforce directions

Educational:
- **Tape recorders** (preferably with head-set)
- **Resource room** services for pre-view/review of difficult subject material
- **Homework books with paraphrased instruction sheet** of work to be performed in work sheet form also preferably in a-b-c order with step one, step two etc.
- **Teacher outlines for parents' use**, for instruction clarity

Physical modifications to the environment:
- **Reduce ambient (surrounding) noise:** Acoustic tiles, rugs, drapes, posters on walls
- **Seat away from distracters:** Pencil sharpeners, door, windows, radiators, A-V equipment
- **Reduce noise from other children:** Scraping of chairs, shuffling of feet, known talkers and the *class clown*, etc.

Test modifications:
- **Extended test time**
- **Repetition of oral information**/review of test material before test (can be recorded earlier and child can listen to it privately before test in quiet test room)
- **Separate place for test taking**

Assistive listening devices:
- These are meant to improve the signal to noise ratio
- Variety of devices available: Tote-able sound field

CAP therapy:
- **Computer based program** (these will be discussed later):
  1) Earobics
  2) Fast Forward
- **Taped programs**

1) Phoneme synthesis training
2) Hooked-on-Phonics®
- **Speech** in-noise desensitized

Child's role:
- Child must be an active participant
- Child can learn how to modify listening environment more effectively to reduce background noise (They need to speak up when they need to have some adjustments made.)
- Utilize strategies such as: learning to vocalize; rehearse/study to master; watch faces of others to learn expressions and watch how information is used by others and put into action. Strategies depend on the lesson.
- Learn how to become their own advocate. Ask for clarification when they (the child) do not understand information given and then ask them to repeat it for clarification

<u>N</u>on <u>V</u>erbal <u>L</u>earning <u>D</u>isorder (NVLD)

Let's not forget the non-verbal learning problems that challenge many children. These must be addressed in a special way. There is a book that is highly recommended called **Non-verbal learning at home** by Tanguay, that many professionals say is very good on the NVLD (Non-Verbal Learning Disorder) subject; as well as many websites such as the one below on learning disabilities. **http://www.GreatSchools.net**

Sequencing, attention and psychological issues

It was difficult to know where I should put the information on this subject I am about to address; because it could be with psychological issues as schizoid-affective or anxiety disorder, yet, it also involves development and education, and what you might want to consider when trying to help children with their disorganized thought patterns. Many VCFS children have problems with sequencing, attention and psychiatric problems that also can play a role in how learning in school is handled. Many children also have obsessive thoughts on one or more subject, idea or desire. We need to have discussions with our child about their obsessive thinking process, including what is right and what is wrong about it. The discussion will more than likely *NOT* stop the obsession right away, but it can help. I used to say we need a reset *"word button"* on our son. But instead we would tell T.G., *"Son you're doing it again, please stop."* Then the child may turn to obsessing over something else (too bad it's not over something like doing the dishes) but the discussion might create a window in-between cycles. That is the objective, to try to put distance between the onsets of obsessions. When this is an issue with your child it may be a good time to have your child evaluated. Many times there is a diagnosis of OCD (obsessive compulsive disorder) and your child should be observed by a professional when actually in the obsessive mode.

As we have discussed before, routines in the VCFS child's life work best. It is helpful to use a chart, list or ledger with a scheduled time table. Place this on their wall where they can readily see it. Remind them to look at it everyday. A smaller version for school is nice. This helps the child to organize and memorize their daily routines. This is good practice for the future. It teaches the use of self-discipline and organizational skills to accomplish goals. What I still do with my son at 32, is to write down a list of errands, appointments, or responsibilities for the day, and sometimes in time-line order. It especially helps when he is driving, so he doesn't waste gas or have to back track. Now that he knows the area, I don't have to do it as much. We live in a large city of 3+ million people so it was important to do this initially.

IQ test scores

IQ tests are recommended *only* to be used as a guide in planning for education and vocational goals; they are not to be taken to heart. These are *NOT* an exact science that can be used to set expectations for performance in the long run. Some children perform better than their testing indicates, while others perform lower than expected in certain areas. However, IQ testing can be useful to measure your child's intellectual development (at the time of the test) compared to other kids of the same age group. This test can help answer the question, *"Is my child's cognitive development keeping pace with their peers?"* IQ scores are also used to compare different school performances. Scores can help diagnose learning disorders. For example: If the child has an average IQ yet still has failing grades, this will bring up questions to be answered.

I seem to have to word my sentences in an elementary form for my child to understand, is this ok?

Whether you are the teacher of a 22q student or a parent, here are some suggestions that might help in the process by teaching the child to listen better; and it might help you to learn more about the child as well.

In my experience we need to put sentences into a *SIMPLE* and *BASIC* form that the child can understand, even if it is below that of his peers. The key is to communicate with them in an understandable way, whatever it takes; this may explain why verbal rephrased instruction worked better than written instruction in our case. In the beginning of the child's speech instruction, try to speak to your child in as proper a manner as possible, in other words not in baby *"goo-goo"* talk. This will advance their vocabulary in time and will help when they get into school. Of course it shouldn't be so proper that they have to pull out a dictionary every time you say something. When they are old enough you may want to have them repeat and explain what you said back to you. I found that when adults truly communicate with the child on a daily basis, it helps to improve the child's communication skills. Energy and consistency on the parents' part are required for each step. The adults will also need encouragement to maintain effort and patience!

- **Use short simple sentences**. We need to allow time for the child to process the information, before going on to another subject or request. This requires *PATIENCE!*

- **Use consistent language.** We want to be consistent in our use of language for the rules we set and the objects we use. For example: if we use the term sofa, always use sofa for the same item. Don't switch to calling it a couch or divan. On one day we can't allow an action and then not allow the same action on another day. It will confuse the child. Be consistent at all times even though it is not easy.

- **Be aware of the tone of your voice.** Another area that I had to learn to be aware of was the tone of my voice. Using the right intonation of speech teaches the right emotion. I was not very good at this at first. I was a worry wart and if T.G. was out of sight, I would panic. (This was partly because I had a close call with a would-be kidnapper with my daughter when she was 2 years old and I don't think I will ever forget the feeling that gave me.) So, when I would find T.G. safe, I was so excited and relieved that my voice sounded angry. Yet I was not angry but relieved. So, I had to learn to pay attention to the emotions he picked up from the sound of my voice.

- **Be calm.** I also learned that there were times I needed to be as calm and silent as possible, keeping others quiet too. If T.G. looked or acted distressed; being calm helped him. Some VCFS children get anxious easily, and can get overly stimulated by outside noise or movement. This can cause the child to act-out in a negative way at home, at school or at anyplace or anytime without warning.

- **Speak slowly.** Speaking slowly and naturally at an even pace (yet not monotone) seems to work the best. Fast speech is difficult to understand and/or imitate, especially when children are learning speech and sentence structure.

- **Repeat key words.** Repetition of key words helps the child remember a task and ensures that they understand the request. For example: Say, *"Would you please put the mail in the mail box and RAISE THE FLAG?"*. Then as they are walking out the door repeat, *"Please don't forget to RAISE THE FLAG."*

- **Use precise language.** In this day when slang is commonly used in language (and Lord knows, I'm the worst) we need to put forth the effort to use precise language, focused on the activity at hand; being direct, simple and explanatory.

- **Never assume anything**, even if you have had the same conversation a million times before. Unless your child says, *"Gee mom, you've told me this a million times before."*, respond as if you haven't ever said it before and ignore your child's eye rolling. Repetition is the best teacher. We have learned that it instills the information in the brain for recall; it cuts a deeper groove in the brain, as they say.

- **Use clear gestures and facial expressions**. If we don't, it can be easy for the child to get mixed messages or wrong impressions. Our VCFS children don't readily understand when someone is being sarcastic because the words don't fit the tone of voice, so this is a concept that has to be taught. Children need to feel secure in all situations and not to have to wonder, *"Am I in trouble for something?"* by the look on our face. If you are not sure you are sending the right message, tell your child how you're feeling and explain the gestures you have made.

• **Explain gestures**. Tell your child that when you hold up an index finger, it means, *"I'll be with you in a moment, when I hang up the phone."* Explain any other common gestures such as a finger to the lips meaning, *"Be quiet".*

• **Create a need to communicate with your child.** Ask your child a question, then wait and listen. Leave pauses for the child to initiate some input. How many of us in this hasty world end up finishing their sentences for them? I know I was guilty as charged, when my kids were little. I hope I have finally learned more patience with age. Perhaps we could involve the child in a family discussion; asking their preference of locations for family vacations or choice of foods to bring on a picnic. We can watch a movie or read a book together then have our child tell us what it was about, in their words. This is getting to the crux of the cognitive challenges and stimulating language maturity.

• **Give visual recognition.** Support understanding of the child's frustrations with visual recognition. Don't ignore a child who is trying to communicate with you, even if you're busy. They want some attention. They need to know they are being heard and counted worthy to be heard.

Medical Trivia

It is a misconception that women change their minds more often than men. Research has shown that the opposite is true.

"Our greatest glory is not in never failing, but in rising up every time we fail." Ralph Waldo Emerson

GRADE SCHOOL AGES

By the time the children reach this age they are beginning the journey for the next 12 years or so of their formal education. VCFS children usually learn how to read, but of course, like everything else in this syndrome, this is not always the case. The children are usually satisfactory spellers but do not always understand the meaning of the words they spell. This stems from the cognitive problems. (Meaning they are not getting the understanding or connecting the meaning to the word). As we have stated before math is usually not a problem when it is the simple math. Since it is factual, it can be memorized and the answer does not change. Children at this age are usually happy and work well in a classroom setting. This is because it is structured, organized, everyone is doing the same things, and they know what is expected of them. They also have many children to imitate. At this stage the step-by-step direction in school is the key to their understanding. This will more than likely be the way they will learn best throughout life.

This is the time period that many of their _special_ _needs_ will manifest themselves and there will be tests to take, accommodations to request, special arrangements to set up, future plans to make, and more.

What approaches should I use to develop language and life skills?

• **VCFS children are visual learners.** They tend to be more aware of, and respond better to visual information than to spoken or auditory information (especially if they have hearing problems). Written instruction vs. oral instruction for things like work sheets and certain tests is an issue that can be debated.

Find what works best for *YOUR* child. However, although <u>*visual*</u> is the best form of teaching information, it can sometimes be misleading, distracting and stressful to a child, if the topic is not totally clear. For example: in a safety *"stranger danger"* film if the child doesn't understand what is going on and sees a child abuse an adult by kicking them and running away; it will be confusing. This can also happen with visual information such as sex education given in health classes when they are older. Therefore, the form of visual support chosen should be easily recognized by the child for the reason intended. It is not a good learning tool if the subject is not understood; or if there is not someone present to teach, answer questions, question what lesson was to be learned, and determine if it did its job.

- **Objects of reference:** A visual chart or timetable of daily activities helps in understanding of language, organization, and memory difficulties. This can be a very important tool as most of these children find it difficult to organize their thinking in many facets of their lives. This also helps teach the child to do ALL that is required of them on the list; and not just to do the things they want or like to do. They learn that life is full of things that are not fun or desirable (like washing dishes, cleaning the toilet or raking leaves), but jobs must be done, that is a fact of life. Most children seem to have a difficult time doing entire multi-tasked projects like *"cleaning their room"*. It becomes too overwhelming for them, not knowing where to start or what steps to take. A good way to teach them this chore, in my opinion, is to sit in the room and dictate what needs to be done, in order. (Perhaps even video tape them or someone else doing it, and replay it so they can learn from it.) Remember it is easier to take time to show them the right way, than it is to let them become frustrated doing a hit and miss cleaning job. Then, they end up either having to do it over or perhaps you might have to redo it yourself. If children are not taught how to do things right, it can carry over into the way they view and handle a job in the future as well. As we teach them, we need to tell them why we do certain chores the way we do, for instance: we dust from top to bottom before we vacuum, because dust settles down on the floor and the floors are always the last thing we clean; we shake the wet clothes from the washer to loosen them up because they get fewer wrinkles and dry faster, which saves electricity. This teaches them to clean, to be helpful *AND* to reason. Maybe even ask them, "Why do you think we do (whatever) this way?"

• **Visual Information.** Photographs, sketches and symbols help keep the child on task. These can be used in a demonstration of the work at hand. For example: A teacher might do a step-by-step drawing demonstration for the first problem on a math assignment. Doing it this way, reminds the student what they are doing and *HOW* it is to be done. Written reminders work well once the child knows how to read. Visual information is also an important tool when teaching <u>sequencing</u> of events. This is important because it is a difficult thing for a VCFS child to learn.
<u>Sequencing example</u>:
  What comes first?
    A). The completed snowman
    B). Rolling of the snowballs
    C.) The snow falling
The example I gave about watching a movie or program and then having the child tell you what it was about is an activity that will help with sequencing. It also helps them when they need to write book reports.

Another great and inexpensive tool is a large yearly calendar. Put a special occasion on the child's calendar and have them mark off the days, as they pass. This not only teaches them the concept of the calendar and how to count, but if they are obsessive about an event, they can see for themselves when it is coming, and know how many days are left to wait. This saves you from getting a few extra gray hairs. (From: **Is it Time Yet?**)

Education gets harder as the child ages

Starting at about 8-10 years of age, schoolwork becomes much harder and more abstract. You may find that a child with VCFS starts to have more difficulty with these advanced concepts. Teachers start looking at English and grammar more critically; math becomes more complicated with decimals, powers, story problems and algebra formulas. Sciences are harder to comprehend. We may have to utilize different aids to help our children, but still use the same type of visual help because that is what they respond to best. For instance, if you buy apples and quarter them this can teach fractions. It is the same with a pie. Better yet, let the child help make the pie, learning what is involved, measuring the ingredients, and seeing that measurements come in fractions wet and dry. This also helps in comprehending story problems. Monopoly® money is helpful to teach about money and values, to use in some story problems and in practicing making change. I have also found that taking real money, making a mini-cashier's drawer, and pretending to buy things; teaches the child to make change. Getting a job when they are older, buying things at the store, or just needing to know how much money they got back in change, are serious concerns for VCFS children and a real concern for parents too. I have found that allowing the child to watch game shows like *The Price is Right*® and *Wheel of Fortune*® has been an excellent teaching aid, they use their brain and have fun. When you play against them, they try harder and are eager to show you their skill.

By this time in the child's life we have worked out an organized plan that works for us and the child. This includes the daily planner concepts, having the child write down assignments and go over them with us each day. This reminds the child what he needs to do, but also alerts us to what is expected of him at school and if help is needed. For example: a science project is due this week. This daily planner can be included in the child's IEP. The teacher can help the child to remember to use the planner, reinforcing their understanding of what they are to do in the form of homework.

Having the child do everything on a schedule to help in organization is key. Lay out what they will wear the next morning just before bed, do homework at a regular time each day, feed the animals or take out the garbage at a specified time. It sounds like military school, but it works. Parents might even tape a hygiene schedule on the bathroom mirror. Some children don't like taking baths or brushing their teeth, so they conveniently forget. This note helps them to form a habit for the future, especially if they are eventually able to live on their own. Many find it is better for their children to be in a group home which is a scheduled environment, monitored and structured with people their own age. *(See Housing and group homes, page 315.)*

We want to really get to know our child. Periodically ask your child the questions below and take note of the answers. The answers will help *YOU* to help *THEM,* one age at a time.
• Do you enjoy school?
• What is it that you like about it?
• What do you find easy or difficult?
There is a more complete list elsewhere in this book to help you. *(See "What is your child like now" on page 123.)*

What can teachers do to help the parents?

- **Written and verbal advice:** The teacher should provide written and verbal advice to the parents. Teachers can give instructions to help parents work with their child directly. They can also provide written guidelines and ideas to help TAs (teacher assistants) or assigned tutors who work directly with your child. Information should be about whatever is needed to complete the assignment.
- **Feedback:** Teachers should provide follow-up feedback about progress in the therapy sessions. Provide ideas to parents about any language tricks (ex: alliteration, rhyme, silly word rhythms) or games they can practice at home. These might include things like a rhythm that can help in articulation or reading out loud into a recorder. Hearing themselves might lead to improvement.
- **Social scenario coaching:** Teachers can write different social scenarios and coach students in these to enhance their social success and awareness. This helps to teach appropriate manners and behavior.
- **Written goals and objectives:** Write out objectives (what we want to accomplish) and discuss them with your child's teacher. In this way, they can back you when you submit the objectives for the IEP or they can be a liaison with other agencies for a united team effort. Parents and therapist should form a partnership and work together toward goals and objectives that will benefit your child in all areas of life including: home, school, job and interactions with others. Let your child master one of the goals before hurrying them on to another goal.
- **Break large tasks into smaller segments:** If there is more than one task to be performed by your child, the teacher should break the tasks up into increments. Perhaps by writing them down and numbering them in order of preference. It will be the best method to use if they are tasks that the child can perform on their own. This teaches independence, logical thinking about the task, the best way to tackle the job, and it improves self-esteem.
- **Teach ways to approach asking questions:** Concepts like opposites hot/cold, same/different, open/close, nouns (person, place or thing) can be made easier to understand by visual means. Figuring out how to form a question is hard for the VCFS child, so it is beneficial to teach different ways to ask a question. Questions formed by using "who, what, when, where, how, and

why" can help in any given situation now and in the future. For example: Teach how to order something by phone, how to ask if a business is doing any hiring, how to make an appointment or how to ask if a friend can come over for dinner.

Sometimes we have to remind our child to look for information and read something instead of asking for help or giving an answer right away without thought. They need to try to learn how to figure things out by themselves first; especially when they get older. An example of this can be putting something in the oven or microwave. Instead of telling them how to set the time and temperature; have them look at the box, read and follow the directions themselves (if they are not cooking by scratch). Supervise at first to see that they are doing it right, but let them do the steps. Perhaps, have them read it to you out loud as they do the steps. However, in some instances with more complicated tasks, it is a good idea to demonstrate, show them how a certain activity should be done, and then show them what the completed task should look like. Then have them do it on their own. Check periodically to make sure they are still on task.

In adolescence/puberty the children are going through so much within themselves, dealing with more than the average child. Do you remember when you were a teen? As we can imagine it is harder for our children in many ways. They may feel clumsy, not attractive and as if nothing ever goes right. Depression can escalate in this time period so it is wise to try and make their day go as smoothly as possible. This will involve the organizational skills we have already discussed. Look for mood swings (extremely happy times followed by a low depressed critical mode). It is not uncommon for a child in this age group to have mood swings; but these can get out of control (manic or deep depression) if not monitored and/or helped with medication, the child can become persistent with emotional problems.

As we have mentioned, a computer is a beneficial tool, especially for this age group. A computer provides help for class work, personal reminder calendars, games, hand-to-eye coordination with keyboarding, and is even a means for them to have friends and communicate with them at their pace, and when they feel like it. This reminds me, my son, T.G. learned to type 35-40 words per minute on the keyboard and I didn't even know he was practicing. He has been able to work once in a while and one of the jobs was doing data entry for a couple of days at a bank through a temporary service. It goes to show that we never know what talents might surface in our children.

A hopeful future

Many children with 22q can progress in school at a reasonable pace if they are given special assistance and accommodations. To re-cap, some can function in a regular classroom environment with added individual help, tutoring at home, and perhaps a modified curriculum. Others will need a more intensive program in the small group setting. Most teachers will not know a great deal about this syndrome and will need to be educated about your child's special needs.

Parents must become informed and learn to advocate for their child, since many 22q children have a very difficult time expressing their needs themselves. Above all, parents and professionals must be flexible, supportive and encouraging to children with this syndrome.

Parents must understand that schools have limited resources and can only do so much. Parents will need to oversee homework completion and work with teachers to ensure a positive outcome for their child. Schools must understand that many 22q children have undergone multiple surgeries, have frequent illnesses and need many years of therapy to address all of their issues.

Deficits in problem solving skills and *common sense* could put these children at risk for school failure. As a rule, learning is not easy for these children; faced with rigid teachers and unbending rules, it can be insurmountable. 22q children may need multiple opportunities to attempt a skill, since most will not progress at the same rate as their peers.

With structure, realistic expectations, and praise, however, VCFS children can be very successful. Many of them graduate from high school, complete training programs, become employed, married and live productive lives. With advances in medication and a better understanding of the chemistry of the brain, it is not unrealistic to expect that things will become easier for VCFS children in the future. Current research, worldwide, on this syndrome is progressing rapidly and is offering new hope. In the meantime, both parents and professionals will need to work together to make school a positive experience for children affected by velo-cardio-facial syndrome/22q.

What is neuropsychology?

It is the study of brain-behavior relationships. It is primarily used in the pediatric population to determine a child's assets (strengths) and deficits (weaknesses). This covers areas such as language, visual spatial (the ability to comprehend a visual presentation and its relationships); non-verbal processing; memory; attention; executing a function or task; sensory and motor ability; academic achievement; and social-emotional functioning. This is very important information to shed light on the impact and the extent of the condition (deletion) and to help design a plan for education as well as development purposes.

Since 22q is not predictable and is so varied, it is wise to work with a neurophysiologist who is familiar with VCFS/22q or other genetic disorder characteristics. The performances in non-verbal learning, attention deficit disorders and in other social-emotional deficits are the focus for the exams that a neurophysiologist performs. The results of these tests help to pinpoint what specialists might be needed; such as, in areas of developmental behavior, speech and language, psychiatry, pathology, audiology, or occupational therapy.

Children with VCFS/22q have unique strengths and weaknesses or as the Children's Hospital of Philadelphia (CHOP) put it *"assets and deficits"*. They are listed as follows:

Assets of a VCFS child

• Verbal IQ and verbal comprehension:
Verbal IQ and verbal comprehension: 22q11 children show a unique ability to learn and retain verbal information that has been gathered through their life's experiences, if the experience makes a real lasting impression on them. They remember general information and comprehend practical verbal content. However, can they repeat the story back in a smooth flowing way? Probably not . . .

- **Rote verbal learning and rote verbal memory:**
One of the most intriguing. This is the ability to remember verbatim and repeat by rote verbal information such as lines from their favorite movies or songs. There are exercises that you can use to help them with this gift to stimulate memory in other areas; strengthening their retention capacity.

- **Initial auditory attention:**
Unlike other attention deficits, the VCFS child has a much more intense auditory (sounds found, for example, in movies and music) attention span. This leads us to believe that utilizing rote (meaning memorizing/repeating without full comprehension) and auditory teaching methods might prove useful.

- **Simple focused attention:**
Children with VCFS seem to be able to focus on a task and do it within normal limits if it is brief and structured. They are often able to maintain or increase their level of performance in time and with experience. (This is important to keep in mind when they are selecting a job in adulthood.)

- **Word reading and word decoding:**
Contrary to what their IQ tests may suggest, VCFS children seem to exhibit a very strong ability to read words and apply phonics in spelling. This does not mean however, that they comprehend the word they can spell or write. It also does not mean that they will be able to stay at the same level of competency if they do not keep active in spelling and reading exercises.

Deficits of a VCFS/22q child

Since many of these children have a more difficult time in school. It is wise, as we mentioned before, to speak with a special education teacher as well as a speech therapist as to what software you can purchase for the child to practice and memorize before the school year begins in that grade. Just this simple task can give your child an edge.

Non-verbal processing

Non-verbal processing involves the ability to relate to and understand what is going on around you. VCFS children have a hard time with skills like visual analysis (interpreting what they see, if it is not obvious), visual and time sequences (memory games or sequencing games, where you put the cards in order, the steps involved in "building a snow man" and determining what comes first in the sequences). This seems to contradict what we said earlier about visual being the best teaching method, but this is different. The child in this instance has to figure out what it means; not just look and learn by visual instruction.

VCFS children have a difficult time understanding relationships, (like snow boots and skis have a relationship with snow or a ski lodge). Looking at it in a literal sense of *relationship* is a good example also: *"How did Uncle Jeff become my uncle?", "How did Lynda become my cousin?", "How can Jackie be my aunt when she is your aunt too?"*. It can become very complicated for them to understand, and requires that they figure out and reason.

These children struggle with what is meant by the non-verbal expression or gesture, etc. This lack of understanding in the non-verbal world can lead to difficulty in the social and pragmatic community. I personally know of times my son would misunderstand or misinterpret intent on the part of another and think that the other person was being rude; when they weren't. This misreading of nonverbal cues and signs can limit their associations with others. It can also leave them open to predators, so we must be forever watchful in their behalf, and train them about the world and how to keep themselves safe.

Visual-spatial skills

• Complex verbal memory and executing the function:
Even though a VCFS child has the gift of remembering quotations or memorization as a whole, they have difficulty in comprehending more complex meaningful dialogue or information. They cannot hold in their memory or organize thought, any information containing long sequences of directions, sentences or stories accurately.

• Motor function, facial processing and recall:
The child may seem clumsy in motor skills and not in total control of their limbs. They also may not be able to process what facial

expressions are being showed by others or to exhibit the proper expressions in themselves. The difficult thing for parents is that the child may not be able to recall the appropriate expressions once learned the first or second time, thus the VCFS child has to learn things over and over again for it to become concrete in their minds and memory.

- **Phonological processing and language processing:**

VCFS preschooler's speech and language delays are fairly consistent if they have them. As they become school-aged and learn more language and verbal expressions, the difficulties often become more apparent. Some of the most common language deficits are verbal fluency, recall of paragraphs, sentence recall, expressive language, and receptive language.

- **Math & Reading Comprehension:**

Most children with VCFS exhibit a consistent deficit in math reasoning and reading comprehension. In the area of math it appears they have difficulty understanding quantitative logic which is needed to process the concept to solve the problems, especially, story problems. Reading comprehension deficit is the inability to process the language and organize their thoughts to form the logic needed for a deeper level of comprehension in order to understand the concept of the text they just read. Abstract thinking, as one would need if reading a mystery or who-done-it story, is difficult.

Social skills

- **Emotional functioning and adaptive functioning:**

Because of their non-verbal processing deficits, children and adolescents with VCFS are at greater risk for experiencing social emotional and mood disorders, such as social anxiety. These children have a tendency to internalize emotions that can bring on depression, anxiety, and some signs of OCD (obsessive compulsive disorders). Some show significant attention deficit behavior or have been diagnosed as autistic or having aspergers. Many of the children are very impulsive and this makes it hard for them to control their tendencies even if it is inappropriate behavior. For example: There was a time when my son was 12 years old, and he was so mad at a boy in middle school for calling him names, that instead of hitting him (which he had been taught not to do) he de-pants him right in the lunch room. His urge was just too great to do nothing. Many of these children also do not seem to comprehend a reward for good behavior system. A parent or teacher may put a reward system for good behavior into practice by giving special stars on the paper or a treat for not talking too much during class and these children may not see what is happening in this situation. Some VCFS children don't seem to understand why they can't do *ALL* they want to do *WHEN* they want to do it (like talking out-of-turn). *"Why is it is wrong?"*, they may wonder? It seems OK to them but, *"It must not be okay, because I didn't get a star and no one is smiling at me."* This is also the case for the concept of <u>consequences</u> of their actions. I have seen this more in the older children when it comes to money issues or choices that did not turn out well. They get frustrated when things don't turn out the way they had imagined or dreamed that they would. They start blaming other things, rather than accepting a *consequence* that resulted from their choice of action. Unfortunately, they tend to repeat the same action. (This is not the case for *ALL* VCFS children.)

Another interesting observation is that many children with VCFS do not seem to do well with perceived major changes in their lives. This can be in everyday community life outside the home, work or in daily living routines. This can even apply to working at communication to interact with others in school. This might answer the question as to why so many VCFS children like doing things by themselves, or will choose to stay home rather than go with a group or crowd to some place new. You may see this more as they get close to adulthood. It's easier for them not to have to change or learn a different behavior. So they may balk at having to learn things, and convince themselves that they mastered everything on the first attempt so they don't have to do it again.

- **Attention, working memory, and executive functioning:**

These areas can cause frequent problems to the parents and teachers of a VCFS child. One area is testing. The child has to shift their attention from what they were doing to now compete with the demand of a test. Many VCFS children have more short term memory than long term memory when it comes to studying for tests. *(I never have figured out exactly what the cut-off time period is between short and long term memory.)*

It seems that when it comes to memory it depends on what the situation is. Tests can be a difficult and stressful time for them. Retaining information is a challenge as well as understanding and remembering the instructions.

I wish I could say I was a saint when it comes to patience, but I was not, and this lack of memory retention can be very frustrating for a parent, teacher or a boss on a job as well. It is tough to know if you can rely on the child to *DO* the requested demand or *REMEMBER* to do the requested task. And if they don't remember, you can never be certain if it is a result of the human trait called _selective hearing_ and/or _procrastination_ or if it is actually due to the syndrome?

Should we discipline on school matters?

I, as well as other parents, believe that one should try not to punish at home for something that happened at school; unless of course, it is a matter that requires more severe punishment than can be administered at school. Instead, try to work to help the child understand that the school has established a punishment system that they use to handle what goes on there and if the child gets extra homework, has to stay after school, do outside clean up work or gets a tongue lashing; it is a direct consequence of their behavior. The parent can then support or back up the school by helping their child to figure out why this happened, what was on their mind when they did it and how it can be avoided in the future. Then, if they feel they were treated unjustly, you can talk about it and determine what makes them believe it is unjust. If you also believe it might have been unjust, you can take it to the school officials. If they refuse to listen, it is time for another lesson. That lesson might be that sometimes life is unfair—so let's learn to cope and move on. (This applies to minor infractions, of course, not to serious matters.) This opens up a great opportunity to brainstorm with your child on coping skills (within their ability) and to encourage personal responsibility. Remember there are times when cause-and-effect events that seem perfectly obvious to everybody else will escape the VCFS child entirely. This can happen regarding their behavior with teachers and/or with other kids.

Even though you have taught them well about one example, the knowledge might not transfer to the next incident! So then you teach it again and again when each problem happens. The lesson is the same, the scenario is different.

Parents Speak Out

It has been determined that many VCFS kids have (NVLD) non-verbal learning disabilities. If anyone has a child that has been diagnosed with this disorder the Web site listed here may be of some help.
http://www.drdeanmooney.com

The *"different learning style"* seems to generally include what's called non-verbal learning disabilities (NVLD). In other words, their ability to read and to speak is better developed than their ability to do math and read maps. (The term also includes having some social difficulties because they don't make the inferences that other kids do. You have to teach them explicitly about the meanings and effects of their own and others' behavior).

All 22q kids are different, but here's an idea of what NVLD (Non-verbal learning disability) is like: math concepts such as 1/2 vs. 1/4 are ideally taught before schooling, using visual and tactile materials. Get them used to hearing and seeing things, for example: *"Here's HALF the pizza. Do you want that much? How about ONE QUARTER of the pizza?"* The same idea applies to inches/cm and yards/meters. Kindergarten and some first grade math teaching concepts do the same thing. It's just that kids with NVLD need a *LOT* more help and repetition.

My son is 10, and he clearly has a non-verbal learning disorder. School has always been difficult, but we've managed up until now. This school year has been a disaster. Initially, he just couldn't remember to bring things home, or turn in his papers. We have called for an unscheduled meeting with his teachers and the school psychiatrist; I'm hoping to convince them that he needs a more hands-on approach, perhaps more one-on-one. It is hard to find short articles or educate the teachers on the concept of non-verbal learning. As an example: You can't teach someone about planetary rotation without starting with the concept of the axis that the planet spins on; but my child has no clue as to what an axis is. Our son requires so much background and context before he can understand something that he spends most of his time lost and frustrated. The frustration is catching.

There is a book I would recommend to anyone who wants to get involved with their child's IEP or is home schooled it is called **Non-verbal Learning Disabilities at School**. This book was published in 2002 and is a excellent book!

My daughter was always having embarrassing accidents at school when she had to go to the restroom more frequently than her peers. I finally made the point of letting her teachers know that it was a problem and had it included in her IEP so there would be no problem any more. I did have to supply the IEP team with a doctor's letter and explain as best I could about VCFS. It has helped.

I just wanted to pass along that my daughter (14-VCFS) made the A/B Honor Roll at her middle school last term. They had a school awards ceremony, and presented a certificate to each honor roll student. It has been a great boost to her confidence (and to ours). As we battle the health issues, emotional issues, insurance issues, life's issues and general worries about our kids, it's nice once in a while to have a *"win"* for our side!

My son's diagnosis came late. He was 15 before someone clued in that perhaps the many symptoms he was experiencing were actually part of a whole, of a syndrome. (I have heard this is very common) He was treated as if these were individual symptoms.

As a result, he has always been part of the regular school, with an aide to help. At first the aide was shared and eventually, as his health deteriorated and his learning disabilities became more pronounced, it became a one-on-one situation.

Which is where he is now. I am sure the fact that he requires oxygen is a major factor in the decision to have the aide. Plus, when he gets into distress. he does not tell anyone. He just becomes very quiet. It just takes one time of a quiet blue (to purple) boy to have the teachers and principal realize that we mean it when we say he can get very sick, very quickly. All his energy goes to breathing at times like this. He cannot spare the breath to try to get someone's attention so he is very quiet. He is in what they call a 100% modified program; meaning he is nowhere near his grade level. We do one hour of homework a night. If we don't complete the assignments in that time, we don't complete them.

I had a meeting with a person from the school board recently and this lady sat there and asked us what we want our son to do when he is done with high school. She insisted that *"We needed to plan."* I asked her if she had ever even met our son. She said, *"No,"* she hadn't. I responded that in 2 years I would be happy if he was still breathing. To answer her question: *"What is the one thing I want the school to do for him?"* I want them to let him have fun with kids of his own age, and to be allowed to be a kid. I can teach him life skills; I could even teach him basics; but I have no way of socializing him or letting him be with kids of his own age. We live out in the country and I feel school, at this point, is to provide the friends my home cannot.

I got this on one of my subscriptions and thought it might be helpful for someone . . . **The Anne Ford Scholarship Funds**. Did you know that a high school student with learning disabilities who needs money for college can get a scholarship? The Anne Ford Scholarship is a $10,000 award given to a high school senior with an identified learning disability (LD) who is pursuing an undergraduate degree. I know there must be more options out there if we knew where to look. For more information on this scholarship, go to: **http://www.ld.org**. (Look under "College.")

Medical Trivia

A cold does not turn into pneumonia as many believe. One is caused by a virus and the other by bacteria.

Blank page for personal data:

"Live the questions now. Perhaps then, someday far in the future, you will gradually, without ever noticing it, live your way into the answer." Rainer Maria Rilke

CHAPTER 6: SPEECH

SPEECH

Speech, language and feeding
Contributed by: Debra Leach M.H.A.
(CRS) St. Joseph's Hospital, AZ

Being a pediatric speech language pathologist for 20 years, I have been exposed to a wide variety of syndromes, disorders and variances in children. One of the most interesting and challenging conditions I have encountered is 22q11/velo-cardio-facial syndrome (VCFS). I have had the opportunity of watching the evolution of the syndrome in our craniofacial clinic from a rare diagnosis, to a frequent finding. Children in our facility are now routinely screened for the syndrome when they present with a cleft or cardiac findings. I have learned, along with parents that the diagnosis of VCFS is not one dimensional, and treatment of our children requires looking at many aspects of the child.

At the time of our first meetings, I usually tell parents that we will get to be good friends. I say this because frequent follow-up visits with the speech pathologist will be necessary as the child grows. In addition to monitoring feeding, the speech pathologist will monitor speech articulation, language and resonance.

I usually meet the families of patients with VCFS in early infancy. Many times parents look puzzled when I enter the room and introduce myself. I know what they are thinking. *"My baby is six days old. Why on earth would we be worried about talking now?"* When I explain that I am there to help with feeding, I am usually met with a big sigh of relief.

Many infants with VCFS have feeding problems. If the baby has a palate anomaly he may have difficulty creating an oral seal and sucking. I meet parents who say that the baby eats all the time. It is taking one to one and one-half hours to drink 2 to 3 ounces of formula. When this happens, it may be that the baby is working too hard to try to suck the formula from the bottle. Changing to a cleft palate bottle and helping the baby extract the formula from the bottle is most helpful for this. The baby then has to burn fewer calories and feeding time is quicker.

The baby may also experience gastro-esophageal reflux and nasal reflux in feeding. Nasal reflux is common in children later diagnosed with a submucus cleft palate. Feeding difficulty and slow growth may occur due to cardiac anomalies. Developmental or neurological problems may also impact upon the infant's ability to feed. They may have difficulty sucking, swallowing, and coordinating the tongue. They might have difficulty transitioning from one food texture to another. There are techniques for dealing with all these issues and the speech pathologist can help you work through them.

In infancy, the speech pathologist may meet with the family regularly to address oral-motor and feeding issues. As the child begins to grow, the role of the speech pathologist will also include monitoring speech and language development. Because children with VCFS are prone to ear infections, language may be negatively impacted. For this reason, the speech pathologist will watch to make sure that pre-speech activities such as cooing and babbling occur and that language emerges along appropriate time lines. If language is not developing, the speech pathologist might recommend the initiation of therapy to stimulate language development.

As the child begins to say words at about 12 months of age, we begin to get a feel for what speech might be like. At about two years of age, we can plan a baseline evaluation to more closely look at speech development. We can look at what sounds the child has, how he is using them, and how well he can be understood. If he is having problems in developing speech sound productions therapy might be recommended at this time.

As the child develops more speech and language and is beginning to speak in phrases and sentences, we will have a better feel for what his resonance is like. Resonance is characterized by airflow and resonation through the nose or mouth. The speech pathologist will want to evaluate the child between the ages of 2 1/2 and 3 to determine resonance function and come up with a follow-up plan of care.

It is necessary for there to be a good seal between the nose and mouth for speech production. If the child has insufficient amount of tissue or weakness in the muscle function in the velar pharyngeal mechanism, his speech may take on the characteristics of nasality. Some children may only be able to produce the sounds /m/ and /n/. Others may give the quality of talking through their nose.

If your child is very hypernasal and is receiving ongoing therapy with minimal improvement in resonance skills, your speech pathologist might recommend further evaluation of the velopharyngeal mechanism. This is achieved by either nasoendoscopic or videofluoroscopic study. Nasoendoscopy is performed by placing a small scope into the child's nose so that the palate and back of the throat (posterior pharyngeal wall) can be viewed. The child is asked to repeat a series of word phrases and sounds so that the speech pathologist can assess what the muscles are doing. From this information, the plastic surgeon and speech pathologist can determine whether to continue with therapy or whether the child would need surgery on the palate (posterior pharyngeal flap or sphincteroplasty).

Some children will not tolerate the nasoendoscopic procedure. In these cases, a videofluoroscopic evaluation might be recommended. This procedure is done in the radiology department of the hospital with the speech pathologist and radiologist present. A small amount of barium contrast might be inserted through the nose to allow for a more accurate viewing. The child is positioned in front of the radiographic camera with a microphone and asked to repeat a series of sounds, words and phrases. Again, the team can use this information to come up with a treatment plan for the child.

If a surgical procedure to correct velo-pharyngeal incompetence is warranted, the speech pathologist will want to see the child for a re-evaluation approximately four to six weeks after surgery. Waiting this long allows the swelling to go down and gives a better picture of what speech and resonance truly are. Most children will require intensive therapy post-operatively to learn how to use the new mechanism and eliminate any compensatory habits still in place.

Along with concerns about language and resonance issues, the child with VCFS may also have articulation disorders stemming from their hypotonia, fluctuating hearing levels, compensation for resonance issues and cognitive delays. Articulation disorders are defined as speech production problems related to inappropriate tongue, lip and jaw placement and movement for sound production. For these problems, speech therapy would be appropriate. The clinician works closely with the child and family to learn oral strength and placement to make sounds more clearly.

Poor social interaction or behavioral difficulties are sometimes associated with VCFS. These factors can impact upon the progress a child makes in therapy. When a child presents with difficulties of attention and focusing, the very structured activities required for articulation and resonance therapy are difficult to accomplish.

How to help your child

Having a new baby brings many new stresses into your life. Knowing the right things to do with the new baby, figuring out a schedule, learning your child's personality, and providing the necessary care your child needs are all challenging. Having a baby with a medical condition adds to these stresses greatly. One early thing that makes us feel successful, as a parent, is being able to care for our baby by feeding them. When a child is not gaining weight, taking a very long time to feed, or struggling through feedings, it is natural for a parent to feel that they are doing something wrong. Feeding may well be one of your biggest challenges early on and there are professionals who can help you in this process. The speech pathologist on the craniofacial team can assist you with finding the appropriate nipple and bottle for your child and help with positioning and oral-motor development. Along with the speech pathologist, you may have to try several tactics before feeding becomes regulated. Remember that the clinician is there to help you and understands how difficult this process is for you.

Babies learn to speak and understand by listening and practicing. Talk to your baby and provide verbal stimulation to him. When your baby coos and babbles, be responsive by cooing and babbling back to him. Engage in vocal play with your child even when he is very young. Label things in your environment to him. Children are never too young for you to talk to them and stimulate them verbally.

If you notice that your child is not responding to sounds in the environment or to your voice it may be that the child is not hearing adequately. Have your doctor check to make sure that the child does not have fluid in his ears and that hearing is tested to assure hearing acuity.

Children will generally say their first words at about 12 to 18 months of age. It is likely that these first words may not be very clear and sound only approximately close to the word intended. Listen closely to your child and pay attention to the environment they are in. If they approximate a word, give them credit for it, acknowledge to them that you understand what it is they are attempting to communicate. Label things in your child's environment for them over and over again and encourage them to try to repeat the words with you. Label foods, items of clothing, toys, and names of family members. If your child first attempts by gesturing, give them credit for this as well. Most children will gesture before they actually say words and some children will learn the signs for things before actually saying words. Signs and gestures can facilitate language development.

When your child is about 2 years old, he should have about 50 words in his vocabulary and start to combine two words together. You can help this process along by modeling two word phrases for your child. When the child says one word to you, repeat the utterance back to him with two words. If for example the child asks for *"Milk"*. You could reply with, *"More milk?"* and then give the child the requested milk. By this age your child should also have several different sounds in his repertoire. You can provide vocal play by modeling and stimulating sounds including /p/, /t/, and /k/ during social play and nonsense productions. By listening to your models and watching your productions, the child may be encouraged to produce the sounds.

At three years of age your child should be expressing himself in 3-4 word sentences and using sounds including /m/, /b/, /g/, /n/, /w/, /h/. He should also be able to follow two part commands at this age. An example of this might be, *"Go into your room and bring me a pillow."* Your child should have a name for about everything and you should be able to understand most of what he is trying to say. If you notice concerns in any of these areas, contact your speech pathologist. She will be able to help you, and determine if there is a problem and what the best strategies for dealing with it might be.

When your child is five, he should be talking in at least 5 word sentences. He should be able to listen to a story and answer simple questions about it. He should also be able to say most sounds correctly when speaking. At this age, there may still be some distortion of sounds including /r/, /th/, /v/, and /s/.

If you notice that your child has trouble blowing air through his mouth, has a nasal tone to his voice or continues to leak food or fluid from his nose when eating these may be signs that the palatal muscles are not working appropriately. If these issues cannot be corrected with therapy, it may be indicated to proceed with additional testing. Contact the speech pathologist on your craniofacial team for further assessment.

Conclusion

The road for development of speech and language may be a long and sometimes trying one. There are many factors, which may influence how well your child develops speech and language. Some children will fair remarkably well and never need the services of a speech pathologist, while others may be involved with speech intervention for many years. Some children may never have completely normal speech and language functioning. As parents, we take the tools we have available to us and use them to the best of our potential. We expose our children to the resources available to them. We teach them to compensate for weakness. We nurture the gifts that they have.

Will my child have speech problems?

After being assessed for feeding problems as babies, our children should then have regular follow-ups to monitor her speech and language, especially if the child is not talking, or not making many sounds by the time they should or if the sounds are different in quality, like grunts, hyper-nasal or high pitched. Your child may, in fact, have some hearing problems which need to be explored for a proper diagnosis. Rule that out first. Every child with VCFS should be considered for early intervention, which is screening, testing, and evaluating as early as possible to determine what types of special services the child might need to enhance their learning and development. This is found to have a positive effect on the outcome of future language and learning skills. It also touches on their social interactions for the future.

Although some children will need surgery eventually, it is very important that they have speech therapy early, to improve the chances of success. It is recommended by some surgeons that the adenoids not be removed in VCFS patients with VPI as this would create a _LARGER_ gap in the throat. This makes speech more difficult to correct due to air escaping through the nose, increasing the hyper-nasal sound of the voice. Yet other surgeons will strongly recommend that both tonsils and adenoids be removed prior to any soft palate surgery. VPI testing cannot be done if the child cannot speak at all. As they are requested to make sounds of letters of the alphabet to move the appropriate muscles in the throat. These muscles are behind the uvula. (Remember, it's the little mini punching bag at the back of the throat.)

My child has a hard time getting people to understand him. What we can do?

Communication takes place in many different ways such as verbal and non-verbal, as we have learned earlier, which can come at many different levels.

The answer to this question is complex, but in my opinion it boils down to teaching and instructing the child to slow down and speak as clearly as possible. Perhaps even teach the child that if the person they are talking to doesn't seem to understand, they can say it again in a different way as we sometimes have to do with them, rephrasing. Your teacher or speech therapist will be helpful in this area.

VCFS children have great difficulty with oral expressive language. A simple thing like asking a question can be exhausting and frustrating on both sides. They may say one thing and mean another. It is a process of groping for the right words to say, and then trying to put them in the right order to form a comprehensive thought. This causes the greatest challenges. These expressive difficulties may be from poor articulation, delays, difficulty with abstract thinking vs. concrete (to the point) communication, and simple concepts of the world.

These communication difficulties, as we mentioned before, can cause the child to have a difficult time finding friends who are willing to be patient enough to get to know the child as a person. It can limit social interactions with others.

The VCFS child might say, *"I can do that."*, when they may mean, they *WANT* to do that. This can cause problems that might get them into trouble at school or on a job when they get older. Sometimes they use the wrong tense or they only hear half of the directions or instructions.

As mentioned before, many VCFS children have a delightful sense of humor, yet they may show not only lack of facial expression but lack of verbal expression. This flat expression of physical and verbal enthusiasm can affect their social life; they find it difficult to look and sound excited or interested. Socially this can be disastrous. It may be these factors that dictate whether or not they get and keep certain jobs or lasting relationships. They may not show it verbally or emotionally, but they feel the same way that we do. This is another area where the speech therapist may help; by having the child practice voice pitch, tone and volume to rehearsed occasions.

Teach VCFS babies sign language
Babies with VCFS can be taught sign language
Contributed by Margi Stenson, M.S.W., Phoenix, AZ

Temper tantrums, frustration and stress can be reduced significantly if you teach your VCFS baby some sign language. Babies who know just a few basic signs can communicate their needs months before they utter their first words.

Babies as young as nine months understand spoken language and have the desire to express themselves; yet, developmentally they do not have the ability to speak intelligibly. They are able, however, to gesture using their hands and fingers because these tasks are not as sophisticated as those required for speech.

Teach your baby a few basic signs such as eat, thirsty, thank you, more and all done. The best place to begin teaching is at the table during meal time because:
• you have your baby's undivided attention, your child is hungry and thinking about food
• you are there together 3-4 times every day
• you are within close proximity to one another so your baby can easily see how and what you are doing
• many of the most useful signs center around food and eating
• A quick online search provided millions of options for "baby signing." There are different signs for America, Canada, Australia and other countries, as well as various systems designed specifically for young children by professionals. For many simple needs, parents can devise their own obvious signs to use repeatedly in daily situations. Here are some examples of signs that can be used with babies and young children who are not communicating with language.

Add more signs as you wish. You will expand your baby's signing vocabulary with each additional gesture you teach. When you

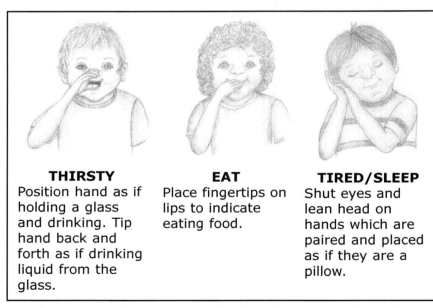

THIRSTY
Position hand as if holding a glass and drinking. Tip hand back and forth as if drinking liquid from the glass.

EAT
Place fingertips on lips to indicate eating food.

TIRED/SLEEP
Shut eyes and lean head on hands which are paired and placed as if they are a pillow.

say the word each time you make the sign, your baby will associate the gesture with the word and its meaning.

After babies learn some sign language they will generalize what they've learned to other situations. For instance, I frequently used the *"all done"* sign to mean we were all finished eating, yet my baby used it on the playground! Here's how it happened: I was pushing my son in the swing. After a very short time, 30 seconds maybe, he signed *"all done"*. I was surprised and asked him if he really was all done. He repeated the sign and seemed content knowing that I understood him. I stopped the swing, lifted him out and brought him over to the sand to play. No screaming. No guessing. No temper tantrum. Just a happy and understood baby.

Here's another example: my husband, Dan, was bouncing our son on his knee. After two minutes of bouncing him Dan stopped. My son looked at Dan with an expression of confusion and then immediately did the sign for *"more"*. Dan smiled, nodded and resumed bouncing him.

Get your spouse, your baby's grandparents and caregivers involved in the teaching process. The more your baby is exposed to sign language the sooner she or he will learn it.

Some children pick up signing quickly while others take a lot longer to learn it. Babies with VCFS can have a speech delay which would mean that they might not speak until two or three years of age. Teach them sign language to bridge the gap between nonverbal behavior and articulated speech.

You and your baby will appreciate the gift of sign language as a way to express needs, wants and feelings.

I am concerned about my child's language skills, should I be?

Language is the means by which our children understand what we are saying, learn from our verbal in-put and try to get us to understand them. When our 22q children get frustrated because their language is not being understood, the frustration sometimes comes out in behaviors. Should you be concerned? Yes, but do not become overanxious. There are things that can be done. Language is broken up into several parts, such as phonetics (sound), grammar, cognitive (meaning) and pragmatics. It is a means by which they learn to understand abstract expressions such as we see in philosophy. For 22q children this type of thinking is a very challenging concept to comprehend because they take things literally. However, there are ways to help children learn these expressions. One is to simply explain what something means immediately after you have said it. It might take several times, but they will learn.

Language becomes even more critical as we mature. It is the basis of our behavior, our success in relationships and how we learn in school and throughout life.

Language difficulties can also affect our communication with the outside world as a whole. The 22q child may not understand different facial expressions, body language or *"take hints."* It is not uncommon for a parent of a 22q child to say, *"Can't you see I am busy?"* or *"Do I look like I am in the mood for that?".* The fact is, they can't see it. With literal eyes yes, but not with eyes of understanding. These social aspects of language are called <u>*pragmatics*</u> (the relationship between signs and expressions in society). A speech therapist should be able to help explain what is necessary to help your child learn in this area.

By learning techniques from your speech therapist or speech pathologist it will show your child that he can successfully communicate his needs, thoughts, and feelings. This may be by a combination of words, gestures, and sign language. Your child's teacher can also help by making sure your child understands exactly what is required of him either by explaining through visual means or by taking a step-by-step approach. Please don't assume that if your child seems to understand something today, that he will understand it tomorrow, without a little refresher course.

There will be some 22q children who never learn to speak in complete sentences or articulate a complete understandable dialogue. They do, however, seem to get their point across. Many children with 22q will have some kind of language delay which can be helped by showing them the different ways of communicating. This may involve gesturing and/or sign language combined with speech as mentioned. Use of these methods might help to prevent frustration, until eventually, speech takes over. This will motivate the child to carry on with his learning, because he will be pleased when he realizes his attempt to communicate with you has been understood.

> *It is important to <u>rule out hearing loss</u> before thinking it is a delay. Consult with your doctor to arrange for a thorough hearing evaluation. Or, contact the* American Speech, Language and Hearing Association *in your state.*

According to the time line of the Children's Hospital of Philadelphia called CHOP, your child should be evaluated for the following at these stages:

- **Newborn:** to evaluate feeding difficulties
- **Follow-up in 6 months to a year:** to monitor emerging speech and language skills
- **Evaluate prior to entering school:** To determine how language is developing and what interventions may optimize your child's learning
- **Older Children:** Follow-up an initial evaluation, follow-up appointments should be tailored to the child's special needs

This reminds me of a quote by Charles R. Swindoll who said, *"Each day of our lives we make deposits in the memory banks of our children."* I hope all of our children's memory bank accounts will be full of wisdom someday.

Parents Speak Out

 I wanted to comment on how we can become confused when our kids say one thing and mean another. My son has had this problem ever since he started talking. When he was little, I just thought it was confusion over words as he was learning them. For example, he used to say, "*I did it on purpose.*" when he did something by accident. Now that he's 11, he's more aware and sometimes catches himself saying the opposite of what he meant and it can be extremely frustrating for him. If we ask him to explain what he means, he just says, "*never mind*" and gets upset with everyone for pressing him to explain what he meant. Another case in point; a child at school tried to help him open his bag of chips. When he couldn't open them, my son took the bag to the teacher and said his friend WOULDN'T open the bag for him. He missed that the friend was TRYING to open the bag. I have read a lot about 22q11 recently because of my son's cognitive regression, and I have come across MANY references to our kids saying one thing and MEANING another. This happens to us all the time. He'll mean to say, "*I DON'T want the sour lemon Popsicle®.*" but what he says is, "*I WANT the sour lemon Popsicle®.*" Now imagine how upset he gets when he ends up with the sour lemon Popsicle® *all* the time. Some of the articles I want to share have stated things like the child says, "*I can't do this.*" when they mean, "*I need help with this.*" This is all very frustrating for the whole family. We have to guess half the time what he means.

I know there has been a controversy over teaching sign language to our children with VCFS so I wanted to add this: A set of researchers at the University of California, Davis, studied a group of *typical* developing children who were taught sign language as infants. They have compared them over a number of years with a control group and found that learning sign language prior to developing verbal speech is NOT associated with delays in speech acquisition. There was more advanced language later on, and higher IQ scores down the road for those who were taught sign language. They believe this is due to the brain development that is facilitated by early language, any language. They also found that the parents really enjoyed being able to communicate with their children during those first three years. I think they wrote a book for the general public in addition to the academic articles.

A personal note from this parent: We taught our son sign language beginning at 15 months of age, and as a toddler/preschooler his language was advanced for his age, despite his inability to speak. Once he was able, the speech replaced the signings. In the transition period, it was wonderfully helpful in enabling us to decipher his attempts at speech, before they were actually intelligible. I believe this encouraged him to keep talking, by lowering his frustration level. We always spoke as we signed—total communication—and therefore he did also.

I have a son who at 2 1/2, was still nonverbal. He seems to be incredibly bright and I was told he had a severe speech output disorder, meaning, *"He couldn't talk!"* At about 3 he started to get very frustrated, so much so that he invented ways to make his own sign language. It was decided then to start teaching him *Makaton* which is simple signing. It proved to be a real benefit and eased his mounting frustration! We always voiced the signs at the same time and it hasn't affected him in any negative way, like causing him not to want to ever talk. As a matter of fact he is now 12 years old and never shuts up! He was a late bloomer in speech, words at 6, and sentences at 8. He has had continuous speech therapy and is doing well.

Medical Trivia

Rain contains vitamin B-12

"Life loves to be taken by the lapel and told: 'I'm with you kid. Let's go.'" Maya Angelou

CENTRAL AUDITORY PROCESSING

Contributed by Debra A. Lightfoot, CCC-A
Syracuse, NY

What is <u>C</u>entral <u>A</u>uditory <u>P</u>rocessing (CAP)?

Simply put it is, *"What the brain does with what the ears hear."* (Katz).

To be more thorough with the definition, it is a deficiency in one or more or the following abilities:
- Attend, discriminate and/or identity as acoustic signal
- Filter, sort and combine information appropriately
- Store and retrieve information efficiently
- Segment and decode acoustic stimuli
- Attach meaning to multiple acoustic signals

There are five basic areas:
- **Auditory memory:** ability to recall sequences
- **Auditory discrimination:** ability to note phonic differences in words
- **Auditory figure-ground:** ability to screen out noises and distractions
- **Auditory cohesion:** ability to organize, interpret and process
- **Auditory attention:** ability to maintain a purposeful auditory focus over an extended period of time

Key concepts involved in auditory processing of speech

- That spoken words are a rapidly decaying event
    1) Speech is a component of the acoustic (sound or hearing)

signal and is also based on predictions of context and familiarity
- Even if auditory stimuli is repeated, it is seldom exactly the same
- Differences may be in context, inflections, and/or intonations within same speakers saying the same message

Top-down approach
- Anticipate what is said using our knowledge of the world and understanding of language
    1) Therefore, we are not dependant on receiving entire acoustic signals (hearing) to understand speech
    2) Use rules of language to *"fill-in"* the missing gaps

Bottom-up Approach
- States we are totally dependent on receiving and interpreting each of the components of spoken sound in order to comprehend speech
- Unrealistic, as it is too slow a method

Your speech pathologist will be able to explain this in depth.

Why might my child be referred for testing?

Most common reasons for referral are:
- Suspect the presence of a peripheral hearing loss based upon behaviors exhibited
- Child is not working to expected potential

Behavior at risk for CAPD
- Says *"huh?"* or *"what?"* a lot
- Gives inconsistent responses to auditory stimuli
- Often misunderstands what is said
- Consistently requests repetition of information presented
- Poor auditory memory
- Easily distracted
- Difficulty listening to speech and noise
- Difficulty with phonics and speech sound discrimination
- Slow or delayed response to verbal stimuli
- Reading, spelling, and academic problems
- Behavioral problems
- Poor receptive and/or expressive language deficits

CAP Test Battery
- Peripheral assessment
- CAP test

The Fisher's auditory problem check list:

Circle each of the numbered items on the list that is of concern.
1) Has a history of hearing loss
2) Has a history of ear infections
3) Does not pay attention (listen) to instruction 30% or more of the time
4) Does not listen carefully to directions often to be able to repeat it back
5) Says *"huh"* or *"what"* 5 or more times a day
6) Can not attend to auditory stimuli for more than a few seconds
7) Has a short attention span
8) Can be easily distracted by background noise
9) Daydreams, attention drifts, not "with-it" at times
10) Has difficulty with phonics
11) Experiences with sound discrimination
12) Forgets what is said in a few minutes
13) Does not remember routine things from day to day
14) Displays problems recalling what was heard last week, last year, etc.
15) Displays problems with recalling a sequence that has been heard
16) Experiences difficulty following auditory directions
17) Promptly misunderstands what is said
18) Does not comprehend many words or verbal concepts for age/grade level
19) Learns poorly through the auditory channel
20) Has a language problem (ex: vocabulary, phonics, etc.)
21) Has an articulation problem
22) Cannot always relate what is heard to what is seen
23) Lacks motivation to learn
24) Displays slow or delayed response to verbal stimuli
25) Demonstrated below average performance in one or more academic area(s)

Peripheral assessment
- Pure tone threshold (air and bone)
- Speech recognition threshold
- Word discrimination (quiet)
- Immittance testing

Purpose of CAP test battery
- Is to address the auditory mechanism at various levels and then to compare results with the age mates

Central testing
- Speech–in-noise
- Phonic synthesis
- Staggered spondaic word (SSW) test
- SCAN-C test screening

What do these mean?

Speech-in-noise
- W-22 words presented at -5 signal to noise ratio (this reflects most class room noise levels) This is the noise level in most classrooms where most children are expected to learn.
- Look at both ears separately
- Assessing the difference in performances in noise as compared to quiet
- Frequency will see a drop of 40%-50%, if not greater

Phonemic synthesis (PS)
- Sound blending task
    4 skills involved (recognize the phoneme, remember them, retain appropriate sequence and combine them to form a word)
- Response is to give word comprised of individual phonemes
- 25 test items
- Most common type of CAPD
- Easiest to remediate
- Qualifiers as important as quantitative score (indicates behavior during test even if answer is correct)

Staggered spondaic word test (SSW)
- 40 spondaic words (2-syllable words) presented which partially overlap
- Response is no repeat stimulus items in order heard

SCAN-C
- Screening test
- Age range is 5 years, 0 months to 11 years, 11 months
- **Comprised of 4 subtests**
  1) Filtered words
  2) Auditory figure-ground
  3) Competing words
  4) Competing sentences

These are:
SCAN: Filtered words
- Repeat words that sound muffled
- Low-pass filtered at 1000mz
- 3 practice, 20 words - in both ears

SCAN: Auditory figure ground
- Speech babbles at a signal to noise ratio of +8 dB
- Repeat stimulus words (mono-syllabic) - 20 words per ear

SCAN: Competing words
- 2 words are heard simultaneously, 1 word to each ear
- 15 words pairs per ear (repeat word heard in right ear first (for initial 15) then word in left ear first (for next 15 words)

SCAN: Competing Sentences
- Hear 2 sentences; repeat sentence heard in primary ear only

Classification of auditory processing disorders

In 1993, the Buffalo Model was proposed.

It is comprised of 4 categories:
1) Decoding
2) Tolerance fading memory
3) Organization
4) Integration

Decoding:
- Break down occurs at the phoneme level
  1) Person may be unsure of what they have heard

2) Deficiency remembering phonemes and manipulating them
• Associated difficulties include poor reading/spelling and receptive language deficits/articulation errors

Tolerance fading memory:
• Poor auditory memory
• Poor figure ground skills
• Difficulty in speech in-noise tasks
• Associated difficulty with poor reading comprehension, a high level of distractibility, poor memory and expressive language deficits
  1) Seems to be as attentive to background noise as they are to primary signals

Organization:
• Very few pure cases
• Tendency to be very disorganized

Integration:
• Experiences significant difficulty in bringing information together
• Tends to have severe reading and spelling deficits (may be identified as dyslexic)

Management - school:
• In the educational setting, students are required to listen approximately 42% of a school day
• Classroom teachers typically spend less than 8% of their time teaching children how to listen
• With advancing grades, frequency of required listening time is increased, while time available to teach listening decreases *(See Education on page 143.)*

Here are some word definitions that you may see in this category and to help you understand the terms:

• **Air conduction:** Response obtained by a person when headphones or insert earphones are placed. Sounds travel through outer ear, middle ear and inner ear.

• **Assisted listening device**: A device that enhances the speech signal to the user; it improves the signal-to-noise ratio by focusing attention on primary speaker and reducing extraneous background noise. Teacher wears a microphone, student wears a receiver coupled to the ears by headphones or ear buds.

- **Auditory attention:** The ability to focus on relevant auditory information and maintain that attention

- **Auditory closure:** The ability to understand words and sentences when a portion of the message is missing

- **Auditory discrimination:** To discriminate between sounds and words

- **Auditory figure-ground:** The ability to isolate primary sound source in the presence of background noise

- **Auditory overload:** When the auditory system is unable to efficiently process incoming acoustic information

- **Auditory Synthesis:** The ability to blend phonemes into words

- **Binaural integration:** Process different information presented to both ears simultaneously and repeat what is heard in both ears.

- **Bone conduction:** Hearing levels obtained by bypassing outer and middle ear. Vibrates skull within thereby stimulating the middle ear directly.

- **Central auditory nervous system:** The pathway to the brain beyond the peripheral mechanism

- **Conductive hearing loss:** Air conduction scores are worse than bone conduction scores.

- **dD-decibel:** A unit to measure hearing sensitivity.

- **Dichotic:** Different stimuli is presented to each ear simultaneously.

- **Diotic:** Same sound presented in both ears.

- **FM system:** An assisted listening device that is comprised of a microphone, transmitter and receiver; used to enhance speech signal and reduce background noise.

- **HL:** Hearing level

- **Peripheral hearing:** Hearing sensitivity as evaluated by air/bone conduction. Comprised of outer, middle and inner ear.

- **Phoneme:** An utterance, to speak; it is the smallest unit of speech. It distinguishes one utterance or word from another.

- **Sensation:** To identify the presence of a sound.

- **Signal-to-noise:** The ratio of the signal to a corresponding noise, tested in the same ear.

- **Speech recognition testing:** A speech identification test.

Parents Speak Out

Since I went to a wonderful workshop last year presented by **Talk Tools Innovative Therapists International**, I have wanted to share this Web site. This is a group of therapists focused on *Oral-Motor Therapies*. Their Web address is: **www.talktools.net**. I got a lot of ideas about helping oral motor work for my son. You can request a catalog on-line. I hope this helps someone.

The inability to blow out birthday candles, blow bubbles or horns is suggestive of either VPI or oral apraxia (or, conceivably, both). It is an accepted practice to do a period of trial therapy to see if the child can learn to direct a stream of air through their mouth. If there is no meaningful progress in a predetermined amount of time (generally not to exceed 6 months) with therapy alone, the standard practice is to recommend assessment for some form of physical management of presumed VPI. Physical management options are basically surgery or a prosthesis. (For those of you new to these terms, VPI stands for velopharyngeal insufficiency or inadequacy. It means that either the physical structures in the soft palate and the top of the throat aren't adequate to allow the muscle to close off the opening between the nose and throat for swallowing and/or speech and/or they don't work well enough, if at all.) If a child is able to pinch his nose and blow through his mouth, chances are that he knows how to blow, but is anatomically and/or physiologically unable to close off the nose when needed. If food and/or liquids come through the nose (so-called nasal regurgitation), the problem is most likely fairly severe. Personally, I would want a recommendation from a reputable cleft palate/craniofacial team before I would attempt therapy to address this aspect. With our VCFS kids, if they are unable to blow, have nasal regurgitation, have a nasal sounding voice, and/or are not developing consonants such as P, B, D, T, K and G, I would suggest a reputable craniofacial team with knowledge of VCFS (or an otolaryngologist with specific knowledge of VCFS) before I would embark on a course of therapy involving directing a stream of air through the mouth. (This is only my opinion of course.)

The incidence of VPI is so high in our kids that my tendency is to presume it is there until demonstrated otherwise; unlike other kids, few of whom will have VPI. That said, following physical management (surgery or prosthesis) or in borderline cases, therapies which incorporate such blowing exercises can have value. I attended a *Talk Tools* workshop and there is a specific set of exercises for improving the function of the velopharyngeal mechanism (presuming that it is in potential working order). The principle is sound, but it is not a replacement for physical management. Rather, it can be seen as one therapeutic tool to use after physical management. Note that the compensatory strategies developed by kids before physical management can get in the way of the development of more normal speech following surgery or the building of a prosthesis. This is one of the reasons the *American Speech Language Hearing Association* takes a strong position on not continuing such therapies if they are not successful in a relatively short period of time.

Lastly, there are some cases where neither prostheses nor surgery are possible. In these cases, teaching of an array of compensatory strategies is the only option. I hope this helps someone. The above is not intended to be a diagnosis or treatment recommendation for any individual, but is offered only as information.

A speech bulb is a painstakingly shaped piece of acrylic, fastened to an acrylic plate that is fitted to the roof of the mouth. It has metal wires that go around the molars, which are fitted with bands upon which there are metal hooks. Think of an orthodontic retainer, but instead of wires around the front teeth it has an arm sticking out the back with a lump sticking up. It comes out to sleep, but is worn all day long. Its purpose is to fill the gap between the soft palate (velum) and the back of the throat (pharynx) in cases of velopharyngeal insufficiency (VPI).

One good thing is that there is no surgery required. It is, then, a conservative treatment for VPI. In some cases a palatal lift is called for instead. This is for a situation where the velum doesn't move, but is long enough. In this case, instead of an arm with a lump of acrylic, there is a shelf that extends out the back. The soft palate rests on this shelf in a higher position than when it's just hanging there.

Medical Trivia

An average human drinks 16,000 gallons of water in a lifetime.

"A happy person is not a person in a certain set of circumstances, but rather a person with a certain set of attitudes." Hugh Downs

CHAPTER 7: BEHAVIOR

BEHAVIORAL PROBLEMS

According to *Webster's Dictionary*, behavior is *"a manner in which one behaves"*. This can change from time to time and it can be altered with teaching. Sometimes there is need for medication. In many instances this will be the case for those children with VCFS/22q.

I've heard that I should watch for behavioral problems. Are they typical for VCFS children?

The answer is yes, they are typical, but this does not mean that YOUR child will have them. My son was one who didn't. There was nothing notable until he was 19 years old. That discussion will come later. *(See T.G.'s story on page 359.)*

Behavioral problems are usually first recognized at home or in school, by either acting out or showing a deficit in social skills. This can occasionally be lessened by talking to a skilled clinical psychologist and/or a speech therapist. (They are both knowledgeable in language and social interaction.) Because language delays can cause the child to be socially immature, proper patterns of conduct must be taught. They can be taught what appropriate social behavior is and what is not, using examples. The child might not understand <u>WHY</u> some actions are considered appropriate, while others are not; but they can learn the behavior and then understand it in time. When my son was a toddler, I personally used a technique when we would go into a market. He would sit in the basket. If there were any children acting up or pulling things off the shelves, I would bend down to his ear and whisper, *"Do you see that? Is that nice?"* Then I would say, *"No, that's NOT nice."*, and then tell him that they are being *"RUDE."* So the word *RUDE* became a word I could use quickly if he was not acting properly, and he knew or understood from having seen it first hand. This was a simple teaching tool, but it worked.

We all need help from time to time, so here are some guidelines to help you.

Guidelines to follow

- **Prepare to back up your words with actions.** Don't just repeat, *"If you do that ONE more time, I'm going to . . ."* and then do nothing but yell. You should not have to ask a child more than twice. We need to keep control as parents or we will have larger problems when they get older.

- **Talk to someone you respect or a professional** if your child's behavior concerns you. Even though we may learn from those with knowledge about how to discipline or how to teach proper instruction; we need to realize that ultimately, we are the ones who will need to follow through. Be resolved to do that, even if it is exhausting; and it will be. But it will be worth it for the greater long term good.

- **Don't jump to conclusions.** We all know that circumstantial evidence can be used to condemn an innocent person. This can also be true of our children. We don't want to lose their trust in us or have them be afraid to tell us anything; so until you know for sure, try not to jump to conclusions. In my opinion parents are right 8 out of 10 times; *BUT* if the training you give your kids is to work as intended, they must see you as reasonable and rational as well as fair. Show that you are willing to give them a chance to prove their statements of innocence. They also need to be willing to listen to why it might look as if they are guilty.

- **Form your statements positively.** This is hard to do at times, especially if you are angry or upset. Try not to always show a negative side to everything. Your child should not have to cringe at the thought of having a conversation with you.

- **Tell your child that you love *them*, but that you do *NOT* like their *behavior*.** Have a plan of action ready and then follow through. There were times I used to tell my children that I loved them *BUT* I didn't like whatever they did. Then I would explain why; and also explain what the punishment was going to be. Then I would ask if they understood all that I had said. I never disciplined without them knowing *WHY* they were being punished. What good would that do? Then, when they would act appropriately I would let them know that I loved them and they were just the kind of person I *LIKED* to be around, and so would others.

• **There have to be rules** but don't make them numerous and complicated. When making the rules involve the child (depending on age of course). These are normal standard rules that most households usually have. By involving the child, it helps the child to think about the purpose of the rules and to remember the rules more concretely.

• **Teach basic right and wrong behaviors.** And as I mentioned before, perhaps have a key trigger word (e.g.: "rude") to use in helping them realize the action is not acceptable.

• **The VCFS child needs to have one-on-one time** with each member of the family. This is not only for the VCFS child but for each family member to really get to know the child, bond and be able to help where help is needed throughout life.

• **Listen to your child patiently.** This can be challenging but it is important as your child struggles to be heard and understood. This could be a time when "signing" may play a positive role. *(Please see "Teach VCFS babies sign language" on page 182.)*

• **Teach independence by offering choices, but not too many.**

• **If you become angry, step away from the situation and cool down.** If you find you are getting angry regularly, it is important for you to have some quality time away from *everybody!* Take some time to rest and get some *quiet* time. As a parent of a child with disabilities and possibly with other children to care for, you need some breathing space to stay focused, healthy and happy. Remember the words of Lawrence J. Peter when he said, *"Speak when you're angry and you'll make the best speech you'll ever REGRET."* So, make plans for regular quiet times for yourself, so all in the family will be happy. This is a PRIORITY.

- **Always be watchful and alert.** Many VCFS children have easy going personalities and are followers. They tend to be people pleasers. Because of this, VCFS children can be easily guided into bad situations. The real problems in this area usually start when they are close to entering middle school and beyond. They want desperately to fit in and be accepted. They might be frightened and believe that if they act tough or join a group of kids, they will be protected and have self esteem. Never fool yourself into thinking *"not my kid."* That can be when we lose them. Always be watchful and alert, but not paranoid.

Many of the VCFS children can have an obsessive, autistic-like or manic behavior pattern. They may even have been diagnosed with Asperger's syndrome. For more information on that, go to **http://www.aspergersyndrome.org**. The child can become totally absorbed in some small unimportant "thing" (ex:object or subject). They can also start fantasizing; thinking their daydreams can come true or believing that they can do wonderful grandiose things. Once they get their mind set, it is difficult to reason with them. You may wonder why I mention it in this chapter. It is because you may have to do more counseling with the addition of medication, if they display any of these behaviors. This will also help them fit in at school and not be labeled disruptive or weird. (See page 220: Autism/Asperger.)

Here is a list of behaviors that might indicate problems. A psychologist or other professional should be informed about them. They will want to know to what degree, if any, these symptoms are appearing in your child and at what age. "0" meaning no symptom to "10" meaning severe symptoms that affect the child and the household. Please use this as a guide and use the blank papers included to keep a diary of the month and date you see any of these symptoms. This will tell you if there is a real problem, a continuous problem or if it is something else that is natural for the age group of the child.

Behavior list

_____Destruction of property
_____Aggressive toward others
_____Consistently irresponsible
_____Lack of impulse control
_____Inappropriately demanding and clingy
_____Stealing

_____Deceitful (lying and conning)
_____Hoarding (ex: food)
_____Inappropriate sexual conduct and attitude
_____Cruelty to animals
_____Sleep disturbance
_____Enuresis (urine incontinence) and encopresis (fecal incontinence)
_____Frequently defies rules – oppositional
_____Hyperactive
_____Abnormal eating habits
_____Preoccupation with fire, gore or evil
_____Persistent non-sense questions and increased chatter
_____Poor hygiene
_____Difficulty with novelty and change
_____Lack of cause-and-effect thinking
_____Learning disorders (when did they appear)
_____Language disorders
_____Perceives self as victim - helpless
_____Grandiose ideas and sense of self-importance
_____Not affectionate on parents' terms
_____Intense display of anger or rage
_____Frequently sad, depressed, or hopeless
_____Inappropriate emotional response
_____Marked mood changes
_____Superficially engaging and charming
_____Lack of eye contact for closeness
_____Indiscriminately affectionate with strangers
_____Lack of stable peer relationships
_____Cannot tolerate limits and external control
_____Blames others for all their mistakes and problems
_____Victimizes others – perpetrator or bully
_____Lacks trust in others
_____Exploitive, manipulative, controlling and bossy
_____Chronic body tension
_____Accident prone
_____High pain tolerance
_____Tactility defensive (pulls back from being touched)
_____Feels lack of meaning and purpose
_____Lack of faith, compassion, and other spiritual values
_____Lack of remorse and conscience
_____Self-defensive
_____Identifies with evil and the dark side of life

Blank page for personal data:

Medical Trivia

A human brain is about 85% water.

*"**P**roblems are only opportunities in work clothes."*
Henry J. Kaiser

PUBERTY

Puberty is a time when, our children are not only dealing with normal hormonal changes, but this might also be the time when emotional disorders flare-up or become exaggerated. Anything traumatic can trigger this if they are predisposed; a move, a severe illness, extreme stress, etc. In my opinion, I believe puberty is the worst time for cognitive problems as well, in as much as it becomes so important for them to know certain things for safety reasons and for moving onto adulthood. Things like understanding tasks, hygiene, self-esteem, social skills, hyperactivity, impulsiveness (hang on to your wallet), gullibility and mood swings. One subject I feel needs to be mentioned in this adolescent time period is that this can be the time when your child might exhibit inappropriate or unusual _sexual urges_. This is usually with the onset of a mania. It can get so severe that they might unwillingly demonstrate improper behavior.

Because high school is a faster moving and more demanding place; it can be very discouraging and intimidating to the VCFS student. Many are already disheartened by not being able to keep up with their peers, and they might feel like they don't belong. So, if we as parents don't get involved and keep a running dialogue, supporting their triumphs; they will find other ways to feel complete. Many times it is in a negative way. Having the benefit of a good speech therapist that understands how hard it is when you have communication difficulties can be a great help and advocate. The therapist can only do this if you inform them and your child's teachers of the difficulty. Explore what they can do to improve their own communication and teaching methods. This could also be brought up when making an IEP.

What are some ways I can promote good behavior?

The word "behavior" can mean good or bad. It would be unreasonable for us as parents to expect our child to be perfect, living under the old dysfunctional line of *"Children should be seen and not heard."* But we can set reasonable standards of conduct; such as: not interrupting when adults are having a conversation, waiting their turn to have the floor, or touching everything in sight when you go places, respecting other's property, etc. If these expectations are met by the child we can have a reward system in place to encourage more good behavior; or verbal praise might be all that is needed. These standards should be the same for all the children in the household. It is much more difficult for a child who has to struggle with impulsiveness or ADD issues, so they deserve special recognition and praise for good behavior. These standards should be appropriate for age, mental capacity, mental maturity and intellect. No matter what standard or rule you set it must be understandable to the child or they will become confused and will have learned nothing.

An encouraging way to show a child that you are proud of him is to do what one parent shared. Start a scrapbook of all their achievements and awards that are commonly given in schools. Include pictures of good times, special events, perhaps when they took music lessons, swimming lessons, dance, etc. When a child feels loved and special they are more excited to please. More recognition can result in more good behavior.

Since it is *Behavior Management* that we are discussing, it is here that we should talk about the task of teaching our child how to act or manage themselves in any given situation. This includes how to act in a situation where: someone might take advantage of them; someone might try to talk them out of their money; they are working and the boss does not want to pay their wages; someone tries to take advantage of them sexually (Explain that people might pretend they are friends or family, and want to touch them inappropriately.); someone is lying to them or conning them to get something from them, etc. I say these things only to help us realize we are human and that any of these things could happen to anyone, at any time. So keep your eyes and ears open when it comes to your children, with or without disabilities.

My child always misunderstands when I am upset.
She thinks I am angry at her. Why?

It is usually beneficial to let a child know what emotion you are feeling. Children learn by this method how to read expressions and to understand that all people, even mom and dad feel the same emotions as they do. This can teach them not to over-react or misread emotions of their parents and others, when there is an unexpected action or expression. I personally learned that this discussion can be helpful. If I were to act out of character, I would sit my son down, apologize for my own behavior, and explain to him what had caused my outburst. I would explain that it is also difficult for grown-ups to act suitably all the time. I would admit to him that this didn't make it right and that I would work on it, just as I expected him to work on improvements. I told him never to be afraid to say that he was sorry and admit his faults. It is a big person who can admit they were at fault and not make excuses or blame anyone else. This also taught him to be truthful and to become empathetic, by showing feelings of kindness toward others. It seemed to have worked. T.G. has always tried to be honest. I am pleased with the adult he has become.

If an outburst on a parent's part is because of the child, an illustration works well. An illustration takes away the blame, shame or finger pointing at the child, while teaching a lesson. After all, we want to teach our children, not humiliate them. In the illustration you choose, it would be wise to show the child how a different action could have produced a different outcome. If it is something that affects the whole family and all should be aware of it, conduct a family gathering. Be upfront and honest on the areas that need to be discussed, without pointing a finger at anyone. This method of dealing with family problems can have another positive influence as well. The VCFS child will also grasp that they are not _different_ but an equal part of the family in all areas; that they are involved in choices, decisions and discussions, both on positive and negative subjects. If you, the parent/guardian think your child may become too distressed on some subjects, you can choose a different approach. Use discretion. No one knows your children as well as you do.

Discipline

I can't take the constant bad behavior, what can I do?

To paraphrase a section of the book ***No more Misbehavin': Different Behaviors and How to Stop Them*** by Michele Borba Ed.D. an educational psychologist and author: *"If your child is disrupting your family life, it is time to take control."* (This goes for disabled children as well.) After you have set the rules of the house and boundaries for conduct, the next thing to do is set up a discipline that relates to the broken rule or bad behavior. The first thing is to respond to the child as soon as possible to administer the discipline and not wait *"till dad gets home"*. You must sit down with the child and tell them what action they did that was wrong and warrants the discipline and why. If it is, for example: getting into mom's make up and getting it on the floor; give the child a rag and bucket and have them wash the floor. I have found this discipline works well; if you have the child clean not only the affected area on the floor, but much more of the floor to emphasize the punishment. And if it is not done to your satisfaction; have them do it over. Do not let your word lose its power of impact by not following through, have them do it right away. The same is true at any age, toddler through teen. Many parents say, *"Oh it's just a phase they are going through."* or, *"They're too young."* I disagree, if they are 2 years and up and were old enough to make the mess, then they are old enough to learn to clean it up. Some may be going through a phase, but most are developing a pattern that will affect behavior at another age. Don't fool yourself. Don't lose control. After the punishment has been completed, make your child sit down and go over the rules once again. Make sure they are understood by the child.

Bullying

How can I assure that my child has appropriate behavior at school?

All schools have policies and rules. The child should be taught what these rules are and that they are expected to abide by them. So that you also know the school's rules, read the school handbook with your child at the beginning of each year. All people who have a share in your child's education should also be aware of the policy and aware of VCFS and what may be "normal" for your child as well as their rights and limitations. Let me give you an example: The first day of high school a child was walking home from school, going through a park across the street. On the way home in the park, a group of kids from the same school jumped the child and she fought back. Do you think she should have gotten kicked out of school and punished the same as the perpetrators? Well, it happened, even after the kids admitted that they jumped her. She got punished because she dared to defend herself. It was in the school policy book that the school had jurisdiction over all students one hour before school and also one hour after school, to extend punishment as they deemed fit; even if the child was not on school property. She lost so many points/credits that she was unable to catch up that semester and her family felt forced to remove her from school. However, if the parents had known that this child had VCFS and had informed the school about the child's VCFS, *and* the school authorities were aware of it, *and* still treated her in this way; there may have been another outcome. As it was, letters to the school board fell on deaf ears. The child was pulled out of school and home-schooled to graduation.

Would the parents have done things differently had they known of her VCFS? Probably, but who can say? I personally believe that if that child had been accurately diagnosed and the school had been informed of the disorder, there may have been some leniency given in the harsh discipline that was imposed. If not, legal action could have been taken. We do what we think is best at the time for our children, and we also need to keep informed in order to be a good advocate for them.

My child has been coming home from school frightened, with torn dirty papers; should I suspect he is being picked on? How can I tell and what shall I do?

As we all know, in the world today, there are those who are unkind and do mean things. When this happens to our child it can rip them up inside and it can affect us as parents as well. I told the story earlier of the first day of high school experience that is a classic example. You wonder sometimes, *"WHY?"* Is it because our child may sound a little different, act a little different or look a little different? What values have these bully's parents taught them? All children deserve to feel secure, free from fear and feel confident that they will be safe. This intimidation is not inducive to productivity at school, work or home. Do YOU know what the policy on bullying is at your child's school? Is it zero tolerance? What is the punishment?

There are different types of bullying just as there are differences in the terms _Assault_ and _Battery_. (ASSAULT is any _deliberate threat_ to inflict bodily harm with the apparent ability to do so. BATTERY is any _unauthorized touching_ of another person.) Isn't bullying just that? There is the deliberate, hurtful behavior and there is the consistent and persistent intent to hurt or upset the individual. Bullying can be physical, verbal or psychological in nature or a combination of all three.

What can the child do? The child can be taught not to just accept this bullying behavior but to tell someone in authority. If at school, the school has the duty to stop the bullying behavior and take appropriate action. Make sure it is documented and get a copy of the documentation.

Signs of bullying or abuse

Here are some signs of bullying or abuse. If your child doesn't tell you, ask them if you see any of the following:

- Bruises on a regular basis
- Supposed accidents continuously
- Doesn't want to go to school consistently and the work is suffering (more than usual).
- Abnormal lack of enthusiasm and energy for school
- The child's personal possessions are lost, damaged or dirty
- Doesn't want to walk home or ride the bus any more with his friends
- Seems unusually unhappy, withdraws, or wants to isolate. (These could also be signs of depression with no bullying involved.)
- Some children develop a stammer or stutters in their speech out of nervousness (rule out normal occurrences)
- Angry outbursts, if out of character
- Uncharacteristic fears, phobias, or negative actions (ex: being mean to an animal or starting to wet the bed—if that is something that normally doesn't happen.

If you believe that something is going on, ask your child directly, but don't do it when he first gets home from school. Children are already wound up from a full day of stress. Give your child some time to unwind, perhaps with a snack. Write down a few specific questions to ask them beforehand. Then, if you don't get any satisfying answers, ask their teacher to investigate. The teacher may know who your child hangs around with at school and they can ask friends if they know of anything that is happening. This is how I found out that my son was being bullied at the bike rack after school in grade school.

Sometimes it might even surprise us to learn that our child was actually fighting their smaller or younger friend's fights for them. I have known my own son to do this by trying to rescue a smaller child who was getting bullied. If that is the case we have to teach them that it was a kind gesture on their part, *BUT* that the fighting will never stop if it's handled in that way. It is best to let an adult handle it.

It is difficult sometimes in this area to determine what are normal VCFS behaviors and what are signs of abuse. As a parent you have to become very discerning.

If the answer you get is that they are being bullied, then what?

First, stay calm and don't go charging in like a raging bull. (I have done this and regretted it.) Still try to stay calm when you find your school may not be able to do much. Stay clear minded yourself and figure out how to handle the situation. First, inform the school. The school may have a protocol that works and that would be great.

Listen intently to your child as they tell you their story. Perhaps take notes and re-ask some questions of the child for clarification, especially if the puzzle is not coming together. Ask who was around so you can interview them to see if the story is as the VCFS child saw it. Eventually the story will unfold and the pieces will fit.

Now, what if YOUR child is the aggressor?

First determine if your child understood what they did and why it was wrong. Try to find out what prompted the action. Then put them in the other person's shoes, so to speak. Help them to mentally feel what the other person may have felt from the results of their actions. In story form, try to think of an appropriate scenario to illustrate their actions, to further explain what actions were not appropriate. Let them know, in no uncertain terms, that this kind of behavior will not be tolerated. This kind of illustrative teaching has proven to be productive. As an added incentive for the child to behave; have the school set an appointment for all involved to come together. Have your child apologize to the other child and to the child's parents. This works remarkably well for bullying. (If your child ever steals anything, have them return it; admitting the error to the store manager with an apology. The embarrassment makes it a quickly learned lesson that lasts.)

In today's world, behavioral problems have become an epidemic and many families are in crisis; not just families with disabled children. Children for the most part are not taught to be considerate, empathetic, moral, or to show an ability to love. So to turn this trend around and have happy families, we as parents must learn good parenting skills and this will be reflected in our children and family unit.

There is a a book, titled **Gorp's Dream: A Tale of Diversity, Tolerance and Love in Pumpernickel Park** which has an anti-bullying message for younger children. Written by Gorp, as told to Sherri Chessen, it is a memorable story told with full color illustrations. *Gorp's Dream* is available from Winmark Communications at: **http://www.winmarkcom.com/gorp.htm**.

Medical Trivia

According to German researchers, the risk of a heart attack is greater on Monday than any other day of the week.

"A teacher affects eternity; he never can tell where his influences stop." Henry Brooks Adams

CHAPTER 8: ADHD/ADD/AUTISM

WHAT IS ADHD/ADD/AUTISM?

Many VCFS/22q children have been diagnosed with ADD or ADHD. Even if your child is not one of those who has been diagnosed, you may see some familiar symptoms that were addressed in other areas in this book that are associated with VCFS. It is of great significance to look at the whole picture of the disorder and to be aware.

ADHD is <u>Attention-Deficit/Hyperactivity Disorder</u>. It is said to be the most common behavioral disorder in childhood. In the past it was thought that people who exhibited certain behavioral characteristics as seen in ADHD were products of head injury, complications at birth, infections in infancy, excess sugar, artificial dyes or poor parenting. It is now known that the majority of ADHD patients have a neurobiological physical brain problem. It is <u>NOT</u> caused by something done to the child or by any trauma in the family, such as divorce. However, even though ADHD has a biological cause, there is evidence that positive treatment which deliberately alters the home and school environment can positively affect the child in how they act or react and function.

This section provides some tips and recommendations about how to deal with an ADHD child. These are suggestions that I have taken from parents with children with ADHD and professional literature such as booklets written for education from **Shire US Inc.**, Florence, Kentucky. **1-800-828-2088**. These ideas are not meant to replace any of your doctor's suggestions. So please ask your doctor if it is right for your child. No matter what suggested treatment you choose to take when dealing with a child with ADHD it is going to be a demanding 24/7 job, requiring long suffering TLC (tender loving care) on your part. But the payoff in the future will be worth it.

First let us look at some of the symptoms and how ADHD is described and diagnosed.

Signs or symptoms related to ADHD
We should observe and take note

(Remember not all children have all symptoms)
- Needs constant attention
- Always on the go, never seems to stop
- Cannot get interested in one toy, object or location and goes from one thing to another
- At school age the teacher complains of fidgeting and excessive talking, blurting out of the first thing on their mind, even if it doesn't fit the subject
- Some children daydream so deeply that they lose themselves, tuning out the outside world, even when being called by name
- Has always had a problem with paying attention
- Struggles to concentrate on homework
- Has difficulty following directions
- Misses details, and is forgetful and disorganized in thought
- Easily bored
- Constantly distracted – with the least little unimportant sight or sound
- Difficulty in relationships
- Suffers from low self-esteem

If you suspect that your child may have ADHD do the following to help your doctor:
Record any and all specific behavior you have observed each time the child does it. Write down what the child did, what you did about it at the time and with what result? (As we mentioned before there are blank pages provided for the purpose of taking notes throughout this book.)

If ADHD is not treated or goes undiagnosed, it can have a profound negative impact on a person's life in adolescence and adulthood. Not only do they continue to struggle with some of the same symptoms they had as children, but now they are dealing with emotional/mental health issues from all the pain, rejection, misunderstood frustrations and failures in their lives without knowing why. As with most of us it is easier to blame something or someone else than to be able to see the real cause within ourselves. They may have problems supporting themselves, losing jobs, failing in relationships etc.

So that your child does not fall through the cracks, and gets diagnosed properly; if you suspect that your child has ADHD and you have documented the behaviors according to the list, tell your child's teacher about your suspicions. The teacher can confirm what you have observed and arrange a meeting with the school psychologist, special education teacher and anyone else who is appropriate in your state. Then they can get you more information on the subject. This will provide an exchange of information from home and school. Then your doctor can arrange the evaluations for a proper diagnosis.

Is there a test that my child can take to know for sure if they have ADHD?

There is no test like a simple blood test or MRI that can give you the answer to this question. There is only a comprehensive evaluation that can be done. This will include a test on developmental levels, an appraisal of the child's academic level, social skills and emotional functioning. Then parents and/or caregivers will be interviewed to listen to their observations and get their input. (This is when all the data and documentation you collected along the way will prove to be especially valuable.) Caregivers will be asked about behavior and any stress factors in the family or environment. This all is used to determine a treatment plan.

There are three main areas of concern that are reviewed:
- **Attention**
- **Hyperactivity**
- **Impulsiveness**

Not all children will have all areas affected, even if they do have ADHD. With ADHD it is the frequency and severity of the symptoms which occur that is of importance to the clinician.

How does my child get a diagnosis of ADHD?
Just like there are three areas of concern, there are three types of ADHD, and with the evaluation the clinician will determine your child's type and treatment plan.

The three types are:
- **Predominately hyperactive-impulsive type:**
The child must have 6 or more symptoms of hyperactivity and impulsiveness.
- **Predominately inattentive type:**
The child must have 6 or more symptoms of inattention problems.
- **Combined type:**
The child must show 6 or more symptoms of attention and hyper-activity problems.

What are the symptoms of each type?

Symptoms of hyperactivity
- Fidgets with hands and feet or squirms in seat
- Leaves seat in situations where staying in the seat is expected
- Runs or climbs excessively when inappropriate
- Has difficulty playing or doing any leisure activity quietly
- Is always on the go as if operating on Ever-Ready® batteries
- Can talk excessively
- Adolescence: may exhibit feeling of restlessness and impatience

(Please note, we are not taking about toddlers who are just energetic, curious and perhaps not disciplined.)

Symptoms of impulsiveness
- Blurts out answers before questions have been completed
- Has difficulty waiting their turn
- Interrupts or intrudes on others' conversations *(Again, we are not talking about children who have not been taught manners, but those who have been taught and still can't seem to control themselves.)*

Symptoms of inattention
- Fails to pay close attention, therefore makes careless mistakes
- Has difficulty maintaining their attention
- Does not seem to be listening when spoken to
- Does not follow directions and therefore fails to complete work assignments at school and/or employment, chores, etc.
- Has a difficult time organizing activities, chores or tasks
- Avoids or dislikes doing anything that requires sustained mental effort
- Loses things that are necessary to complete the task, assignment etc. (ex: the paper assignment from school)

- Is easily distracted
- Is often forgetful of daily activities, even if they have gone over them a million times

Because there are other conditions that can imitate the symptoms of ADHD the clinician must rule out things like vision impairment, attention lapses from possible epileptic seizures, middle ear infections that can cause intermediate hearing loss, psychiatric disorders or a combination of the above. All would need to be treated and/or addressed.

As a parent, I would have appreciated knowing early that my son had this disorder, and could have gotten help for him. For so long I just thought he was rebellious and lazy. Now that I understand that this is a disorder, I am much more patient with him and I try to work with the disorder to help him feel less stress—and me too!

Can my child also have coexisting disorders?

Some of the coexisting disorders that can accompany ADHD are:
- **Depressive disorders**
- **Anxiety disorders**
- **Learning disorders**
- **Conduct disorder**
- **Bipolar disorder**
- **Oppositional defiant disorder**

These disorders can go hand in hand with ADHD and can be quite disabling for the child.

Is there beneficial treatment available?
Effective treatment of ADHD requires a multi-system approach of medication, psychology and education.

Interventions for training in development of social skills, improvment of study habits, organizational skills, behavioral therapy and medication are normally addressed by your doctor. This requires a team of experienced clinicians with each working in their specialty.

What are some tips I can use to build confidence and help my child with ADHD?

Anyone can help to boost self-esteem in the ADHD child, whether it is at home or school. Here are some tips about what to do and how to do it:

Confidence building tips
• Recognize and build on existing strengths. Perhaps in art, sports, spelling, etc.
• Teach responsibilities that the child can surely meet and build on those.
• Teach decision-making skills by allowing your child to make simple, concrete decisions.
• Reinforce self-discipline by allowing the child to receive a consequence for wrong behavior that fits the behavior. For example: time-outs, loss of privileges, scraping and cleaning their own shoes after walking in mud, etc.
• Develop problem solving skills by taking a specific problem with which your child is familiar and reviewing a list of options that can resolve the problem. Discuss pros and cons of each suggestion.
• Offer encouraging and positive feedback by being attentive to good behavior. Attention and praise can motivate the child to further good/positive behavior.
• To encourage homework, create a proper homework space that is quiet; establish a regularly scheduled time each day for homework to be done; provide all necessary supplies and have them readily available; use little mini-breaks at intervals. (Many kids will want to study while listening to music and/or the TV, but this should be discouraged.)
• Make sure all the teacher's instructions and requirements are understood, as well as the schedule for when the work is due to be turned in to the teacher.
• Organize the school assignments. Put one book and assignment on the desk at a time, and then go on to the next assignment after each one is completed. The child may get overwhelmed to see all the books and assignments at once, and might get the assignments mixed up as well.
• Keep in regular contact with your child's teachers at school and inform them of any areas of difficulty that your child is having. Then the teacher can give your child more attention or assign an aid or tutor.

• Find ways to keep your child motivated so that they will want to exhibit good behavior and correct their wrong behavior, to do well in school and other areas in life. This is not an easy task, but can be done step-by-step once you *know* your child.

Other areas that need to be addressed for a child with ADHD include peer relationships and dealing with adults and others who know nothing about ADHD and VCFS. This disorder often reflects on the whole family socially, so it is important to get this condition diagnosed as soon as possible and get interventions in place.

There are numerous books on the subject that you can find at many libraries.

Here is a list of a few support groups:
Children and Adults with ADHD (CHADD)-
    Phone: 301-306-7070 Web site: **www.chadd.org**
Learning disabilities Association of America (LDA)
    Phone: 412-341-1515 Web site: **www.ldanatl.org**
National Attention Deficit Disorder Association (ADDA)
    Phone: 847-432-2332 Web site: **www.add.org**
National center for Learning Disabilities
    Phone: 212-545-7510 Website: **www.ncld.org**
National Institute of Mental Health (NIMH)
    Phone: 301-443-4513 Web site:
    www.nimh.nih.gov (Use on-site search for "ADD.")

Any of these organizations above can help you negotiate with the school system for an ADHD friendly environment. They can educate you about your child's rights under the law, and on subjects such as *classroom accommodations*, *physical education* and *activities*. (See Negotiations on page 261.)

Here is a web site for the ADD warehouse that claims to be *"the leading resource for understanding and treatment of ALL developmental disorders, including ADHD and related problems."* It seems to have a wealth of information for parents and educators who can browse through information on different subjects.
http://addwarehouse.com/shopsite_sc/store/html/index.html

Autism: It's Not All in the Mind

The Background

The last century, particularly during the last 30 years, has seen a dramatic change in modern Western society. As we have moved into being an industrialized society living predominantly in cities, we have also moved from extended family groups to nuclear families and increasingly single-parent families with little social support. The explosion of technology has resulted in a dramatic change in environment, lifestyles, and ways of eating.

Along with these societal changes have come dramatic changes in patterns of illness, seen particularly in the changing patterns of health in children over the past 30 years. Over this timeframe, asthma, allergies and atopy, ADHD, and autism "the Big Four" have all increased significantly. Allergies have increased in incidence up to seven times, now affecting up to 50% of children; asthma has increased from affecting some 10% of children in 1982 to 36% of children in 2002; ADHD is now the most common diagnosed psychiatric condition in school-age children, affecting an estimated 11%; and autism has increased from an incidence rate of four in 10,000 in 1985 to greater than one in 100 school-aged children as reported in The Lancet in 2006. Food allergies, in particular, have increased over six times, affecting 30% of children; peanut allergy has gone from being a rarity to a common problem now affecting one in 50 children. The concern with food allergies is the correlation between early food allergies and later asthma, food allergies, and eczema, and the now proven association between eczema and later asthma. Eczema as an early marker of an atopic tendency has itself increased in incidence nine times over the past 20 years. At the same time, autoimmune diseases have been increasing in children, with conditions such as autoimmune thyroiditis, previously typically only seen in mature females, occurring increasingly in children. Autoimmune diabetes, for example, has been increasing at the rate of 3% a year over the past 15 years, and is occurring at increasingly younger ages. What all of these diseases indicate is an underlying immune system dysregulation.

In the same time period, there has been a concomitant rise in various infections, which had not been vaccinated against, including otitis media, sinusitis, gastrointestinal infections, and respiratory tract infections, as reflected in the 40% increase in the rate of hospital admissions for upper respiratory tract infections and gastroenteritis over the 1990s. The increase in the incidence and changed pattern of otitis media is particularly noteworthy. Prior to 1970, it was an uncommon diagnosis and seen usually in children aged four to six years; by 2000, the peak incidence had shifted to six to 12-month-old babies, with 48% having at least two infections by the age of six months and 91% by 24 months, and infants being affected by otitis media with effusion (glue ear) for 30% of their first year of life. The insertion of tympanostomy tubes (grommets) is now the most common operation in young children. Since the 1980s, childhood sinusitis has also become an increasing clinical problem and is frequently a missed diagnosis since sinusitis was previously a disease of adulthood. Along with these increasing rates of infection, there is an increasing use of antibiotics, with most of the antibiotics used in childhood being used in boys and for the treatment of otitis media; and paracetamol is increasingly used as an analgesic and antipyretic.

This rising rate of infection and antibiotic and paracetomol use is not without other effects. There is a direct correlation between respiratory tract infections, particularly pneumonia, otitis media, and the development of asthma as the risk of asthma increases with the increasing number of infections. Recently, correlation with the use of paracetamol and later development of asthma has also been noted. There is also an apparent association between infections in childhood and neurodevelopmental disorders: speech difficulties in boys correlate with asthma, allergies, ear infections, and other mental health disorders; recurrent otitis media in infancy correlates with a greater incidence of ADHD and later difficulties in school behavior and performance. A high correlation between the prevalence of otitis media and use of antibiotics and the development of autism has been noted, with earlier incidence and more frequent infection correlating with increasing severity of autism. Asthma itself is also correlated with behavioral disorders in preschool boys.

Given these correlations, it is not surprising that the rates of neurodevelopmental abnormalities have increased dramatically over this time period. The impairment of speech and language has become the most common neurodevelopmental condition among children under five, particularly in boys, affecting up to 8% of the population. These boys are also more likely to have other behavioral issues, including social and emotional adjustment problems and dysfunction. A large study reports that one child in five in the United States has been found to have a diagnosable "mental health" problem, with 62% being boys; in Australia, this was one in six children, with 11% diagnosable with ADHD. Learning disorders affect some 20% of modern children. Other neurological conditions, including childhood epilepsy and cerebral palsy, have also increased in incidence, along with increasing severity and atypical presentations.

What would at first sight appear to be unrelated is that, in the same timeframe, children are increasingly suffering from various gastrointestinal diseases. Gastroesophageal reflux disease, previously the domain of adults, has increased in incidence in infants in recent years and is now a major cause of illness and failure to thrive, particularly in neurologically impaired children. This reflux disease is part of the spectrum of cow's milk allergy and is linked to the development of otitis media, sinusitis, and asthma. The treatment of this condition with the use of gastric acid inhibitors has been shown to increase the rates of gastroenteritis and pneumonia as a result of an interaction with white blood cell function and gastrointestinal microflora. Anecdotally, there appears to be a correlation between reflux disease in infancy, the use of gastric acid inhibitors, and the later development of autism. As mentioned previously, children are more often suffering from gastroenteritis, and Crohn's disease is also being increasingly diagnosed in young children.

Autism

It is against this background that we see autism, a previously rare disease, increasing in incidence such that it now affects up to one in 100 children. That autism is a largely environmental disease is becoming increasingly apparent, as no single clear genetic predisposition can be found, although there are many genetic polymorphisms that confer a degree of increased risk. It has been said that genes load the gun but the environment pulls the trigger. Autism occurs in a much higher incidence in increasingly industrialized societies and also in pockets with higher levels of environmental pollution.

The predominant symptoms in behavior. It is classified as a psychiatric disease but it is a neurodevelopmental disorder. A number of neurological abnormalities have been verified in autism, including abnormal head size, reduced cerebral blood flow, loss of cerebellar Purkinje fibers, and reductions in neurotransmitters and neuropeptides. These abnormalities result in functional abnormalities, with reduced synchronicity and connectivity in the brain. Autism is also an autoimmune disease, with high levels of autoantibodies against various neuronal tissue being documented and in conjunction with antibodies to cross-reactive peptides, including streptococcus and milk.

In addition to autoimmune dysfunction, abnormal gut-based immune responses are well documented in autism. There are increases in all Th2 cytokines, activation of both Th1 and Th2 systems, and a Th2 predominance but no compensatory increase in the regulatory Th3 cytokines. IgA produced by the Gut Associated Lymphoid Tissue (GALT) is typically reduced, while antibody-producing B cells are increased.

The developing picture of abnormal brain function, neuronal autoantibodies, cross-reacting antibodies to streptococcus and milk, Th2 shift, and abnormal GALT-associated antibodies indicates abnormalities in the gut microflora in autism as the gut microflora is increasingly being recognized as the major determinant of immunological function.

Gastrointestinal symptoms are now well recognized in autism, with abnormal bowel movements and food sensitivities and reactions being the norm. Numerous gastrointestinal abnormalities have been documented, including gastric hypochlorhydria with resultant raised pH (Horvath); reflux esophagitis; chronic gastritis and duodenitis (Horvath); reduced intestinal disaccharidase enzyme function (Horvath, Kushak) and dipeptidase function; enhanced pancreatic bicarbonate response to secretin, indicating lowered secretin stimulus to the pancreas (Horvath); colitis and lymphoid hyperplasia (Wakefield, Krigsman, Kushak); and increased intestinal permeability (Eufemia). Such a wide-ranging gastrointestinal disruption would indicate a significant change in the underlying gastrointestinal flora as the function of healthy symbiotic flora includes being a physical barrier to gut pathogens, the production of numerous immune-acting substances, the neutralization of bacterial toxic by-products, and the enhancement and regulation of mucosal immunity.

The healthy gut will contain an average of one-half to 2 kg of bacteria, comprising some 10 to 50% of the total number of body cells. There are over 500 species, with the predominant symbiotic bacteria being Bifidobacteria, Lactobacilli, Propionobacteria, physiological strains of E. coli, Peptostretococci and Enterococci. Other bacteria that frequently present opportunistically include Streptococci, Clostridia, and yeasts, especially Candida. In a series of studies involving stool analysis in a total of 124 cases and with 117 controls, a significant pattern of colonic dysbiosis emerged. Overall, there is a reduction in symbiotic E. coli and Lactobacilli and a corresponding increase in Enterococcal (formerly Strep faecalis) and Streptococcal species, with predominant overgrowth of the non-pneumococcal alpha hemolytic or non-hemolytic Streptococcus. This pattern is similar to that seen in small intestinal bacterial overgrowth (SIBO), which correlates with malabsorption syndrome (MAS). Studies in SIBO with MAS have grown bacteria in most jejunal aspirates, the predominant gram positive organism being Streptococcal species, with anaerobes, infrequent.

There are <u>several mechanisms</u> by which disturbances in gut function can affect the brain, the so-called gut-brain axis. <u>The first</u> is via the enteric nervous system and the gut based neurotransmitters: there is a great deal of overlap between enteric neurotransmitters and those of the brain; indeed the gut has a nervous system second only to the brain, and it has been called the "second brain."

The <u>second mechanism</u> is increased intestinal permeability; this allows increased passage of toxins and antigens, which may also cross the blood-brain barrier (which frequently is also hyperpermeable). Poor food digestion and resultant abnormal food reactions can contribute to increased intestinal permeability. Gluten and casein reactions are well documented in children with autism. Gluten and casein are proline and glutamine rich and resist degradation by proteolytic enzymes. The enzyme DPP-IV is specific for the proline bonds in gluten and casein, producing dipeptide fragments. It is a brush border enzyme of the duodenum, and it is also an immune modulator. Limited production of the enzyme or failure of the enzyme to function correctly will result in the formation of exorphins, short-chain, proline-rich peptides that can interact with opioid receptors in the brain (Reichelt). Interestingly, DPP-IV binds more strongly to streptozyme (Streptococcal enzymes) and mercury than to gluten and casein (Vodjani, Panghorn) thus providing a mechanism of action by which Streptococcal overgrowth can result in gluten and casein sensitivity.

The third mechanism is via abnormal activation of the gut associated lymphoid tissue, which could occur in a small bowel overgrowth with increased intestinal permeability; this could result in antibody formation and possible cross-reacting antibodies with brain tissue. Streptococcal species have long been known to be associated with neuropsychiatric disorders, abnormal movements, and autoimmune phenomenon, as in scarlet fever, rheumatic fever and Sydenham's chorea. Dr. Susan Swedo's work in the recognition of the association between the development of obsessive compulsive disorder following Streptococcal infections (Pediatric Autoimmune Neuropsychiatric Disorders Associated with Streptococcal Infections) has further developed this link, and the correlation with other abnormal movements and psychiatric disturbances has also been recognized. Raised Streptococcal titers have also been found to correlate with abnormal basal ganglia volumes in the brain in children with ADHD. A particular antibody, monoclonal antibody D 8/17, has been found to be expressed in increased amounts in association with a Streptococcal antibodies in these conditions. This antibody has also been found high in patients with autism correlating with the severity of their compulsive symptoms, which indicates a likely connection with Streptococcus and autism. A personal review of 29 children with ASD who were tested for GABH Streptococcal antibodies found nearly 50% positive while only one quarter had any past any history of GABHS disease (scarlet fever or tonsillitis), two thirds had a known history of other Streptococcal diseases such as otitis media, sinusitis, or pneumonia (Strep pneumoniae), and one-third had positive serology with no known history of any Streptococcal infection. While confirming a correlation with Streptococcal antibodies in autism, this indicates a connection with other Streptococcal diseases and the possibility of a state of Streptococcal carriage with antibody reaction but often no apparent clinical disease.

The fourth mechanism is the disruption of normal gut bacterial metabolites. Numerous metabolic abnormalities have been demonstrated in children with autism. As Dr. Andrew Wakefield has stated, "often manifest are complex biochemical, metabolic, and immunological abnormalities that a primary genetic cause cannot readily account for." A significant disruption in gut microflora has been described and could readily account for many of the gut brain associations seen in autism. The reduction in E. coli results in a loss of various vitamins, antibacterial substances, and the precursors for folic acid, Coenzyme Q10, tryptophan, tyrosine, phenylalanine and vitamin K, with possible disruption in methylation pathways, cellular respiration, and various neurotransmitters. Low E. coli with high Enterococcus and Streptococcus also correlates with malabsorption, deconjugation of bile acids, and increased ammonia production. The deconjugation of bile acids and increased ammonia production in conjunction with increased intestinal permeability results in increased entero-hepatic circulation and, therefore, increased toxin load to the liver.

Streptococcus and Enterococcus are also predominant producers of D-lactic acid. This results in an acidic pH of the stool and changed bacterial metabolism in the other gut microflora, including increased amine production, which is a possible mechanism for some of the food-related reactions seen in autism. High levels of Streptococcus and Enterococcus in the stool correlate with neurocognitive symptoms in adults. The symptoms are remarkably similar to those seen in short bowel syndrome in adults, where headaches, weakness, cognitive impairment, fatigue, pain, and severe lethargy are related to D-lactic acidosis, presumably as a result of increased colonisation by lactic acid producing gram positive bacteria. D-lactate is poorly metabolized in humans, as we lack the enzyme D-lactate dehydrogenase. In a retrospective analysis of 34 children seen in clinic, one-third had a frankly reduced serum bicarb and two thirds had a raised anion gap consistent with a chronic metabolic organic acidosis, which could be caused by raised D-lactate. 95% of those children with metabolic acidosis had an overgrowth of Enterococcal and/or Streptococcal species in the stool.

There is, therefore, a developing picture in which underlying dysbiosis with predominant overgrowth of Streptococcal and Enterococcal species can result in gastrointestinal dysfunction in autism, with increased intestinal permeability, immunological dysfunction, cross-reacting antibody production, and disruption to digestion and normal gastrointestinal flora function and metabolites, with metabolic sequelae and organic acidosis. If there is a concomitant overgrowth of Clostridial species, as is commonly noted post heavy antibiotic use, the potential metabolic and neurologic disruption is even greater. (Bolte)

The cause of overgrowth of these bacteria is usually apparent from the history. These children frequently have a history of reflux disease with proton pump inhibitor treatment and resultant gastric hypochlorhydria. As mentioned previously, otitis media has also been noted to occur frequently in the history of these children as well as sinusitis, which is frequently silent. Nasopharyngeal carriage of Strep pneumonia is very common in young children and correlates with the development of otitis media. Antibiotic treatment of this otitis media, however, results in an alteration in the pattern of these bacteria to non-pneumococcal alpha-hemolytic Streptococcus (as seen in the fecal analyses), which predisposes the individual to further otitis media and sinusitis and reduces the levels of symbiotic bacteria in the gut. In both these conditions, Streptococcus exists in a biofilm, a matrix of mucus slime that protects the organism. This results in a chronic carrier state of the organism that is resistant to antibiotics and provides a constant source of bacteria and bacterial toxin-laden mucus from the Eustachian tubes and sinuses down through the stomach and into the duodenum. Once in the duodenum, the organism is able to establish a biofilm in the gut, which causes disruption of enzyme function, digestion, and absorption as well as an altered duodenal pH. Modern diets are typically also high in cow's milk dairy, and several large scale studies have demonstrated that residual Streptococcus and Enterococcal bacteria are in the milk; this provides a further source of the bacteria and bacterial toxin. Modern diets are also deficient in the traditional lactose fermented foods that were historically part of everyday fare and a natural dietary source of symbiotic bacteria, natural competitors of the Streptococcus and Enterococcus bacteria.

<u>In summary</u>, autism can be viewed as an autoimmune and toxin-mediated disease in conjunction with malnutrition and metabolic derangement secondary to gastrointestinal dysfunction and dysbiosis in genetically susceptible individuals. A typical scenario would be where the normal early infancy Th2 responses fail to mature to Th1 responses as a consequence of early dysbiosis from such factors as maternal dysbiosis, Caesarean section, loss of breastfeeding, recurrent infections, early antibiotic and paracetamol use, vaccination, and other toxin exposure. The resultant predominant Streptococcal and Enterococcal overgrowth in the gut increases intestinal permeability, which further skews Th2

responses with overproduction of IgG antigen-specific responses. The toxins are those associated with Streptococcal infections, Streptococcal and food-derived toxins traversing the intestinal barrier, other bacterial toxins and other environmental toxins notably mercury. The D-lactate produced by excessive Streptococcus and Enterococcus disrupts duodenal enzyme function, the metabolic function of other gut-based bacteria, and produces a chronic metabolic acidosis. The Streptococcal enzymes disrupt DPP-IV function, with increased immunogenicity and neurological effects from gluten and casein, and the Streptococcal and milk antigens activate gut associated lymphoid tissue and cross-react with neuronal antigens.

Management of such a complicated disorder can never be as simple as a single pill or supplement. There is a need to restore diets to those traditionally eaten, restore symbiotic bacteria, heal the gut wall, modulate immune responses, improve hepatic detoxification function, support the metabolism and nutritional deficiencies, clear biofilms, treat occult infection, and remove heavy metals and other toxins. The growing number of successfully treated children indicates that it is possible to modify and, to some degree, reverse the underlying factors that have combined to result in the condition we know as autism.

Dr. Robyn Cosford, MBBS(Hons), FACNEM
Conjoint Lecturer,
University of Newcastle School of Biological Sciences.
Director, Northern Beaches Care Centre
Australia

ADD, can you have it WITHOUT the hyperactivity?

Yes, you sure can! I came across a wonderful article not long ago on www.addchoices.com. This was the best article I have found that spells out the symptoms of ADD without hyperactivity accurately. Very well done! I don't believe I could have written it any better than BJ, so I would like to share it with you and encourage you to go to her site for other information. This article will either validate your suspicions or rule it out.

Written by BJ Madewell
Masters Degree: Educational Psychology, with emphasis in Learning disabilities
Veteran Special Education Teacher for 33+ years

ADD without Hyperactivity

It's always been pretty easy to recognize hyperactive children. They whirl through life bouncing off teachers, parents and friends alike in an active, never stopping frenzy of activity. There's usually the tell-tale sign of wiggly feet, drumming fingers, non-stop talking and so on.

The really difficult students to recognize and diagnose, are those with ADD without Hyperactivity. Literally thousands of these ADD without hyperactivity children sit in our classrooms and homes. These students are particularly frustrating. They appear LAZY, SHY, QUIET, "SPACEY," INATTENTIVE, "LOST," UNABLE TO SUSTAIN A TASK, CAN'T FIGURE OUT WHAT TO DO, and INCONSISTENT.

(Be aware that having an ADHD child does not make you immune to having an ADD child. In fact, some researchers claim ADD and ADHD are inherited syndromes.)

Below is a "checklist" of the ADDer without hyperactivity.
- Below level of body activity
- Reticent to get involved in group activities
- Does not volunteer information
- Would rather be alone

- Does not volunteer to answer questions
- Appears to be in a private fantasy world
- Needs oral information repeated
- Uses very few words when required to speak
- Looses train of thought easily
- Needs frequent reminders to finish tasks
- Mental images disorganized
- Continually looses important details or essential facts
- Cannot do a series of tasks without starting to make mistakes
- Stays out of group discussions
- Can't remember a series of instructions
- Can't remember what the assignment is
- Forgets names of people and things
- Forgets the rules of games
- Can't stay on a schedule
- Unorganized
- Often loses personal things
- Needs supervision to complete homework
- Forgets daily routines
- Lives in a "pile of clutter" and can't seem to straighten it up or keep it neat once it is organized
- Erases a lot
- Thinks others are playing tricks as mental images shift
- Continually surprised or startled
- Misunderstands others often
- Whispers to self = trying to remember
- Says "What?" or What do you mean?" often
- Claims "You didn't say that!" or "I didn't hear you say that."
- Long pauses before responding
- Needs prodding to start assignments
- Long periods of time go by with nothing accomplished
- Can't keep up with pace of group activities
- Tends not to finish tasks without supervision
- Has numerous unfinished tasks to do
- Thinks the task is done when it is not
- Whines
- Wants to quit before others normally would give up
- Has difficulty with "small talk" in social situations
- Avoids personal involvement
- Easily distracted by sounds, odors, movement etc.
- Fiddles with things
- Behavior appears immature
- Prefers to be with younger persons

- Does not try to accept responsibility
- Impulsive, does not think ahead
- Easily bored
- Conversation jumps around

If this resembles someone you know, consider ADD without Hyperactivity as a possible cause of school/home related problems.

Parents Speak Out

Our daughter, with VCFS, is nearly 24 years old. I know exactly what it means to live with "a walking time bomb and you never know when it is going to explode." In my view, this is part of her rapid cycling bipolar disorder that afflicts so many kids who have VCFS.

For us, it was like walking on eggshells all day every day, never knowing what might set her off. For her, it was day after weary day of misery. For about four years she's been wonderfully stable (and happy!) on a combination of medications that includes a mood stabilizer. Without that miracle drug, she wouldn't be able to take an antidepressant (she'd be at risk of mania) or a stimulant to aid her ADHD (she'd be at risk of psychosis). My husband and I didn't fully realize, until it was over, just how exhausting and demoralizing it had been to live with chronic fear of her meltdowns. We are so grateful for research and medications that can help lives.

Medical trivia

The average person inadvertently swallows 8 spiders a year.

Blank page for personal data:

"What lies behind us and what lies before us are tiny matters compared to what lies within us"
Ralph Waldo Emerson

 CHAPTER 9: PSYCHOLOGICIAL ISSUES

VCFS PSYCHOLOGICAL ISSUES

We need more research and understanding of the brain!

Since the brain is the captain of our ship the body, it directs the body's actions and responses; and dictates how we learn, act, speak, interract, and become the very person we are. It's important to understand how the brain works, to understand how to heal the brain, both physically (as in surgery), emotionally, chemically (as in drug therapy), and in couseling for guidance. The brain is what directs our life.

The talking network within the brain is done by synapse energy; much like zaps of electrical current which is limited at birth, but grows more complex as we grow older and have experiences. Directly connected with the CNS (Central Nervous System) and it's nerve partners, the brain responds to commands as directed by hormones and other chemicals secreted from various parts of the body, such as the endocrine glands.

Genes, made up of amino-proteins in DNA, are housed in the chromosomes, and dictate how we react to our environment. Individual sections of chromosomes are not dedicated to just one specific part of the body. All the DNA in each gene is intertwined with others, with each gene affecting several parts of the body, including the brain, all at once. This is why missing even a small amount of DNA can affect so many areas of the body as we see in 22q.

Our sensations and perceptions team together to produce the information we need to intrepret the world around us. One depends on the other to fully complete us. They work together, allowing our different senses to work properly; by reporting or transmitting needed information to the brain to respond accordingly. Much of our perception is inborn or inherited; it is also defined by the psychological influences of our environment as we grow and mature.

Intelligence is very difficult to define. However, research has been developed to make the process more accurate. (The cognitive approach is more accurate than the old I.Q. tests, most of us experienced in school.)

Babies continue to develop the pre-frontal cortex and Amygdala of the brain. Babies are born with "mirror neurons" that develop and mature in the brain by using them to imitate action and language; by mirroring they interpret what they observe in action, emotion, and intent.

I have learned that the discovery of "mirror neurons" may someday empact our future for finding cures in areas such as Autism, mental disorders, motivation, language, child development, and perhaps anti-social or sociopathic behaviors. I hope this helps you to understand a little more about our development.

What kinds of psychological problems are associated with VCFS/22q

I have heard that children with VCFS might have psychological difficulties which contribute to behavioral problems. This scares me. What can I expect?
Not all children with VCFS have _mental_ _illnesses_ that bring on abnormal behavioral problems; but it is estimated that 1 in 3 VCFS children will have some behavioral problems, at one time or another. These _behavioral_ _problems_ are not just the normal behavioral difficulties that all children experience while growing up. This behavior is unsettling, and more than just a little difficult. *(See Behavior list on page 202.)*

Behavioral or psychological problems can appear at any age. But more often than not it is in pre-puberty and puberty that many behavioral/psychiatric problems begin or are noticed to increase in severity. Many believe it is the hormonal changes in adolescence that trigger these symptoms, but it can happen earlier. For the sake of argument as to when the onset begins, I will address the adolescent years. Many of these children have a tendency toward developing _manic/depressant_ _disorder_ or _bipolar_; with or without schizoaffective disorder. These schizoaffective symptoms are sometimes called psychoses. Severe mania or depression can bring about psychosis, also known as psychotic episodes. Common symptoms of psychosis include hallucinating and hearing voices, thinking people are reading their minds, and seeing things that aren't there. Many fixate on religion (as my son did, carrying his Bible around clutched in his arms). Sometimes they feel a presence that is not there (as my son did, thinking Jesus was always standing behind me) or that people want to harm them (paranoia). These can be imaginary people or they can be family members. Bipolar individuals can become delusional, having false but strong beliefs that they are right and no one else is thinking straight. Psychotic symptoms in bipolar disorder tend to reflect severe mood states (whatever the state is at the time) either the manic elation or depression. Thus, it is called _polar_ because they can be at opposite poles in moods within minutes and this can be within a wide spectrum of severity. In many cases there are delusions of _grandiose ideas_ of one's self. As an example: they may believe they have special powers or abilities, typical of mania; or they may suffer from the delusion they are homeless and have committed a crime for which to be punished, and are in a state of depression. Because of the similarities in the mania of bipolar with the schizoaffective aspect and schizophrenia, a person is oftentimes misdiagnosed. (_NOT_ all VCFS children have this psychosis with bipolar.) My son was one of them who did develop bipolar with psychosis, but that story is later. _(See T.G.'s story on page 359.)_

It has been said, by professionals, that many people tested in their research who had previously been diagnosed with schizophrenia (with no knowledge of VCFS at the time) were found to have the 22q11 deletion when tested. This is not surprising because the field of genetics is in its infancy and VCFS is not well known yet. Just think of all the older people who might now be helped when their doctors become well informed about VCFS. They can make a patient's true diagnosis, determine what might have caused it; connect it with a name; and treat them accordingly with this information in mind.

It is noted that many with VCFS also suffer from ADD or ADHD. *(See ADHD on page 213.)* Some VCFS experts will caution against using common stimulant medications, because bipolar symptoms are at risk.

In all honesty, in my opinion and experience, I believe it is the psychological and emotional symptoms that are the most difficult to cope with and understand. Bipolar and schizoaffective symptoms can be challenging to manage; however, in nearly all cases, they can be successfully managed over time. We were fortunate enough to find a good psychiatric nurse practitioner in Phoenix, who was willing to learn and listen to us.

When should I take my child to see a professional?

Both children and adults can develop bipolar disorder. It is more prevalent if one of the parents has bipolar or some mental illness. Children with bipolar seem to be less common than adults, or it might be the lack of diagnoses. Children usually show a very fast onset with each episode. But, in children, it is difficult to tell if the symptoms or behavior are age related or if it is, in fact a disorder. Younger children seem to exhibit more of the irritation and destructive mode than the euphoric mood of mania. Many of the symptoms of bipolar resemble symptoms for ADHD. For more information on kids with bipolar go to the web site: **www.bpkids.org** which is the *Child & Adolescent Bipolar Foundation*. The site is filled with pertinent information on this subject. You could also read the book: **The Bipolar Child: The Definitive and Reassuring Guide to Childhood's Most Misunderstood Disorder,** by Dr. Demetri Papolos. *(See page 250 for more information.)*

A child is not bipolar if they are merely having the blues for a few days. This might be simply due to a problem at school, or a short-lived sadness which is nothing that a little love and praise can't fix. Give the child a task they can successfully complete and then praise the efforts. They should perk up. This type of child needs to build their self confidence. One way to do that might be to allow them to do more things that you know they can do successfully, balancing their limits with their strengths. Maybe this can be accomplished by allowing them to make some decisions. For example: give your child cooking lessons once a week and then let them choose what will be served on the menu (even if the choices are not ones you would have made). This aids a child in developing abilities they will need for life. Professionals tell us that if we *"baby"* the child too much for too long they will become so dependant on us that they will have a tough time merging into the world of adulthood. Stress and responsibilities can become enormous to these kids and to us. By taking the sorts of steps we mentioned; we will be able to see how our child reacts. It will give us a good gauge to know whether the child needs to see a doctor or not. Like all other issues, we learn as we go along. *BUT*, if you are concerned about your child's behavior because it has changed to a degree that it is noticed by all who are around them, for an extended period of time; make notes on the list of symptoms below. Put the times and observational dates on the calendar or in the blank pages that have been provided for you in this book. This will assist you and your doctor in knowing how long it has been going on and seeing if there is a pattern. Dealing with depression, bipolar or schizoaffective disorders requires first and foremost an understanding of the depressive, bipolar and schizophrenia symptoms. These are common, treatable illnesses; the symptoms can include some of the following:

Manic episodes:
- Increased energy, activity and restlessness
- Excessive *"highs"*, an over-happy euphoric mood
- Extreme irritability
- Racing thoughts; talks too fast; jumps from one thought or idea to another
- Distractible and can't concentrate
- Needs little sleep
- Unrealistic of one's self, abilities, and powers
- Poor judgment
- Spending sprees
- A lasting duration of these symptoms that is out of the norm
- Increased sexual drive
- Abuse of alcohol and/or drugs
- Provocative, disturbing, aggressive behaviors
- Denial that anything is wrong

A manic episode is suspected if an elevated mood exists with at least 3 or more of the symptoms mentioned most of the day, nearly everyday for a week or longer.

Depressant episodes:
- A persistent sad, *"empty"* or anxious mood/feeling
- Self-absorbed, selfish, unaware or unconcerned about the needs or feelings of others
- Loss of interest or pleasure in activities previously enjoyed
- Unresponsive, uncommunicative, indifferent
- Sleeping and eating disturbances (either too much or too little)
- Changeable and unpredictable; illogical and unreasonable (in thought and/or action)
- Difficulty concentrating, remembering, making decisions
- Manipulative, pleasant and charming in public, the opposite at home
- Feelings of hopelessness, pessimistic (nothing will change), wants to die
- Thoughts of death or suicide; suicide attempts
- Mean, belittling and critical of self and others
- Irritability, excessive crying
- Makes unexplainable and sudden references to leaving
- Chronic aches and pains that don't respond to treatment
- Increased use of alcohol and drugs
- Feelings of guilt
- Sluggish, fatigue no energy

(See Behavior List on page 202.)
A depressant episode is diagnosed if 5 or more of these symptoms last most of a day for 2 weeks or longer.

Blank page for personal data:

Blank page for personal data:

After viewing this list of symptoms, it should explain why depressive illnesses can destroy relationships and disrupt families, if not treated. We need to identify, understand, and learn how to cope with the affected child or adult as the first step in helping both our loved ones and ourselves.

If you see any of these symptoms in your child or yourself you may be dealing with an emotional problem that requires the help of a professional for diagnosis and treatment.

Psychological stories from parents with VCFS children

I have included a few extreme case studies of bipolar disorder that I believe explain what a varied pattern this disorder can have, as well as show how it can affect families. Names were changed, as requested, to shield the identity of the families.

Please note that these are examples of what some parents are dealing with in the mental illness arena of VCFS. They are not typical and do not happen with all children. Bipolar, for the most part, is very treatable. Most of these children have multiple diagnoses and need more time to find a combination of medications that will work for them, with more medications becoming available all the time.

Adam

My son Adam is 22 years old now. He appeared to be a normal, relatively healthy baby at birth. However, we soon found that he was not gaining weight as he should, seemed to vomit formula out of his nose, had lots of respiratory infections and was irritable and restless as an infant. It was hard to get him to sleep and for him to stay asleep. Therefore WE got little sleep, and this caused difficulty in our marriage because we were tired and irritable ourselves. As time went on, he seemed to level out, and, except for his speech delay, he seemed to be developing in his age group normally according to the chart the doctor shared with us. It wasn't until he was about 5 years of age that we became more concerned about his voice quality and speech problems. This is when he started to see doctors on a regular basis. When the doctors looked more carefully they decided that Adam had some facial characteristics that concerned them. The suspicions started then. On further examination it was found that Adam had VPI (velopharyngeal insufficiency) so a surgery was performed and he had speech therapy in school. He was also showing signs of heart problems. Tests found that he had a small ventricular septal cardiac defect (which is a congenital defect in the walls separating two chambers of the heart). After finding these, the FISH test was done and it was confirmed Adam had VCFS.

Adam was a timid and clumsy boy growing up, with feelings of inferiority during his early years of childhood. He had few friends and tended to stay to himself a lot. At age 15, after changing schools several times, Adam began to hide and isolate himself from people. He seemed to get his days and nights mixed up and rarely slept. His appetite dwindled to nothing and he lost considerable weight. Over several months he continued to stay depressed and we didn't know what to do. It seemed like he had come full circle and he was acting like he did when he was a baby. Adam then started to say odd things like, *"his teachers were all spying on him, writing notes about him and didn't like him"*. We took him to a psychologist and then a psychiatrist who was recommended by our doctor. Adam was diagnosed with paranoia and prescribed medication. Within a month Adam was acting-out with violent outbursts, argumentative and, on occasion, would try to hit his father. Within a year his medications were reduced and his depression seemed to be better. Although the depression was better, his ability to focus on any task and his over-all attention span were still limited. By the time Adam was 20 he was periodically expressing grandiose ideas about himself. He initially wanted to be a famous athlete or rock star. Later he started to think of himself as a tycoon and wanted to invest in money-making schemes and gamble at casinos. He envisioned himself rich and famous. During these obsessive times he would be very talkative, restless and could go without sleep much of the time. It was very scary.

Adam was diagnosed with: *bipolar* or *manic-depressant disorder* (grandiose ideas and paranoia).

Betty
Betty is now 17 yrs old. She was first seen by the doctor for a small cleft in her palate when she was 4 months old. I had toxemia when pregnant with Betty so the delivery was by C-section. After Betty had her palate repaired she seemed to grow very

slowly and had poor weight gain. By the time Betty was 4 years old it was decided she would be given growth hormone. She continued to be socially immature and we saw no real positive results from the hormone therapy that she was given.

At age 11, the time had come when Betty needed to see a professional. At this time she was living with me, her father and her 2 sisters and 2 brothers. Her father had just been diagnosed with a terminal illness; and with the cancer treatments came the stress that weighs upon the whole family when dealing with such situations. The family was always talking and wondering about what might transpire when the inevitable happened.

During Betty's growing-up years she had always had some behavioral problems, such as rebellion against parents, teachers, and even other adults who tried to guide her. She was extremely impulsive. She would steal; skip school, and show disrespect with no fear for any authority figure, not even the police. As a matter of fact, Betty physically struck one of her teachers during an argument about dressing-down in physical education.

By the time Betty was 13 she had become obsessed with make-up, colorful false eye lashes and colorful stains on her hair. She started to steal products from stores and take any items she desired.

Betty dropped out of school in 10th grade when she was 16 years old. She said she just couldn't make it to class anymore. She was too tired and could not concentrate, had a hard time understanding what it all meant and felt confused all the time. This was not really surprising because when Betty was 11 she attended group therapy for depression. She would say now (I'm sure) that she had been having mood swings since she was about 10 or earlier. (She started menstruation at 9, and we wonder if this contributed to anything.) Betty's father and I agreed that she seemed to get worse about 2-4 times a year. During these periods Betty would get extremely talkative; then in just a short period of time would get sulky, weepy and withdraw, not wanting to communicate or talk to anyone. Betty would say that, at those times when she was talking all the time, she also felt real excited and full of super amounts of energy. Her personality would get playful and she would exhibit a jovial joking persona; at times trying to make up jokes, many that didn't make sense, but she would make herself laugh. Also during these high-energy moods she felt hungry and would eat like there was no tomorrow, I am told it is called binging. The scariest times for us were when she would go on an excessive shopping spree with the credit card that we had given her for emergencies, buying anything and everything that caught her fancy. Many of the items she didn't need, but they were *"on sale"* so that justified her purchase. This created a financial burden on us because she would run the credit card up to maximum if we didn't recognize the symptom and stop it in time.

On the other side of the mood spectrum Betty would then suffer from bouts of irritation, annoyance, rejection and find it hard to get along with anyone. This made it difficult to deal with having her at home.

Betty recalls her first noticeably serious episode with depression was at 15. She said she had suicidal thoughts, didn't feel like eating and lost a lot of weight. She would lock herself in her room and sleep day and night. She felt wasted and had no energy. The next year, at 16, she had another severe episode. This one seemed even worse to me than the first (if that is possible). The first episode lasted about 2 weeks but this second one lasted about a month or two. She remembers only wanting to dress in black and think morbid thoughts.

With her second bout of depression, instead of sleeping day and night she experienced insomnia. Her depression got worse and she felt useless and worthless and had no positive thoughts about anything and she felt she was doomed and a burden to everyone.

Right now, Betty has been in a year-long depression with intermediate bouts of elation with energy, then back to irritability which lasts for days. We have not yet found any medication that has stabilizes her year round.

She was diagnosed: _Bipolar II with rapid cycling_ and tested positive for VCFS

Charlie
My son Charlie is now 15 yrs old and has been diagnosed with _Bipolar II with rapid cycling_.

He was first taken to see the doctor at 4 years of age with speech and language problems that we now know were related to VCFS. Charlie was a product of a high risk pregnancy. I had polyhydramnios disorder (meaning too much amniotic fluid) from toxemia.

Charlie suffered from pulmonic stenosis (cardiac condition), and chronic infections throughout his whole infancy and early childhood. Along with these symptoms there were learning and emotional problems. Charlie was given a genetic test at about 9 and it was affirmed he had VCFS.

Charlie is the younger of two children. He attends a school for learning disabled children.

Charlie was very docile and content until he was about 4. He was perfectly happy playing by himself and content with occupying himself, and didn't seem to want or need to mingle in preschool with the other kids.

Charlie started to experience what was called _separation anxiety_ between the ages of 4-9 even at home. It was not uncommon for someone to have to go and sit at his bed side or lay with him until he fell asleep at night. He experienced a type of night fears and couldn't fall asleep easily, afraid he would have more. If we tried to sneak out early he would start to panic and say he had bad dreams. Many people thought he was just spoiled, so we tried the *tough love* theory and it did not work. It was something real to him, not an act for attention. In kindergarten his teacher said she noticed he was very restless and distracted, of course I thought it was because of all the anxiety of not being at home. By 7 it was obvious he was easily distracted and he was diagnosed as learning disabled. The school psychologist described Charlie as *"emotionally immature and impulsive"*. Charlie could get very angry under stressful conditions and lash out at anyone in his way. It became more noticeable between 7-8 years of age. This is when he started to show he was consistently unstable in his moods. From age 7 on, the irritability became a constant feature of his emotional state. Charlie could not accept "NO" for an answer. If he was prevented from doing something he wanted to do, he became angry and he would throw tantrums. At this point you might wonder if we disciplined. The answer is *"yes"*, we did have discipline and structure in his life. Every time he acted up, I thought it HAD to be a "disorder"; one that had to be dealt with and not just a lack of parenting skills. Charlie started to become destructive, breaking things, putting holes in the walls and throwing food.

When Charlie was 10 years old he was taken to another child psychologist for his learning and behavioral problems. This doctor diagnosed him with ADHD and he was put on medication. This seemed to help at first but within a year Charlie began to exhibit high-energy states that would fluctuate over a period of hours up and down. Charlie would exhibit characteristics that were uncommon for him, stay up all night to do chores like cleaning his room, rearranging things on his shelves, sorting his clothes in his drawers, emptying out the closet and getting carried away with making absurd plans/goals for himself.

He would become so excited and determined that what he was doing was so important that he would have to be told to calm down and be quiet so the rest of us could sleep. This would happen on the average of 2-3 times a month.

We are presently working to get this under control and there is a lot of promise.

Dena
Since I was asked to talk about Dena's mental health history, I will not go into all the other medical problems that most of our VCFS children have had to endure. I just hope that what we have gone through can help another family feel they are not alone and find hope.

It was almost unbearable to think that my darling little princess, Dena, needed to see a psychiatrist when she was 6 years old and entering kindergarten. We knew she would be a little anxious when having to leave her mother's side. She was a timid and shy little girl and had been sheltered, and was never left alone at any time. Dena tended to worry a lot for her age, but we were surprised to the degree that this went. Dena became extremely fearful and anxious about the separation from home when she had to go to school and needed to engage in activities with others. The doctor diagnosed her with *"separation anxiety with panic attacks"* and put her on a medication.

We were elated when she was able to return to school. She appeared to actually be happy, outgoing and started to play well with other children in her class. This lasted for several months. Then in the following year, things started to change. She started saying (after getting into trouble) that it was hard for her to "listen or pay attention" to the teacher. She started to dwell on "death and doom." She was scared we were going to die and leave her alone, had all kinds of negative thoughts, got anxious in the mornings again and did not want to go to school. She used every trick in the book and then some, not to go. We think back now and realize that the lack of concentration may have been her preoccupation with death and negative thoughts, which is the effect of depression, we were told. We took her to another doctor. She was mis-diagnosed with ADHD and was given another medication to use. Within the first 24 hours of this stimulant drug we noticed she became extremely talkative, exuberant about everything, then oppositional and argumentative, and downright sassy. One of the most unnerving things was the extremely wild hyper-sexual tendency that arose in her at this age.

She was so out of control that she started chasing her brother around the house and tried being physical with him. (We now understand that this activity was done while in a manic state.) But soon after that she would become downcast in her demeanor. She would become very emotional, tearful, obsessing about death again. Back and forth these extremes would go, all within hours of each other and this lasted for several weeks. Needless to say, the medication was discontinued.

Dena is now 13 and she is mostly in a depressant mental state. We have not found the perfect medication for her as of yet. The diagnosis in her chart says *Bipolar II disorder (rapid cycling)*. Her doctor has said he has *"many other medications up his sleeve"* to try. I am sure this is to reassure us, so we don't lose faith. We are determined that one or more of them will work and we will have our sweet little girl back to normal and happy.

Eddie
Eddie has always been an impulsive child. It became noticeable at about 2 years old. He was a bed wetter until he was 8. (Even though this is not a *"typical"* VCFS symptom I have corresponded with many parents who have had this same problem with bed-wetting.)

Every minor change in life seems to disturb Eddie and he found it difficult to adjust. Even something as elementary or uncontrollable as a plan being cancelled or altered or a store he wanted to go to being closed on Sunday. He had no self control and would become very upset. It could take hours for him to forget about it. Eddie was very distractible, and unable to concentrate on things that were important. His moods would fluctuate between very low self esteem and a very inflated self image.

His temper flared up about twice a month. These tantrums were full blown fits. They could last 30-60 minutes at a time and would consist of yelling, screaming, kicking, pounding on the walls, spitting and anything else he could do. During a couple of his worst episodes he got physical with all of us. He hit his father, choked his sister and pulled her hair, and he scratched me (his mother). The funny thing is, these tantrums could be triggered by something minor and silly. Usually as minor as having to share or give up something like the dice on a board game, when it was someone else's turn. As Eddie got older he started to see that his behavior was not normal (compared to his peers) and so he would feel remorse and apologize profusely. He would tell us it wasn't the real him that was acting like this, but that *"his brain was telling him or making him do it"*. He declared he was not able to control the impulses. We often asked him if he heard voices and he would say, *"No."*

Eddie is now 14 and was just suspended from school. He could not control his anger and he threw a tray of food on a student in the cafeteria because he believed the student was making fun of him. Also, we all have noticed that his mouth has been getting more and more vulgar, both at home and at school.

I think one of the most difficult things for us to deal with right now, is his stealing and lying. I believe the impulsiveness has increased lately as a result of the suspension and his frustration has hit a high. He feels that he cannot control himself.

Eddie wants to be funny so people will like him, but his antics almost always turn into mischief or vandalism. As an example: he will put something slippery or gooey on door handles when we are all in a hurry. He doesn't seem to see that it is *NOT* funny, or why it is especially not funny at this time. I can't seem to teach him that it is all in the timing. He doesn't comprehend that practical jokes can only be funny if they don't hurt someone, don't cause harm to property or don't cause people to be inconvenienced.

When Eddie was 12 we all (including his teachers at school) could tell he was going through a depression. It would escalate at times if he was not having a good day or when he experienced a disappointment. He would dwell on the fact that he had VCFS and felt different and would wonder, *"Why ME?"*. The frustration of knowing there is no immediate *"cure"* would trigger a negative response. He would say he felt different and rejected by his classmates. Sometimes, he would call himself names. When these more severe episodes of depression would happen, Eddie would withdraw from everyone. He would demand to be left alone. We could hear him crying in his room, but he would say he didn't want to talk about it. At these times Eddie would want to do <u>everything</u> in his room: eat, watch TV, play on the computer, play video games, and *SLEEP*. He always seems to be a little depressed unless he is on a mental high and extremely impulsive. In these "high" times he is totally the opposite. Now that I can see this vast difference, it makes me wonder if I would have thought my son was on drugs if we didn't know about the disorder. It makes me wonder how many kids are not on drugs but who are getting blamed for it, and don't get help?

Over the past year Eddie has had more and more of the impulsive excited "high" spells. These seem to escalate if there is a special event in his life. Eddie loves going on dirt bike rides and fishing with his uncle. At these times he becomes overly talkative, uptight and pressured in his speech and repetitive in his thoughts. He can stay on a subject for days and weeks until the event is over. This is one of the hardest things to live with at times. We find that it is almost impossible to get his mind onto another subject.

Eddie receives speech therapy 3 times a week and he has a regular appointment with a psychotherapist once a week. We hoped that this would change his thinking pattern but it has not so far. We are now looking into biofeedback as an option. *(See Biofeedback on page 121.)* This is supposed to help a person gain control over impulses.

In the meantime we took Eddie to yet another child psychiatrist and he said Eddie has ADHD and he put him on a medication.

Although there has been a slight improvement in his impulsive behavior and attention, it has increased the mood swings with the irritability and restlessness. Eddie is almost always unbearable in the mornings; he is rude and obnoxious to everyone. He does seem to be better in the afternoons after school. We wonder if the bad morning moods are a direct result of the medication or is he just *"not a morning person"*.

The latest condition we are dealing with is a new form of obsession. He is concerned all the time now with germs and becoming sick and dying. He refuses to eat any food we order in a restaurant if he can't see it being cooked. He says he can visualize the people in the kitchen having dirty germy hands or spiting in his food and he just can't eat it.

Eddie has always been a little afraid of the dark but this too has now gotten out of control. With some help from the therapist he is able to go into the garage at night without freaking out.

He recently started to have an unnatural fear of dogs, both strange and familiar. We are concerned as to how far these phobias are going to reach; only time will tell.

His diagnosis is <u>*Cyclothymic disorder, hypomania, impulsive disorder and depression*</u>.

What is schizophrenia?

Schizophrenia is a chronic and sometimes serious disabling brain disease. It usually appears in someone's late teens or early twenties. People with this disorder often suffer terrifying symptoms. Much like the psychosis state experienced by schizoaffective disorder; people with schizophrenia hear internal voices. Both male and female voices, belonging to people whom they do not know. These voices are not heard by anyone but themselves. They can believe they are controlled in thought and action by people who want to do them harm. This can cause the person to be fearful and withdrawn, not knowing who to trust. Their speech is many times incomprehensible and it causes people who come in contact with them to be frightened. Schizophrenia is much more complicated than bipolar and is not easily contained. Many with this disorder wrestle with this all of their lives. However, we live in a time where new research is being done that gives promise of new medicines for people afflicted with schizophrenia and their families.

The first signs of schizophrenia may be shocking or scary to the family with the changes in behavior. Coping with the symptoms can be very difficult for the adolescent, as well as for the family that is in shock. The onset of schizophrenia usually has a sudden onset of psychotic symptoms or as we explained before *"psychosis."* This is a state of mental impairment, with the hallucinations (loss of perception), delusions (false but strong ideas), and an inability to separate real from unreal. Some people have only one true psychotic episode where some have many throughout their life. Since there are so many symptoms in the mania of bipolar and that of schizophrenia, it is this fact that makes it so important to know a doctor who understands the difference between the two and can diagnose correctly.

Can children have schizophrenia?

According to the *National Institute of Mental Health:* Although it is very rare before adolescence, it is possible for children 5 years of age and older to have schizophrenia. It is just difficult to diagnose when they are that young. For more information go to: **www.nimh.nih.gov** (Use on site search for "schizoprenia.")

Information on bipolar disorder:
When I was faced with the diagnosis of bipolar disorder for my son T.G. and contacted the VCFS Education Foundation I was given a phone number to contact Dr. D. Papolos. At the time, he was doing the case studies, that I now know were for a book that he and his wife were writing. My son was too old for the study, but he gave me some good advice and some areas to research. Now his book is complete and I would recommend reading it because Dr. Demitri Papolos is one of the leading experts in the area of children with bipolar disorder. He is an Associate Professor of Psychiatry at Albert Einstein College of Medicine and the Medical Advisor for Parents of Bipolar Children, Dr. Papolos has published the book **The Bipolar Child: The Definitive and Reassuring Guide to Children's Most Misunderstood Disorder**, which he co-authored with his wife, Janice.

In this book the doctors touch on many areas that come to mind when we, as parents, wonder if our child has a bipolar disorder.

One of the major questions raised is *"What are the symptoms?"*. He states that bipolar disorder in childhood has a number of different symptoms. Some of the chief symptoms include: extreme temper tantrums; out-of-control behavior that is related to difficulty controlling aggression; rapidly fluctuating mood swings which may occur many times within a day; problems with the sleep/wake cycle; carbohydrate cravings; changes in appetite; problems with concentration including distractibility, and racing thoughts; and often oppositional behavior. Many parents might describe the cycles as a *Dr. Jekyll and Mr. Hyde phenomenon*. There are other features as well, but these are the most remarkable.

One commonly finds other disorders paired with bipolar such as ADHD.

Dr. Papolos has stated: "In these children with bipolar disorder their sleep/wake cycle is often disturbed. Parents can help by trying to set regular limits to the sleep/wake cycle—meaning having their children try to sleep at a reasonable time at night and getting them up in the morning—preventing the tendency that exists with the condition to have a sleep/wake reversal. This, in many cases, can help to prevent the development of symptoms.

"This condition is hereditary and in fact many studies of adult bipolar disorder have demonstrated convincingly the genetic basis of this condition. However, the environment can alter the predisposition that is there on a genetic basis. Both sides of the gene pool have an equal chance of passing down the gene and there is currently no genetic test to determine if one has a bipolar disorder.

"It is important for people to know that there are new medications that are just starting to be used, and they are approved. When someone is properly medicated and takes care of herself, such as taking care of the sleep cycle, and not taking drugs and alcohol; she can enjoy a full, creative and meaningful life. As more attention is being given to this disorder in children and adults by the medical research community, there is every reason to hope there will be better medication that will follow."

I found it interesting that in this book there is also an (IEP) Individual Education Plan for children with bipolar disorders that takes into account their seasonal mood swings, shifts in energy and attention. It seems very detailed and would help a school-system to plan for a child with this disorder.

Medical Trivia

7-UP® once contained lithium carbonate, a powerful and effective drug used in psychiatry and administered as a treatment for manic-depression.

"Confidence, like art, never comes from having all the answers; It comes from being open to all the questions" Earl Gray Stevens

CHAPTER 10: COMT GENE

WHAT IS THE COMT GENE?

The COMT gene was first introduced to me at the 9th Annual VCFSEF (Velo-Cardio-Facial-Syndrome Education Foundation) Conference in San Diego in August of 2003. At that time there was a discussion about genes in reference to what is known as the COMT genes, and how they can affect VCFS/22q children. The symposium was called **"Effects of a functioning COMT polymorphism on neurocognitive function in the 22q deletion syndrome."** The study was done by Carrie E. Bearden, T.J. Simon, P.P. Wang, A.F. Jawad, D.R. Lynch, D. McDonald McGinn, and E.H. Zackai, B.S. Emanuel.

So it piqued my interest when I would find other articles regarding the study of the COMT genes. From my research I have found that the COMT genes are believed to be responsible for more than just the tolerance of pain, but have connections with other disabilities such as OCD, schizophrenia, and even breast cancer.

First let's get some general information on the COMT genes before we start discussing a few of the areas that are associated with 22q.

Everyone inherits two copies of the catecho-O-methyltransferase (COMT) gene, one from each parent. It codes for the enzyme that metabolizes neurotransmitters like dopamine and norepinephrine and comes in two common versions. One version, the *met,* contains the amino acid methionine (*met*) at a point in its chemical sequence where the other version, (*val*) contains a valine. Depending on the mix from heredity, a person's COMT genes can be typed as *met-met, val-val,* or *val-met.* Since all versions of the COMT gene are common in the population at large, it makes sense that each would present some advantages and disadvantages.

Do the COMT genes linked to pain tolerance, differ in VCFS?

The answer is, *"Yes they can."* Health Day News shared some articles that help us understand more about pain. People with one of the common versions of the gene, in association with more efficient working memory (Researchers have found that neurons *form memories* of past pain.) and frontal lobe information processing; may have adverse responses to amphetamines (mood elevators) in heightened anxiety and sensitivity to pain. In the study on *pain memories*, researchers concluded, from previous research, that when a certain group of neurons in the spinal cord is stimulated by pain-related matter, it enhances sensitivity to pain.

It is suggested that for those with a double *"low active"* version *met-met* of the COMT gene; antidepressants (which regulate dopamine) may be effective in reducing pain levels. This makes sense since dopamine produces endorphins, which not only give the *"up-lift"* in mood but are the body's natural painkillers; that is what antidepressants do as well.

People with the *val-val* combination can naturally absorb more enzymes and thus have a naturally higher pain tolerance. *Val* is considered the *"high activity"* of the two and the *met,* the *"low activity."* So one can conclude that a person with both *met-met* may experience more natural pain, with the least amount of pain-making stimuli.

Since researchers have identified that neurons may form *"memories"* of past pain, it might explain why some people experience on-going or chronic/persistent pain, even after the source of the pain is gone. The Discoveries, which appeared in the February 21, 2002 issue of Science Magazine, added to a growing body of research finding that *"An individual's perception of pain is largely dependent on his genes and brain chemistry."* (as quoted in Health Day News). This research explains why, when given the same experiences of stimulus to pain, some people will react like babies while the others will barely sense it.

Researchers focused on the COMT gene because it contains enzymes that control the metabolism of the neurotransmitters dopamine and noradrenaline, which affect pain tolerance by influencing the release of endorphins, the body's natural pain killer. Those with one copy of each gene, *met-val* combination, fall somewhere in the middle with pain tolerance; therefore it is important to note that the *met-val* combination is the most common. Researchers found that people with the *val-val* combination were able to activate the brain's painkilling system better than those with the *met-met* combination. Therefore, people with the *val-val* combination were able to tolerate the most pain. Isn't it amazing that a single gene can dictate how the body responds to pain?

In research studies the clinicians found some study participants who had a <u>single</u> variation of the COMT gene, like the variation we see in VCFS. They found that when pain was stimulated by pain induced mechanisms, persons with only <u>one</u> COMT gene of either *val* or *met* were more troubled by the experience than a person with two COMT genes. Hence, it would be reasonable to conclude that between two people, who only have <u>one</u> COMT gene, that the person with only the *Val* gene would be able to tolerate more pain, over the one with only a *met* gene. This might explain why some VCFS children have a variation in pain tolerance, even though both only have one COMT gene each.

For more information on pain check out the **American Pain Foundation***

What about the COMT gene and schizophrenia?

Genes don't directly program humans for psychopathology, such as hallucinations, delusions, and panic attacks, etc.

Schizophrenia is a serious brain disorder that alters a person's perceptions of reality, emotions, and thought processes.

---

*SOURCES: Jon-Kar Zubieta, M.D., Associate Professor of Radiology and Psychiatry, University of Michigan, Ann Arbor; Charles Argoff, M.D., Director, Cohn Pain Management Center, Syosset, N.Y., and Assistant Professor of Neurology, New York University School of Medicine; Feb. 21, 2003, Science.

Researchers have found that those with schizophrenia did worse on tests that evaluated their mental abilities. This group was more likely to have the variant in the COMT genes.

According to the *National Institute of Mental Health*, scientists have discovered a certain variation in a gene that appears to be associated with schizophrenia. This gene turned out to be the COMT gene. It affects the *brain chemistry function* and thereby *increases* the risk for schizophrenia. Studies of this COMT gene, that affects how efficiently the brain can process information, have revealed some differences in its molecular structure and how it increased susceptibility to schizophrenia. Researchers at the *National Institute of Health (NIH)* report that *"A common version of the COMT gene may predispose the brain toward a pattern of neurochemical activity associated with psychosis."* The alternate version of the COMT gene slightly reduces this mental ability. Events of the prefrontal brain function guided researchers to study the COMT gene and showed how COMT worked to increase risk for schizophrenia.

It is important to support the study of the COMT gene variants because they provide insight into how a gene affects the way the brain processes information. Moreover, this may ultimately reveal how there may be an increased susceptibility to schizophrenia. By studying brain chemistry, and through this study, there will be more medications formulated that can pin point where to focus for the best results of management of this disorder.

The NIMH researchers reported that people with reduced dopamine activity had less efficient information processing of the brain, along with increased risk for schizophrenia. People with *val-met* had more efficient information processing function, and people with *met-met* the most efficient of all. Since amphetamine (anti-depressants) boosts dopamine activity, researchers believe that this type of drug, should in effect, correct a deficiency in people with *val-val*. (Refer to Dr. Doron Gothelf's study of 22q11 which is available online.)

Mayada Akil, M.D., Joel Kleinman, M.D., and colleagues in the NIMH Clinical Brain Disorders Branch, examined expression of the gene that codes for tyrosine hydroxylase, the enzyme that makes dopamine, in the brains of 23 deceased normal subjects. In the March 15, 2003 *Journal of Neuroscience*, they reported that the gene turns-on more neurons projecting to the striatum (an area in the middle of the brain) in people who inherited two copies of the COMT *val* variant than in those with only one copy of the *val* variant. The higher expression of the tyrosine hydroxylase gene reflects higher dopamine synthesis, and presumably, higher activity of dopamine neurons."

Researchers found that people with *met-met* on amphetamines (used as antidepressants) performed worse. People with COMT genes *val-val* on amphetamines showed more efficient brainpower and didn't have to work as hard. The brain performs most efficiently when dopamine activity is at a *moderate level*. So this tells us that dopamine activity needs to be neither too low nor too high for optimal prefrontal brain functioning. Amphetamines and other drugs that affect prefrontal dopamine systems are used to treat attention deficit hyperactivity disorder (ADHD), and other psychiatric illnesses. Some people respond better than others on these medications. COMT gene typing may become a relevant factor when considering treatment with certain medications and their dosage.

Preliminary evidence suggests that having even one *met* gene predicts a better response on working memory tests in schizophrenia patients after treatment with an antipsychotic medication. Moreover, there is new evidence about how the *val* variant may increase risk for such psychotic illnesses.

"Many mysteries will continue to surround the COMT gene's possible role in schizophrenia and other psychiatric illness. Probably the most important and direct implication of this finding is that currently available medications that inhibit COMT may improve cognition and perhaps reduce illness severity in patients with one (as seen in VCFS) or two copies of this gene," Egan told *Reuters Health*.*

---

*Source: Daniel Weinberger, M.D., National Institute of Mental Health (NIMH), whose research team, headed by Venkata Mattay, M.D., reports on how the variants affect the brain's response to amphetamine in the May 13, 2003 *Proceedings of the National Academy of Sciences*, already published online.
SOURCE: Proceedings of the National Academy of Sciences 2001

OCD and the COMT gene, is there a connection?

<u>The Rockefeller University</u> did a study on the COMT gene in reference to OCD and found that an altered COMT gene *"increased men's risk for obsessive-compulsive disorder. The altered form of the gene involved the communication between the brain's nerve cells that may put certain men at greater risk of developing obsessive-compulsive disorder (OCD), report scientists from <u>The Rockefeller University</u> and four other institutions in the April 29 <u>Proceedings of the National Academy of Sciences</u>. The discovery, was the first susceptibility gene isolated for OCD, it offers a possible target for developing treatments for the disorder, which affects 1 to 3 percent of the U.S. population."* Although this was not a study that directly used participants with only one COMT gene (as we find in VCFS), it does give us a hint as to the direction research is going and gives us hope for discoveries for new medications to control this disorder.

PARENTS SPEAK OUT

One mother writes: "My daughter has what two different neurologists have dubbed *congenital* *indifference* *to* *pain*. She's had nerve conduction studies done and so forth, and they have found no neuropathy. She does not feel pain normally and in her case she just doesn't *feel* things that other people find painful. She has been stung on the lip by a hornet and had no reaction, dislocated her elbow never cried, and gashed open her lip side to side without a whimper!!! We frequently track her in the house by the trail of blood she leaves because of her foot getting cut; she doesn't feel it until it's pointed out to her. If she scrapes her knee, or does something that she knows is usually painful from watching other people, she will cry. It's not an, *'Ouch I am hurt!'* cry it's an, *'Oh, this is supposed to be painful and crying should be my response.'* type of cry. One neurologist said he feels sure she *feels* the pain but does not process it as pain due to her extreme processing difficulties. It gets lost in the wiring, so to speak. She has horrible ear infections and never says a word and has never needed pain medication after minor surgeries. Now that she is older she's done really dangerous things, such as putting her hand into a pan on the stove to have a taste of something hot, rummaging in the knife drawer to look for something (she cut herself), touching the burner on the stove, and so forth. We have to watch her very carefully!!!"

A parent shares: "I have 2 kids with VCFS. My daughter (age 7 1/2) has an extremely high tolerance to pain. I could go through an amazing list of things that she's gone through without *EVER* complaining about being in pain, but it's way too long. It includes getting her finger stuck in a door and not crying out, and many visits to the E.R. for stitches, with *NO* anesthesia except for the topical kind.

"My son (age 6) had major dental work done recently. When the dentist finally got him under general anesthesia (done in the hospital), he couldn't believe that this child didn't complain about the deep infection and obvious cause for pain in his mouth, ever. He had so much dental work done that it took 2 1/2 hours to complete. This would have put me out of commission for a week. I would not have been able to eat or talk or anything, but not this little guy. He wanted to go back to school the next day! No pain . . . no complaints . . . nothing."

For you parents who have kids with an unusually high tolerance of pain: Ever since my *two VCFS kids* could crawl, walk or get around independently, I have had a difficult time with them because they seem to have no fear of anything. They climb up bookcases, fences, run in front of cars, get out of cars seats and swing the car door open while I'm driving, swing way too high on swings, pet strange dogs . . . the list is infinite. I'm not talking about usual things that normally developing kids do; I'm talking about risk-taking to the extreme. My theory is that since they don't really know pain, or maybe just a small part of what other people would feel, they have no way to judge whether or not they can get hurt by doing something dangerous. They have no concept of what *getting hurt* is. This causes them to have behavior which is very scary to me. The risks they take are dangerous and cause them to get hurt because they have absolutely no concept of pain and danger. I take them to the park, or out in public and my son walks in front of swings while other kids are swinging. I yell out, *"Danger!"* but he just doesn't really understand. He's been knocked down many times, and it just doesn't faze him in the least. My daughter likes to climb up on walls and walk along them like a tightrope. Even if she falls, which she does, she wants to get back up and do it again. It's difficult to keep her off these things. It must be that because they don't feel pain when they get an injury, they are not afraid to do these things. I have 3 other kids and have never had this type of problem with them.

I hear many people talking about *pain tolerance*. Most discussions are on kids with high tolerance of pain. But my daughter like many other VCFS children, is just the opposite. She is constantly complaining about something hurting, such as leg pain. Just like most things with VCFS you have some that swing one way and some that swing in the opposite direction.

Medical Trivia

Every human spent about a half hour as a single cell.

*"**S**nowflakes are one of nature's most fragile things, but just look at what they can do when they stick together."* Author unknown

CHAPTER 11: NEGOTIATIONS

HOW TO NEGOTIATE

I have to admit from the beginning; negotiating is not easy for many of us. We sometimes know what we want for our children and sometimes we just *THINK* we know. Nothing is worse than going into a negotiation without enough knowledge to hold our own and feeling intimidated or losing our confidence. So in this chapter I have included things I have learned and suggestions from professionals with whom I have spoken or whose information I have read and found useful.

The first step is to learn <u>IF</u> you *HAVE TO NEGOTIATE* for what you want for your child; or if it is already in place in your area and you just need to find the information, such as special programs or software for learning difficulties, etc.

Decide what issues deserve the most merit and attention. It is best to prioritize them, not just in your head but on paper and then brainstorm with someone to make sure they are in the right order. After that, do your homework and educate yourself on the subject(s) about which you want to negotiate. Know your subject well but learn to separate the people you will be negotiating with from the issue. Learn what steps you need to take to make a successful plan, according to the rules of the institution involved in the negotiation. (If it is the school system; know <u>your</u> rights and <u>their</u> policies.) It also helps if you have a note book with you to keep you on course; have the facts you might need on the subject handy, as well as a picture of our child to show from the onset of the discussion. The photograph enables the *"other side"* to put a face with the child for whom you are negotiating. It also gives the cause some realistic value; a name, a face and a reason for the visit.

When starting the negotiation meeting, focus on common interests of the issue and discuss them first. (This sets the positive mood.) Come up with options that have a mutual gain to all involved. In other words, negotiate for something that you know will be agreeable to all.

This starts the pre-negotiation phase by mentally claiming the right to negotiate as well as showing you can be flexible and workable. Try not to pre-judge the people or the outcome before you go into the meeting, or your face and the tone of your voice may show this.

Negotiation begins by making the informed decision that you have the right to negotiate and that what you want or need for your child is worth negotiation. Claim your lawful right, but try not to be confrontational. Rather, be a collaborator, a team player. Select a subject that starts you out on common ground.

As an example: *"I have concerns about (whatever). How can we meet Bobby's needs? How can we perhaps make (your suggestion) work?"* Then listen for input from the other person. Remember, *"Listening is more than just waiting for your turn to talk."* It may be that easy.

If you are going to be a successful negotiator you will have to overcome being intimidated, being emotional or having a dislike of negotiating (what we might view as a confrontation). Not many of us like it; but we have learned to understand that's how the system often works. To get what we want, we must demand our rights for our children—tactfully!

If the people with whom you are dealing, whether a school, doctor, clinic, health insurance, etc. are not accommodating at first, say something like, *"You need to help me here, we all want what is best for Bobby, how can we achieve that goal?"*

If we are good at stroking egos and don't come off as phony, it doesn't hurt to stroke the person a little by giving a compliment or showing appreciation in words for their efforts to help (which is sincere in most cases). But nothing will kill a negotiation faster than a person who comes off as a phony, insincere or manipulative. No one likes to be perceived as a gullible fool.

One of the hardest steps in negotiating is not taking anything personally; this is especially hard when we are dealing with our own child. If the negotiation is not successful the first time; there is always another time and another person. It may be that the person we encountered is not having a good day, or is not familiar with what we are asking them to do or why it is important. Remember there's nothing wrong with _US_ if the person isn't giving us what we want. People nowadays like the easy, familiar way of doing things and are not prone to go out of their way if they can avoid it. In other words they get into a comfortable rut of how things are done; and it is hard for them to get out of that rut, to be creative or to make waves for others who are also in that *"don't rock the boat"* state of mind.

At this point it may be to your advantage to state what you are willing to do to acquire the services (maybe even put them in writing) and set another appointment. Then the other person knows what you are requesting and knows that you have taken notes about the grounds for the refusal. Leave them to process the information.

If there is a successful negotiation, it might be a good time to paraphrase what you understand to be the final outcome or decision from the notes you have taken (I hope) to make sure you are on the same page (so to speak). It confirms to them that you were listening. Just make sure it is written down on paper and it is specific with no room for error. This will minimize any misunderstandings later, which could slow down progress.

Body language in negotiations

When we go anywhere that is important, especially to a meeting, we need to be aware of our body language. Body language has been studied for years. It is obvious that it has some merit when numerous books, seminars and classes have been taught to people in business, showing them how to use it to their advantage and how to read it in others.

Since we are talking here about negotiation or meetings that can affect our children we should want to know how to give ourselves an edge. It helps to understand how we are perceived to procure services for our child.

Body language can be an affective way to communicate without saying a word. It can deliver a positive or negative response, send a positive or negative message to another, *or* it can say the wrong thing at the proverbial *"first impression."* Research has shown that the combination of voice tone and body language together account for 60-80% of what is communicated.

I hope the simple guide below will help in preparing you for any negotiations you may have to perform by giving yourself the successful edge that you will need to go forward in helping your children or those you love to get the services and recognition of 22q. This is what this book is intended to do. The information should prove to be valuable as we read the chapter on *Advocacy* and the need to negotiate. *(See Advocacy, page 269.)*

Guide to positive body language

• Be responsive, show eagerness:
This is done by slightly sitting forward in your seat. Have your arms open, perhaps resting on your legs. Nodding in the affirmative, will assure the other party you are listening and that you have some agreeable points upon which to build. If an agreement has been made, is about to be made, or if they believe the meeting is coming to a close; there are sign(s) you will see like, the closing of a folder, paperwork being gathered, putting down a pen and putting hands flat on the table. At this point in time, if you are confused about what has been accomplished or what the final decision was in their minds; *THIS* is the time to state this and get it clarified. When you leave the meeting, you need to know what the next step might be.

• Listening:
This is an important step in being able to find a meeting of the minds. Most people who are interested in what you have to say will tilt their head slightly. You should make consistent eye contact without looking like you are staring at them. Make verbal acknowledgements when they are on the same line of thought and you agree. This will help the other party to have a more open mind when it is your turn to speak and reason your point. To show a reflection that the content of what is being said is being evaluated, a person might ponder while biting on the earpiece of a pair of glasses, stroking their chin, or looking up and

to the right of the person. Sit with your legs crossed only at the ankle. You can learn more by listening than by talking.

• Be attentive:
SMILE--at appropriate times of course. This will help to relieve tension on both sides and will set a more pleasant stage for negotiations. Smiles do not cost you a dime, but they are priceless and you can cash-in on the proceeds from a well placed and genuine smile. When asking a question make them open-ended questions so that you encourage feed back; not just questions requiring only *yes* and *no* answers. This helps you to understand the mind set of the other person and their personality. Then you will learn how to proceed with them from there.

Guide to negative body language

• Signs of boredom:
People get bored and irritated when someone is a poor speaker so they tend to slump in their posture or sit too casual. Other signs of bordom include: tapping of one's foot or fingers, wandering eyes, inappropriate responses to questions or doodling on paper.

• A sign that what is being said is being rejected:
People who are rejecting information often show it with arms folded in a defensive position; a head down, as though looking at the floor; subconscious frowning as if they are not pleased with what they hear; or a sarcastic smirk, perhaps with biting on the inside of the cheek.

• Perceived aggression:
You come off as agressive if you lean too far forward in your chair. It makes it appear as if you are about to leap from it, if you hear something you don't like. Other aggressive indicators include: finger pointing, grinding of the teeth, tightening of the jaw and shaking of the head.

• Lying:
Lying is perceived if you touch your face while speaking, hold your hand over your mouth when you speak (no, it's not bad breath), avert your eyes, shift and are uneasy in your chair, or stall in answering a direct question. (Try not to get too carried

away with analyzing people. Some people ponder before answering, are shy, or glance around while talking.)

Behaviors that enhance negotiation

Below is a list of behaviors that have a positive influence in not only negotiating but with any interpersonal relationship.
- Convey an attitude of open acceptance and lack of prejudice
- Be honest
- Take initiative and responsibility
- Be reliable
- Demonstrate humility
- Show respect
- Accept accountability
- Be confident and prepared
- Show genuine interest
- Convey an appreciation of the other person's time
- Accept expressions of positive or negative feelings from others
- Take enough time
- Be frank and forthright
- Be able to admit when you are wrong
- Apologize when you have caused distress or inconvenience

Behaviors that can inhibit negotiations

Likewise the list below can affect interpersonal relationships in a harmful way.
- Conveying an attitude of doubt, mistrust or negative judgment
- Giving false information
- Conveying, *"It's not my job.", "It's your job, so do it."* or *"Aren't you supposed to help me? The law says so."* kind of attitude
- Not meeting commitments, only meeting half of your commitments or not being punctual
- Demonstrating self-importance
- Talking down to a person or assuming familiarity without permission like, *"May I call you Sherry?"*
- Making excuses or placing blame where it doesn't belong
- Being unsure and trying to wing it
- Acting like you are only doing something because it is your job, showing a lack of sincerity
- Acting as if others have more time than you do
- Demonstrating annoyance when a negative feeling is being expressed

- Rushing though something without giving it due attention
- Sending mixed messages, only saying things because we think that is what the other person wants to hear, talking behind someone's back or spreading vicious gossip that can ruin another person's reputation
- Denying or ignoring when we have made a error
- Acting like nothing happened or making excuses

With these clues of conduct and how to be prepared for negotiation; we wish you the best in your endeavors. Think Positive!

Medical Trivia

Humans shed about 600,000 particles of skin each hour, about 1.5 pounds per year. By 70 years of age, an average person has lost 105 pounds of skin.

Blank page for personal data:

"The secret of success is consistency of purpose"
Benjamin Disraeli

CHAPTER 12: ADVOCACY

WHAT IS ADVOCACY?

According to *Webster's II New Riverside University Dictionary*, *Advocacy* is *"Active support; as for a cause."* Isn't that what we do for our families? So to be an advocate we would be one who supports or defends this cause. In the case of VCFS we plead on behalf of our child or loved one; just as an attorney would be an advocator in court.

Below you will first hear from a parent whose child is a young adult with VCFS. She shares with you some of her experiences, her ideas, and how she successfully advocated for her child.

Advocacy amongst the turmoil
Contributed by Frances P. (parent and advocate)

It's a devastating thing for a parent to find out that their new baby has health problems that will affect their lives and the lives of their family. Children are a precious gift and even with a disability they have an impact on the future of the family in many interesting ways. Whether children are labeled normal by society or not, parents strive to protect and advocate for their children throughout their lives.

We are blessed with a different focus on parenthood when a child is born with a disability or a disabled child is brought into the picture; perhaps by marriage. It creates more issues; however, the parent's role in that child's life remains the same.

As parents we still focus on providing what the child needs to live a fulfilling life. No matter what we face. The first thing I had to do as the parent of a child living with VCFS/22q, was to face the reality of the fact that things would be different and that I had to maintain a stronger defense around my child.

As a parent, understanding the disability is a critical issue. Obtaining a proper diagnosis for the condition is a strong aspect for future decision-making and creating a good foundation. I don't believe that generalizing a disability into generic labels is sufficient for providing an environment that supports the child's individual needs throughout the early years and during educational settings. (Unfortunately, labels are needed in society for recognition and services in most cases.)

Once the disability is understood, the needs of the entire family must be considered. Especially, when milestones in the development within the family and the child begin to take place.

Special education, health care and service providers have a much better chance of understanding and cooperating when there is information shared and direction given by the parent. As the child grows and moves into the educational setting, if you have awareness of the type of environment that works best for the child, planning is helpful. This way you don't have to function on a crisis basis when dealing with each issue.

Finding a health care professional, who either knows the disability or is willing to learn about it and what it all can entail, can create a tremendous foundation for knowing what direction you need to take. Once the disability or need is pinpointed, the individual's personality can take shape within the structures of care that is needed/given. From this we can realize a healthier future.

There are many helpful organizations all across the country from which you can access a wealth of information.

www.snap4kids.org *(Special Needs Assistance Program)*
e-mail: *pam@ snap4kids.org*
www.rarediseases.org *NORD. (The National Organization for rare disorders) 1-800-999-6673 (Voice Mail)*
55 Kenosia Avenue, P.O. Box 1968, Danbury, CT 06813-1968
e-mail: *RN@rarediseases.org*

There is also a Web site that provides names and addresses for parent-training and information centers by state. They may be able to provide lists of groups and organizations in your state. That Web site is **http://www.taalliance.org** *(Technical Assistance Alliance for Parent Centers)* Find a local advocacy group in your community. Even if they are not dealing with the same disability, they can educate you on the laws pertaining to special education and will help you understand your rights as a parent dealing with the educational system. *(See Education on page 143.)*

Many school district personnel may view parents, who are advocates for their children's needs, as pushy at times. School districts believe they know what is best because they are the professionals, but not all children with special education needs have the same problems. 22q is one area where I have yet to find anyone in the special education arena who will admit to being an expert. Remember, there is no one who knows your child better than you.

As a parent advocate YOU must become an expert on both your child's disability and the educational techniques/programs that need to be accessed for your child's particular disability. Then you will have the tools to acquaint them with programs you would like for your child, or the teaching approach most likely to be successful. Then, they will respect your knowledge. Schools will not usually share the information that they are obligated to afford all children with disabilities the programs necessary for learning, because they are afraid parents *will ask* for the needed services. That costs money, and most schools, as a rule, are on a strict budget. But schools are required under *federal law* to provide whatever the child needs.

Prepare yourself by reading the laws, medical literature, and obtaining an advocate to support you. Free or low cost advocacy is available through the *Federation for Children with Special Needs* **www.fcsn.org** and your *State Department of Education*. Get organized and document every contact made with professionals. Have a ledger specifically for this purpose; not only document who you talked to, but the time, the date, the content and highlights of the conversation and the results. These organizations can usually be found in the phone book, on the Internet (under *Advocacy* or *Disabilities*) through a referral service and, in many cases, through your local Chamber of Commerce. Perhaps speak with a social worker who works in a children's hospital or ask around to get information by word-of-mouth.

Knowing the direction in which you want to go and having goals, are major factors in how you will get there. An _Individual Education Plan_ (IEP) provides the starting point with an evaluation to determine the present levels of performance. From that point you can add in the personal needs of your child and attempt to persuade the school to provide the needed services. (See IEP in Education on page 145.)

As your child progresses through the system you need to maintain a relationship with the school and continue to ask for the services. If you don't ask, they won't offer. When your child approaches the end of their educational journey they will need to have a plan that will take them into their adult life. By LAW, a <u>Transition Plan</u> is required to be in place by the age of 14.

A <u>Transition Plan</u> spells out the goals you want to obtain for your child as they step into the next aspect of life. It should include all the aspects of adult life such as: employment, independent living, community participation, adult services and post secondary education. If these areas are desired to be a part of the child's future, they can be pursued through the transition plan within the school system. The plan will also address which members of the IEP team will be responsible for each of the areas. The IEP team can take part in the support of the child's future in a more productive manner through a detailed plan. When a child becomes the appropriate age for transition plans he or she has the right, by law, to be a part of the IEP meeting and the transition planning. It is never too early to plan for the future. It may change over the years but at least there is a foundation plan in place. You are not alone. This law was not in place for many of us who are parents of older children. So this is a wonderful addition.

When dealing with all the various health, education and community programs and providers, record keeping is imperative to ensure that the situation can be readily understood and relayed to others. This written documentation of contacts provides protection against misunderstandings and miscommunications that can have an adverse affect on the relationships with the professionals with whom you will be dealing. In many cases it would be appropriate to send a follow-up letter after a positive phone conversation. Stating that, "As per our telephone conversation on 00/00/00 at 0:00 a.m. I am writing to thank you for . . . and appreciate your efforts regarding . . ." Then keep a copy on file.

Records to keep on file

It has been recommended that an expanding file with a compartment for each area of the child's life works well. One parent created a file system that holds information in the following areas:

Medical history
- Family history
- Developmental history
- Medical reports, diagnoses
- Recommendations and evaluations
- Progress reports

Educational history
- IFSP/IEP - individual plans for educational goals
- Consents
- Correspondence logs to and from professionals
- Evaluations
- Transition plans and goals
- Progress reports

It is very <u>important to acquire a *medical* diagnosis</u> rather than rely on an educational diagnosis. School districts focus on ability to perform tests, and what the written tests tell them; and a medical diagnosis provides a level of functioning beyond cognitive functioning or a test score. If possible <u>have the medical professionals write a letter to the educational institution</u> and try to get the two systems to work together on the IEP (Individual Education Plan).

Technical Assistance Alliance for Parent Centers

What is the Alliance?
<u>The Technical Assistance Alliance for Parent Centers</u> (the "*Alliance*") is funded by the U.S. Department of Education, Office of Special Education Programs, to serve as the coordinating office for the Technical Assistance to Parent Projects in

the beginning stage. The *Alliance* is an innovative project which focuses on providing technical assistance for developing and funding: parent training, information projects and community parent resource centers under the _Individuals with Disabilities Education Act_ (IDEA).

The Alliance project:
• Provides technical assistance for establishing, developing, and coordination of parent centers
• Offers technology resources and information to lead parent centers in the new millennium
• Informs parent centers about IDEA and other laws that affect families of children with disabilities
• Builds leadership among families and advocates to help them secure services and opportunities for children and young adults with disabilities
• Promotes cultural diversity and cultural competency
• Works with members of the national business and media communities to gain expertise for helping families
• Collaborates with other national organizations serving families of children with disabilities

What are parent centers?
Parent centers affect the lives of children and youth with disabilities across the nation.

Whether it's the family of a Georgia 2-year-old who cannot walk or talk, a Texas 12-year-old squirming through social studies class, or a young Montana woman with mental retardation seeking a job, parent centers are there to help.

Parent centers: _Parent Training and Information Centers_ (PTIs) and _Community Parent Resource Centers_ (CPRCs) serve families of children and young adults from birth to age 22 with *all* disabilities: physical, mental, learning, emotional, and attention deficit disorders. They train and inform parents and professionals and help families obtain appropriate education and services for their children with disabilities.

Parent centers work to improve education results for all children, resolve problems between families and schools or other agencies connected with children with disabilities to community resources that address their needs.

Since the parent centers are funded by the U.S. Department of Education, under the *Individuals with Disabilities Education Act (IDEA):* each state has at least one parent center, and states with large populations may have more. There are approximately 100 parent centers in the United States at this writing.

Parent centers are as individual as the families and communities they serve. Whatever their size or geographic location, commitment is the mark of parent centers. Staff is likely to be parents and siblings of children with disabilities, or adults who have disabilities themselves. The staff personifies the philosophy of: *Parents helping parents.*

How is the project administered?
The Alliance coordinating office, located at PACER Center in Minneapolis, administers the Alliance grant and supports the four regional centers that serve local and statewide parent centers. The coordinating office produces materials on IDEA and other special education or disability subjects. It conducts a national conference and operates four institutes covering specific issues and offering high-quality technical expertise to parent centers nationwide.
Its toll-free number is: **1 (888) 248-0822**.

The regional centers are a parent's first resource for technical assistance. Each regional office conducts an annual conference for parent center staff in the geographic area served by the regional office and facilitates *Individualized Technical Assistance Agreements (ITAGs).* Other work may include providing conferences, meetings, and training; publishing printed and Internet materials; and conducting conference calls, meetings, and site visits among parent centers.

Who is a part of the Alliance?
The Alliance is a partnership comprised of a coordinating office located at PACER Center in Minnesota and four regional centers located in New Hampshire, Ohio, Texas, and California.
Alliance partners are as follows by states.

Alliance coordinating offices:

PACER Center
8161 Normandale Blvd.
Minneapolis, MN 55437-1044
(952) 838-9000 - Voice
(952) 838-0190 - TTY
(952) 838-0199 - Fax
1-888-248-0822 toll free number nationwide
E-mail: **alliance@taalliance.org**
Web Site: **www.taalliance.org** also **www.pacer.org**
Sue Folger, Project Co-Director: **sfolger@pacer.org**
Kelly Lorenz: **klorenz@pacer.org**
Dao Xiong, Multicultural Advisor: **dxiong@pacer.org**
Jesús Villaseñor, Multicultural Advisor: **jvillasenor@pacer.org**
In partnership with: Judith Raskin, National Technical Assistance Consultant, Parent Information Center, New Hampshire

Northeast Regional Center
Parent Information Center of Delaware
700 Barksdale Road, Suite 16
Newark, DE 19711
302-366-0152
302-366-0276 fax
E-Mail: **picofdel@picofdel.org**
Web Site: **www.picofdel.org**
Marie Anne Aghazadian, Regional Director
Kathryn Herel, Technical Assistance Coordinator
CT, DE, DC, ME, MD, MA, NJ, NY, PA, Puerto Rico, RI, US VI, VT

Midwest Regional Center
Ohio Coalition for the Education of Children with Disabilities (OCECD)
Bank One Building
165 West Center Street, Suite 302
Marion, OH 43302-3741
(740) 382-5452 voice
(740) 383-6421 fax
E-mail: **ocecd@gte.net**
Margaret Burley, Regional Co-Director
Lee Ann Derugen, Regional Co-Director

Dena Hook, Technical Assistance Coordinator
CO, IL, IA, IN, KS, KY, MI, MN, MO, NE, ND, OH, SD, WI

South Regional Center
Exceptional Children's Assistance Center (ECAC)
907 Barra Row, Suite 102/103
Davidson, NC 28036
Phone: 704-892-1321
Fax: 704-892-5028
E-mail: **sregionta@aol.com**
Web Site: **www.ecac-parentcenter.org**
Connie Hawkins, Regional Director
Judy Higginbotham, Technical Assistance Coordinator
Johnny Allen, Multicultural TA Coordinator
AL, AR, FL, GA, LA, MS, NC, OK, SC, TN, TX, VA, WV

West Regional Center
Matrix Parent Network and Resource Center
94 Galli Drive, Suite C
Novato, CA 94949
(415) 884-3535
(415) 884-3555 fax
E-mail: **alliance@matrixparents.org**
Web Site: **www.matrixparents.org**
Alan Kerzin, Regional Director
Nora Thompson, Technical Assistance Coordinator
AK, AZ, Department of Defense Dependent Schools (DODDS), CA, HI, ID, MT, NV, NM, OR, Pacific Jurisdiction, UT, WA, WY

National Indian Parent Information Center
Judy Wiley
560-A NE "F" Street, PMB 418
Grants Pass, OR 97526
(541) 472-9467
(541) 472-9611 fax
E-mail: **Indian.Info@nipic.org**
Web Site: **www.nipic.org**
Serving: nationwide resources for Native American Tribes, Nations and Clans

Specialized Training of Military Parent (STOMP)
Heather Hebdon, Adriana Martinez
6316 South 12th Street, Suite B
Tacoma, WA 98465-1900
(253) 565-2266 Voice and TTY
(253) 566-8052 FAX
1-800-5PARENT Voice and TTY
E-mail: **stomp@washingtonpave.com**
Web Site: **www.stompproject.org** & **www.wapave.org**
Serving: Military Families; U.S. Military installations; and as a resource for Parent Centers and others

Contacts by state:

Alabama
Alabama Parent Education Center
Jeana Winter, Director
10520 US Highway 231
Wetumpka, AL 36092
334-567-2252
334-567-9938 FAX
E-mail: **jwinter@alabamaparentcenter.com**
Web Site: **www.alabamaparentcenter.com**
Serving: Statewide

Alaska (CPRC)
LINKS Mat-Su Parent Resource Center
Eric Wade
6177 East Mountain Heather Way, Suite #3
Palmer, AK 99645
907-373-3632
607-373-3620 FAX
E-mail: **eric@linksprc.org**
Web Site: **www.linksprc.org**
Serving: Matanuska-Susitna Borough

Alaska
Stone Soup Group
Pamela Shackelford
307 E. Northern Lights Blvd., Ste. 100
Anchorage, AK 99503
907-561-3701

907-561-3702 FAX
E-mail: **pams@stonesoupgroup.org**
Web Site: **www.stonesoupgroup.org**
Serving: Statewide

American Samoa
CPRC in American Samoa
Elda Najera-Suisala
PO Box 2191
Pago Pago, AS 96799
684-699-6621
684-699-6619 FAX
E-mail: **eldasuisala@yahoo.com**
Web Site: **www.taalliance.org/ptis/amsamoa**
Serving: Statewide

Arizona
Pilot Parents of Southern Arizona
Lynn Kallis
2600 North Wyatt Drive
Tucson, AZ 85712
520-324-3150
520-324-3152 FAX
1-877-365-7220
E-mail: **ppsa@pilotparents.org**
Web Site: **www.pilotparents.org**
Serving: Southern AZ

Arizona
RAISING Special Kids
Joyce Millarde-Hoie
2400 North Central Avenue, Suite 200
Phoenix, AZ 85004
602-242-4366 Voice & TDD
602-242-4306 FAX
1-800-237-3007 (in AZ)
E-mail: **info@raisingspecialkids.org**
Web Site: **www.raisingspecialkids.org**
Serving: Central and Northern AZ

Arkansas
Arkansas Disability Coalition
Wanda Stovall

1123 S. University Avenue, Suite 225
Little Rock, AR 72204-1605
501-614-7020 Voice & TDD
501-614-9082 FAX
1-800-223-1330 (AR only)
E-mail: **adcwstovall@earthlink.net**
Web Site: **www.adcpti.org**
Serving: Statewide

California
Chinese Parents Association for the Disabled
Rachel Chen
P.O. Box 2884
San Gabriel, CA 91778-2884
626-307-3837
E-mail: **chen_Rachel@hotmail.com**
Web Site: **www.cpad.org**
Serving: Los Angeles and Orange Counties

California
DREDF (Disability Rights Education & Defense Fund) and FYEW (Foster Youth Resources for Education)
Susan Henderson
2212 Sixth Street
Berkeley, CA 94710
510-644-2555 (TDD available)
510-841-8645 FAX
1-800-466-4232
E-mail: **shenderson@dredf.org**
Web Site: **www.dredf.org**
Serving: Alameda, Contra Costs and Yolo Counties

California
Exceptional Family Resource Center
9245 Sky Park Court Suite 130
San Diego, CA 92123
619-594-7416 Voice
858-268-4275 FAX
1-800-281-8252
E-mail: **efrc@projects.sdsu.edu**
Web Site: **http://www.efrconline.org**
Serving: San Diego and Imperial County

California
Exceptional Parents Unlimited
Bobbie Coulbourne
4440 N. First St.
Fresno, CA 93726
559-229-2000
559-229-2956 FAX
E-mail: **bcoulboune@exceptionalparents.org**
Web Site: **www.exceptionalparents.org**
Serving: Central California

California
Fiesta Educativa
Irene Martinez
161 S. Avenue 24
Los Angeles, CA 90031
373-221-6696
323-221-6699 FAX
E-mail: **info@fiestaeducativa.org**
Web Site: **www.fiestaeducativa.org**
Serving: East Los Angeles

California
Matrix Parent Network and Resource Center
Nora Thompson
94 Galli Drive, Suite C
Novato, CA 94949
415-884-3535
415-884-3555 FAX
1-800-578-2592
E-mail: **info@matrixparents.org**
Web Site: **www.matrixparents.org**
Serving: Northern California
with Parents Helping parents of Santa Clara

California
Parents Helping Parents
Mary Ellen Peterson
Sobrato Center for Nonprofits-San Jose
1400 Parkmoor Avenue, Suite 100
San Jose, CA 95126

408/727-5775
408-727-7655 TDD
408-727-0182 fax
E-mail: **infor@php.com**
Web Site: **www.php.com**
Serving: Northern California

California (CPRC)
Parents of Watts
10828 Lou Dillon Ave
Los Angeles, CA 90059
323-566-7556
323-569-3982 FAX
E-mail: **pow90059@yahoo.com**
Web Site: **www.cde.ca.gov** (Search for "Parents of Watts")
Serving: Watts Neighborhood of Los Angeles

California
Rowell Family Empowerment of Northern California
Kathleen Lowrance
962 Maraglia Street
Redding, CA 96002
530-226-5129
530-226-5141 FAX
1-877-227-3471
E-mail: **sklowrance@aol.com**
Web Site: **www.rfenc.org**
Serving: Far northern California

California
Support for Families of Children with Disabilities
Juno Duenas
1663 Mission Street, 7th Floor
San Francisco, CA 94103
415-282-7494
415-282-1226 FAX
E-mail: **jduenas@supportforfamilies.org**
Web Site: **www.supportforfamilies.org**
Serving: San Francisco

California

TASK, San Diego (Team of Advocates for Special Kids)
Brenda Smith
3180 University Avenue, Suite 430
San Diego, CA 92104
619-794-2947
619-794-2984 FAX
1-866-609-3218
E-mail: **taskca@yahoo.com**
Web Site: **www.taskca.org**
Serving: City of San Diego and Imperial county

California

Team of Advocates for Special Kids (TASK)
Mata Anchondo
100 West Cerritos Ave.
Anaheim, CA 92805
714-533-8275
714-533-2533 FAX
E-mail: **h8flying@yahoo.com**
Web Site: **www.taskca.org**
Serving: Southern California

California (CPRC)

Vietnamese Parents of Disabled Children Assoc., Inc. (VPDCA)
Hung Nguyen
7526 Syracuse Ave
Stanton, CA 90680
949-724-2359
949-724-2914 FAX
E-mail: **vdpcahung@yahoo.com/hgnguyen@vpdca.org**
Web Site: **www.vpdca.org**
Serving: Los Angeles

Colorado

Denver Metro Community Parent resource Center
Yvette Plummer
14501 E. Alameda Ave., Ste 205
Aurora, CO 8012
303-365-2772
303-365-2778 FAX
E-mail: **info@denvermetrocprc.org**

Web Site: **www.denvermetrocprc.org**
Serving: Denver Metro Area (Denver, Arapahoe, Adams, Jefferson and Douglas Counties)

Colorado
PEAK Parent Center, Inc.
Barbara Buswell
Julie Harmon
611 North Weber, Suite 200
Colorado Springs, CO 80903
719-531-9400 voice
719-531-9403 TDD
719-531-9452 FAX
1-800-284-0251
E-mail: **info@peakparent.org**
Web Site: **www.peakparent.org**
Serving: Statewide

Connecticut
Connecticut Parent Advocacy Center (CPAC)
Nancy Prescott
338 Main Street
Niantic, CT. 06357
860-739-3089 Voice & TDD
860-739-7460 FAX (Call first to dedicate line)
1-800-445-2722 in CT
E-mail: **cpac@cpacinc.org**
Web Site: **www.cpacinc.org**
Serving: Statewide

Delaware
Parent Information Center of Delaware (PIC/DE)
Marie-Anne Aghazadian
5570 Kirkwood Highway
Wilmington, DE 19808-5002
302-999-7394 Voice & TDD
302-999-7637 FAX
1-888-547-4412
E-mail: **maghaz@picofdel.org**
Web Site: **www.picofdel.org**
Serving: Statewide

District of Columbia
Advocates for Justice and Education
Kim Jones
2041 Martin Luther King Ave., SE, Suite 400
Washington, DC 20020
202-678-8060
202-678-8062 FAX
1-888-327-8060
E-mail: **kimjones@aje-DC.org**
Web Site: **www.AJE-DC.org**
Serving: District of Columbia

District of Columbia
Advocates for Justice and Education CPRC
Caru Echenique
4201 Georgia Ave. NW
Washington DC 20011
202-265-1432
202-265-9012 FAX
E-mail: **caru.echenique@aje-dc-org**
Web Site: **www.AJE-DC.org**
Serving: District of Columbia

Florida
Central Florida Parent Center
Eileen Gilley
Assistance with Achieving Results in Education
1021 Delaware Avenue
Palm Harbor, FL 3468
727-789-2400
727-789-2454 FAX
1-888-61A-WARE (612-9273)
E-mail: **cfpc@cflparents.org**
Web Site: **www.cflparents.org**
Serving: 30 counties in central and northeast Florida

Florida
Family Network on Disabilities
Jan LaBelle
2735 Whitney Road
Clearwater, FL 33760-1610
727-523-1130

727-523-8687 FAX
1-800-825-5736 (FL only)
E-mail: **fnd@fndfl.org**
Web Site: **fndfl.org**

Florida
Parent Education Network Project
Wilbur Hawke
Tara Bremer
2535 Whitney Road
Clearwater FL 33760-1610
727-523-1130
727-523-8687 FAX
1-800-825-5736 (FL only)
E-mail: **pen@fndfl.org**
Web Site: **www.fndfl.org/pen/index.htm**
Serving: Dade, Broward, palm Beach, Monroe, Collier, Lee, Hendry, Martin, and Glades Counties

Florida (CPRC)
Parent to Parent of Miami, Inc.
Isabel C. Garcia
7990 SW 117th Ave. Suite 200
Miami, FL 33183
305-271-9797
305-271-6628 FAX
1-800-527-9552
E-mail: **info@ptopmiami.org**
Web Site: **www.ptopmiami.org**
Serving: Miami Dade and Monroe Counties

Florida
Parents of the Panhandle Information Network
Jeanne Boggs
Nicole Brown
541 E. Tennessee Street, Suite 103
Tallahassee, FL 32308
850-847-0010
E-mail: **popin@fndfl.org**
Web Site: **www.fndfl.org/popin.htm**
Serving: Northwest Florida from Escambia Country to Alachua

Georgia
Parent to Parent of Georgia, Inc.
Stephanie Moss
3805 Presidential Parkway, Suite 207
Atlanta, GA 30340
770-451-5484
770-458-4091 FAX
1-800-229-2038 in GA
E-mail: **info@parenttoparentofga.org**
Web Site: **http:/www.parenttoparentofga.org**
Serving: Statewide

Hawaii
Hawaii Parent training and Information Center
Michael Moore
245 N. Kukui St., Ste. 205
Honolulu, HI 96817
808-536-9684
808-536-2280 Voice and TTY
808-537-6780 FAX
1-800-533-9684
E-mail: **mmoore@ldahawaii.org**
Web Site: **www.ldahawaii.org**
(Can be accessed through **www.taalliance.org**)
Serving: Statewide

Idaho (CPRC)
Idaho Parents Unlimited, Inc.
Evelyn Mason/Susie Depew
1878 W. Overland Rd. / P.O. Box 50126
Boise, ID 83705
208-342-5884 Voice & TDD
208-342-1408 FAX
1-800-242-4785
E-mail: **evelyn@ipulidaho.org**
Web Site: **www.ipulidaho.org**

Idaho
Native American Families Together Parent Center
Chris Curry
103 S. Polk
Moscow, ID 83843

208-596-2777
509-335-7339 FAX
1-866-326-4864
E-mail: **ftpd@familiestogether.org**
Web Site: **www.faniliestogether.org**
Serving: Nationwide resource for Native American families, tribes, communities, parent centers, and others.

Illinois
Designs for Change
Donald Moore
814 South Western Ave.
Chicago, IL 60612
312-236-7252 voice / 312-857-1013 TDD
312-236-7927 FAX
E-mail: **info@designsforchange.org**
Web Site: **www.designsforchange.org**

Illinois
Family Matters
Debbie Einhorn
1901 S. 4th Street, Suite 209
Effingham, IL 62401
217-347-5428
217-347-5119 FAX
1-866-436-7842
E-mail: **info@fmptic.org or deinhorn@fmptic.org**
Web Site: **www.fmptic.org**
Serving: 94 counties outside of Chicago Area 1

Illinois
Family Resource Center on Disabilities
Charlotte Des Jardins
Michelle Phillips
20 E. Jackson Blvd., Room 300
Chicago, IL 60604
312-939-3513
312-939-3519 TTY & TDY
312-939-7297 FAX
1-800-952-4199 (in IL0
E-mail: **info@frcd.org**
Web Site: **www.frcd.org**
Serving: Chicago Area 1

Indiana
IN*SOURCE
Richard Burden
1703 S. Ironwood Drive
South Bend, IN 46613
574-234-7101
219-239-7275 TDD
574-234-7279 FAX
1-800-332-4433 (in IN)
E-mail: **insource@insource.org**
Web Site: **www.insource.org**
Serving: Statewide

Iowa
Access for Special Kids (ASK)
Susan Myers
321 E. 6th St
Des Moines, IA 50309
515-243-1713
515-243-1902 FAX
1-800-450-8667
E-mail: **smyers@askresource.org**
Web Site: **www.askresource.org**
Serving: Statewide

Kansas
Families Together, Inc.
Connie Zienkewicz
3303 Wesst Second, Ste. 106
Wichita, KS 67203
316-945-7747
316-945-7795 FAX
1-888-815-6364
E-mail: **connie@familiestogetherinc.org**
Web Site: **www.familiestogetherinc.com**
Serving: Statewide

Kentucky (CPRC)
FIND of Louisville
Sandra Duverge
1151 S. 4th Street, Ste. 101
Louisville, KY 40203

502-587-6500
502-584-1261 FAX
E-mail: **smduverge@findoflouisville.org**
Web Site: **www.findoflouisville.org**
Serving: Jefferson County

Kentucky
Kentucky Special Parent Involvement Network (KY-SPIN)
Paulette Logsdon
10301 B Deering Road
Louisville, KY 40272
502-937-6894
502-937-6464 FAX
1-800-525-7746
E-mail: **spininc@kyspin.com**
Web Site: **www.kyspin.com**
Serving: Statewide

Louisiana
Louisiana Parent Training and Information Center
Cindy Arceneaux
201 Evans Rd., Bldg. 1 Ste 100
Harahan LA 70123
504-888-9111
504-888-0246 FAX
1-800-766-7736
E-mail: **carceneaux@laptic.org**
Web Site: **www.laptic.org**
Serving: Statewide

Louisiana (CPRC)
Pyramid Community Parent Training Program
D.J. Markey
3132 Napoleon Avenue
New Orleans, LA 70125
504-899-1505
504-827-2999 FAX
E-mail: **PyramidCPRC@aol.com**
Web Site: **www.pyramidparentcenter.org**
Serving: Greater New Orleans Metropolitan Area

Maine
Maine Parent Federation
Janice LaChance
12 Shuman Ave. #7,
Augusta, ME 04338-2067
PO Box 2067, Augusta ME 04330
207-623-2144
207-623-2148 FAX
1-800-870-7746 (ME only)
E-mail: **parentconnect@mpf.org**
Web Site: **www.mpf.org**
Serving: Statewide

Maine (CPRC)
Southern Maine Parent Awareness
Sue Henri-MacKanzie
886 Main Street, Suite 303
Sanford, ME 04073
207-324-2337
207-324-2338 Voice and TTY
207-324-5621 FAX
E-Mail: **support@somepa.org**
Web Site: **www.somepa.org**

Maryland Families Together
National Capital Region
Cindi Clark
4417 Great Oak Road
Rockville, MD, 20853-1946
301-460-4417
301-455-5636
E-mail: **mileniumcc@aol.com**
Web Site: **www.familiestogether.org**

Maryland
Parents Place of Maryland, Inc.
Josie Thomas
801 Cromwell Park Drive, Suite 103
Glen Burnie, MD 21061
410-786-9100 Voice & TDD
410-786-0830 FAX

E-mail: **info@ppmd.org**
Web Site: **www.ppmd.org**
Serving: Statewide

Massachusetts
Federation for Children with Special Needs
Richard Robison / Robin Foley
1135 Tremont Street, Suite 420
Boston, MA 02120-2140
617-236-7210 (Voice and TTY)
617-572-2094 FAX
1-800-331-0688 in MA
E-mail: **fcsninfo@fcsn.org**
Web Site: **www.fcsn.org**

Massachusetts (CPRC)
Urban PRIDE
Charlotte R. Spinkston
184 Dudley Street, Suite 104LL
Roxbury, MA 02119
617-989-3929
617-989-3925 FAX
E-mail: **c.spinkston@urbanpride.org**
Web Site: **www.urbanpride.org**
Serving: Inner City Boston

Michigan
CAUSE
Patricia Keller
6412 Centurion Dr. Suite 130
Lansing, MI 48413
517-886-9167 Voice & TDD & TDY
517-886-9633 FAX
1-800-221-9105 (in MI)
1-866-454-4464
E-mail: **info@causeonline.org**
Web Site: **www.causeonline.org**

Michigan
Tri-County Partnership (Division of Cause Online)
Patricia Keller
NW Activity Center

18100 Meyers, Suite 305/307
Detroit, MI 48235
313-863-0813
313-863-8048 FAX
1-800-298-4424
E-mail: **infodetroit@causeonline.org**
Web Site: **www.causeonline.org**
Serving: Wayne, Oakland, Macomb counties

Michigan (CPSA)
Southwest Detroit Family Center
Services to Parents of Children with Disabilities
Malisa Pearson
69 McGraw
Detroit, MI 48210
313-895-2860
313-895-2867 FAX
E-mail: **swdacmh@aol.com**
Web Site: **www.acmh-mi.org/swdetroitfc.html**

Minnesota (CPRC)
Discapacitados Abriendose Caminos
Ana M. Perez de Perez
400 Southview Blvd.
South St. Paul, MN 55075
651-293-1748
651-293-1744 FAX
E-mail: **discapacitados@comcast.net**
Web Site: **www.php.com/discapacitados-abriendose-caminos**

Minnesota
PACER Center, Inc.
Paula Goldberg/Virginia Richardson
8161 Normandale Blvd.
Minneapolis, MN 55437-1044
952-838-9000 (Voice)
952-838-0190 (TTY)
952-838-0199 FAX
1-800-537-2237 (in MN)
E-mail: **pacer@pacer.org**
Web Site: **www.pacer.org**

Mississippi
Parent Partners
3111 North State Street
Jackson, MS 39216
(601) 366-5707
1-800-366-5707 (in MS)
Web Site: **www.autismlink.com/listing/parent-partners-of-mississippi**

Mississippi (CPRC)
Empower Community Resource Center
Agnes Johnson
136 South Poplar Street
Greenville, MS 38701-4025
PO Box 1733, Greenville, MS 38702-1733
662-332-4852
662-332-1622 FAX
1-800-337-4852
E-mail: **empower@suddenlinkmail.com**
Serving six EMPOWER counties

Mississippi MS PTI
Pam Dollar
2 Old River Place, Suite A
Jackson, MS 39202
601-969-0601
601-709-0250 FAX
1-800-721-7255
E-mail: **mspti@mscoalition.com**
Web Site: **www.mspti.org**
Serving: Statewide

Missouri
Missouri Parents Act (MPACT)
Mary Kay Savage
8301 State Line Road, Suite 204
Kansas City, MO 64114
816-531-7070
816-931-2992 TDD
816-531-4777 FAX

E-mail: **msavage@ptimpact.com**
Web Site: **www.ptimpact.com**
Serving: Statewide

Montana
Parents Let's Unite for Kids
Roger Holt
516 N. 32nd Street
Billings, MT 59101
406-255-0540
406-255-0523 FAX
1-800-222-7585 (in MT)
E-mail: **plukinfo@pluk.org**
Web Site: **www.pluk.org**

Nebraska
PTI Nebraska
Glenda Davis
3135 North 93rd Street
Omaha, NE 68134
402-346-0525 Voice & TDD
402-934-1479 FAX
1-800-284-8520
E-mail: **Info@pti-nebraska.org**
Web Site: **www.pti-nebraska.org**

Nevada
Nevada Parents Encouraging Parents (PEP)
Karen Taycher
2810 W. Charleston Blvd., Suite G-68
Quail Park IV
Las Vegas, NV 89102
702-388-8899
702-388-2966 FAX
1-800-216-5188
E-mail: **ktaycher@nvpep.org**
Web Site: **www.nvpep.org**

New Hampshire
Parent Information Center
Heather Thalmeimer

P.O. Box 2405
Concord, NH 03302-2405
603-224-7005 (Voice & TDD)
603-224-4365 FAX
1-800-232-0986 (in NH)
E-mail: **hthalheimer@parentinformationcenter.org**
Web Site: **www.parentinformationcenter.org**
Serving: Statewide

New Jersey
Statewide Parent Advocacy Network (SPAN)
Diana Autin, Debra Jennings
35 Halsey Street, 4th Floor
Newark, NJ 07102
973-642-8100
973-642-8080 FAX
1-800-654-SPAN
E-mail: **diana.autin@spannj.org**
Web Site: **www.spannj.org**
Serving: Statewide

New Mexico
Parents Reaching Out
Larry Fuller
1920 "B" Columbia Drive SE
Albuquerque, NM 87106
505-247-0192 Voice & TDD
505-247-1345 FAX
1-800-524-5176 (in NM)
E-mail: **prodreamaker@msn.com**
Web Site: **www.parentsreachingout.org**

New Mexico (CPRC)
Abrazos Family Support Services
Jeanette Trancosa
P.O. Box 788
Bernalillo, NM 87004
505-867-3396
505-867-3398 FAX
E-mail: **info@abrazosnm.org**
Web Site: **www.abrazosnm.org**
Serving: 22 American Indian Communities

New York
The Advocacy Center
Barb Klein
590 South Avenue, Averill Court
Rochester, NY 1460
585-546-1700
585-546-7069 FAX
1-800-650-4967 (NY only)
E-mail: **klein@advocacycenter.com**
Web Site: **www.advocacycenter.com**
Serving: Statewide except for New York City

New York
Advocates for Children of NY
Ana Espada
151 West 30th Street, 5th Floor
New York, NY 10001
212-947-9779
212-947-9790 FAX
E-mail: **aespada@advocatesforchildren.org**
Web Site: **www.advocatesforchildren.org**
Serving: Five boroughs of New York City

New York
Parent Network of WNY
Susan Barlow / Kim Walek
1000 Main St.
Buffalo, NY 14202
716-332-4170
716-332-4171 FAX
E-mail: **infor@parentnetworkwny.org**
Web Site: **parentnetworkwny.org**
Serving: Erie, Niagara, Orleans, Wyoming, Genesee, Chautauqua, Cattaraugus, and Allegany Counties

New York
Resources for Children with Special Needs, Inc.
Rachel Howard
116 East 16th St., 5th Fl.
New York, NY 10003
212-677-4650
212-254-4070 FAX

E-mail: **info@resourcesnyc.org**
Web Site: **www.resourcesnyc.org**
Serving: New York City (Bronx, Brooklyn, Manhattan, Queens, Staten Island)

New York
Sinergia / Metropolitan Parent Center
Myrta Cuadra-Lash
2082 Lexington Ave, 4th Floor
New York, NY 10035
212-496-1300
212-496-5608
E-Mail: **information@sinergiany.org**
Web Site: **www.sinergiany.org**
Serving five boroughs of New York City: Bronx, Brooklyn, Manhattan, Queens, and Staten Island

New York (CPRC)
United We Stand
Lourdes Rivera-Putz
91 Harrison Avenue
Brooklyn, NY 112006
718-302-4313
718-302-4315 FAX
E-mail: **uwsofny@aol.com**
Web Site: **www.uwsofny.org**
Serving: Brooklyn

North Carolina
Exceptional Children's Assistance Center (ECAC), Inc.
Connie Hawkins / Mary LaCorte
907 Barra Row
Suites 102/103
Davidson, NC 28036
704-892-1321
704-892-5028 FAX
1-800-962-6817 (NC only)
E-mail: **mlacorte@ECACmail.org**
Web Site: **www.ecac-parentcenter.org**
Serving: Statewide

North Carolina (CPRC)
Hope Network
Vickie Dieter
300 Enola Road
Morganton, NC 28655
828-433-2825
828-433-2821 FAX
E-mail: **vbdieter@charter.net**
Web Site: **www.fsnhope.org**
Serving: Burke County

North Dakota
ND Pathfinder Parent Training and Information Center
Cathy Haarstad, Executive Director
Arrowhead Shopping Center
1600 2nd Ave. SW, Suite 19
Minot, ND 58701-3459
701-837-7500 Voice
701-837-7501 TDD
701-837-7548 FAX
1-800-245-5840 (ND only)
E-mail: **info@pathfinder-nd.org**
Web Site: **www.pathfinder-nd.org**
Serving: Statewide

Ohio
OCECD
LeeAnn Derugan (Region1), Margaret Burley (Region 2)
Bank One Building
165 West Center St., Suite 302
Marion, OH 43302-3741
740-382-5452 Voice & TDD
740-383-6421 FAX
1-800-374-2806
E-mail: **ocecd@gte.net**
Web Site: **www.ocecd.org**
Serving: Statewide

Oklahoma
Oklahoma Parents Center, Inc.
Sharon House
700 N. Hinkley (P.O. Box 512)

Holdenville, OK 74848
405-379-6015
405-379-0022 FAX
1-877-553-IDEA (4332)
E-mail: **info@oklahomaparentscenterorg**
Web Site: **www.oklahomaparentscenter.org**

Oregon
Oregon FIRST
Anne Saraceno
830 NE 47th Ave.
Portland, Or 97213
503-215-2268
503-215-2478 FAX
E-Mail: **info@orfirst.org**
Web Site: **www.orfirst.org**
Serving: Multnomah, Washington and Clackamas Counties

Oregon
Oregon PTI
Janice Richard
2288/ Liberty Street NE
Salem, OR 97301
503-581-8156 Voice & TDD
503-391-0429 FAX
1-888-505-COPE (2673) (OR only)
E-mail: **orpti@open.org**
Web Site: **www.open.org**

Pennsylvania (CPRC)
Hispanos Unidos para Niños Excepcionales
(Hispanics United for Exceptional Children)
Luz Hernandez
2200 North Second Street
Philadelphia, PA 19133-3838
215-425-6203
215-425-6204 FAX
E-mail: **huneinc@aol.com**
Web Site: **www.huneinc.org**
American Street Empowerment Zone

Pennsylvania (CPRC)
The Mentor Parent Program
Gail Walker
P.O. Box 47
Pittsfield, PA 16340
814-563-3470
814-563-3445 FAX
1-888-447-1431 (in PA)
E-mail: **gwalker@westpa.net**
Web Site: **www.mentorparent.org**
Serving: Rural NW PA

Pennsylvania
Parent Education Network
Kay Lipsitz
2107 Industrial Hwy
York, PA 17402-2223
717-600-0100 Voice & TTY
717-600-8101 FAX
1-800-522-5827 (in PA)
1-800-441-5028 (Spanish in PA)
E-mail: **mentorprogram@verizon.net**
Web Site: **www.parentednet.org**
Serving: Central and Eastern PA

Puerto Rico
APNI
P.O. Box 21280
San Juan, PR 00928-1280
787-763-4665
787-765-0345 FAX
1-800-981-8492
E-mail: **centralinfo@apnipr.org**
Web Site: **www.apnipr.org**
Island of Puerto Rico

Rhode Island
RI Parent Info Network, Inc. (RIPIN)
Sue Donovan
175 Main Street
Pawtucket, RI 02860-4101
401-727-4144

401-727-4040 FAX
1-800-464-3399 (in RI)
E-mail: **donovan@ripin.org**
Web Site: **http://www.ripin.org**

South Carolina (CPRC)
Parent Training & Resource Center
Beverly McCarty
1575 Savannah Hwy. Suite 6
Charleston, SC 29407
843-266-1318
843-266-1941 FAX
E-mail: **bevmccartyb@frcdsh.org**
Web Site: **www.frcdsn.org**
Serving: Tri-county: Charleston, Berkeley, and Dorchester

South Carolina
PRO-PARENTS
Mary Eaddy
652 Bush River Road, Suite 203
Columbia, SC 29210
803-772-5688 Voice
803-772-5341 FAX
1-800-759-4776 in SC
E-mail: **proparents@proparents.org**
Web Site: **www.proparents.org**
Serving: Statewide

South Dakota
South Dakota Parent Connection
Elaine Roberts
3701 West 49th St., Suite 102
Sioux Falls, SD 57106
605-361-3171 Voice & TDD
605-361-2928 FAX
1-800-640-4553 (in SD)
E-mail: **eroberts@sdparent.org**
Web Site: **www.sdparent.org**
Serving: Statewide

Tennessee (CPRC)
Tennessee Voices for Children
701 Bradford Ave.
Nashville, TN 37201
Phone: 615-269-7751
Fax: 615-269-8914
E-mail: **TVC@tnvoices.org**
Web Site: **www.tnvoices.org**
Serving: Statewide

Tennessee
Support and Training for Exceptional Parents, Inc.
Jenness Roth/Karen Harrison
712 Professional Plaza
Greeneville, TN 37745
423-639-0125 voice / 423-639-8802 TTY/Text
423-636-8217 FAX
1-800-280-STEP (in TN)
E-mail: **information@tnstep.org**
Web Site: **www.tnstep.org**

Texas (CPRC)
The Arc of Texas
Parents Supporting Parents Network
8001 Centre Park Dr., Suite 100
Austin, TX 78754
512-454-6694
512-454-4956 FAX
1-800-252-9729
E-mail: **info@thearcoftexas.org**
Web Site: **www.thearcoftexas.org**

Texas
Partners Resource Network—PATH Project
Gary Ferguson
1090 Longfellow Drive, Suite B
Beaumont, TX 77706
409-898-4684 Voice and TDD
409-898-4869 FAX
1-800-866-4726
E-mail: **partnersresoure@sbcglobal.net**
Web Site: **www.PartnersTX.org**

Serving: Dallas, Fort Worth, Austin, Wichita Falls, Southeast and East Texas

Texas
Partners Resource Network—PEN Project
JoAnn Rodriguez
1001 Main Street, Suite 701
Lubbock, TX, 79401
806-762-1434
806-762-1628 FAX
1-877-762-1435 (in TX)
E-mail: **wtxpen@sbcglobal.net**
Web Site: **www.PartnersTX.org**
Serving: Amarillo, Lubbock, Abilene, San Angelo, and El Paso

Texas
Partners Resource Network—TEAM Project
Amy Woolsey
5005 W. 34th St., Suite 207A
Houston, TX
713-542-2147
713-942-7135 FAX
1-877-832-8945
E-mail: **prnteam@sbcglocal.net**
Web Site: **www.PartnersTX.org**
Serving: San Antonio, Houston, Corpus Christi, Rio Grande Valley, and Laredo

Texas
Special Kids, Inc. (SKI)
Rose Ferguson
P.O. Box 266958
Houston, TX 77207-6958
713-734-5355
713-836-1955 FAX
E-mail: **Specialkidsinc@yahoo.com**
Web Site: **www.specialkidsinc.org**
Serving: Houston Independent School Districts: South, South Central, and Central

Utah
Utah Parent Center

Helen Post
2290 East 4500 S., Suite 110
Salt Lake City, UT 84117-4428
801-272-1051
801-272-8907 FAX
1-800-468-1160 (in UT)
E-mail: **upcinfo@utahparentcenter.org**
Web Site: **www.utahparentcenter.org**
Serving: Statewide

Vermont
Vermont Family Network
Kathleen Kilbourne
600 Blair Park Road, Suite 240
Williston, VT 05495
802-876-5315 Voice and TDD
802-876-6291 FAX
1-800-800-4005
E-mail: **kathleen.kilbourne@vtfn.org**
Web Site: **www.vermonthfamilynetwork.org**
Serving: Statewide

Virgin Islands
V.I. FIND
9500 Wheatley Shopping Center
#2 Nye Gade
St. Thomas, US VI 00802
340-774-1662
340-774-1662 FAX
E-mail: **vifind@islands.vi**
Web Site: **www.taalliance.org**
Serving: Virgin Islands

Virginia
Parent Educational Advocacy Training Center
Suzanne Bowers
100 N. Washington St. Suite 34
Falls Church, VA 22046
703-923-0010
1-800-693-3514 FAX
1-800-869-6782 (VA only)

E-mail: **partners@peatc.org**
Web Site: *www.peatc.org*
Serving: Statewide

Washington
Families Together
Sherry Watson
Smith Gym 213D
Pullman, WA 99164-1410
509-335-2321
509-335-7339 FAX
1-866-326-4864
E-mail: **ftpd@familiestogether.org**
Web Site: *www.familiestogether.org*

Washington (CPRC)
Parent to Parent Power
Yvone Link
1118 S 142nd Street, Suite B
Tacoma, WA 98444
253-531-2022
253-538-1126 FAX
E-mail: **yvone_link@yahoo.com**
Web Site: **www.p2ppower.org**
Serving: Asian families in Western Washington and Canadian border

Washington
PAVE
Tracy Kahlo, Vicky McKimmey
6316 South 12th Street., Suite B
Tacoma, WA 98465-1900
253-565-2266 Voice & TTY
253-566-8052 FAX
1-800-572-7368 (in WA)
E-mail: **tkahlo@washingtonpave.com** or **vmckinney@wapave.org**
Web Site: **www.washingtonpave.org**
Serving: Statewide

West Virginia
West Virginia PTI
Pat Haberbosch
1701 Hamill Ave.
Clarksburg, WV 26301
304-624-1436 Voice & TTY
304-624-1438 FAX
1-800-281-1436 (in WV)
E-mail: **wvpti@aol.com**
Web Site: **www.wvpti.org**
Serving: Statewide

Wisconsin
Alianza Latina Aplicando Soluciones
Becky Medina
1615 S. 22nd Street, Suite 109
Milwaukee, WI 53204
414-643-0022
414-643-0023 FAX
E-mail: **alianza.latina07@yahoo.com**
Web Site: **alianzalatinawi.org**
Serving: City of Milwaukee

Wisconsin
Native American Family Empowerment Center
Great Lakes Intertribal Council
2932 Highway 47N, P.O. Box 9
Lac du Flambeau, WI 54538
715-588-3324
715-588-7900 FAX
1-800-472-7207
E-mail: **glitc@glitc.org**
Web Site: **www.glitc.org**

Wisconsin (CPRC)
Wisconsin FACETS
Jan Serak, Courtney Slazer
2714 North Dr. Martin Luther King Dr., Suite E
Milwaukee, WI 53212
414-374-4645 / 414-374-4635 TTD
414-374-4655 FAX
1-888-374-4677

E-mail: **wifacets@wifacets.org**
Web Site: **www.wifacets.org**
Serving: Milwaukee

Wyoming
Parent Information Center
Terri Dawson
500 West Loft Street, Suite A
Buffalo, WY 82834
307-684-2277 Voice & TDD
307-684-5314 FAX
1-800-660-9742 (WY only)
E-mail: **tdawson@wpic.org**
Web Site: **www.wpic.org**

Without advocates of the past, our children living with disabilities would still be living within institutions of old. We have come a long way in our society but we have a long road ahead. Fortunately we live in a country where we are able to protect our rights as individuals.

Battling school districts can be a hard up-hill battle and can take so long it makes us feel they are trying to wear us down. Many parents have had a struggle when it comes to educational placements. Because many children function or learn way below their grade level, the officials want to place them in special schools. Sometimes this is appropriate, but at other times it can have a traumatizing affect on the child; especially if the child is older, has gone to school with the same group since kindergarten and has grown up with these friends for years. It can be scary to have to go to a strange school, with strange teachers and not have a pal with whom to spend time. As we noted before, VCFS children seem to do better in familiar surroundings.

In this instance, before negotiating for the child, make a list of the good reasons she should stay, or the opposite if you believe a change is better for services. Reasons can be things such as: good behavior (if it is true), whether they have problems socializing and whether or not they already have a support system of friends where they are presently.

Even if the school says they are not prepared for your child's level of disability; they must try to accommodate. If the school authorities adamantly refuse, even though schools are *obligated* to provide accommodations according to federal laws; first, get familiar with the *federal laws* and then file a grievance with your state's department of education. It is not the teachers who are holding things back. (Many teachers are very sympathetic.) It is the political entity that pulls the purse strings.

What is FAPE Solutions?
FAPE Solutions is a unique company whose mission is to help parents, advocates, and others throughout the whole USA, develop and implement better IEP's.

What does FAPE Solutions do?
Provides fee-based personal IEP assistance for individuals. They provide customer's with excellent, actionable information while taking as little time as possible. Usually they start with an existing IEP and do a detailed analysis, adjusting with recommendations. They offer: IEP action plans, IEP progress reports and individual support services.

To learn more visit:
www.fapes.org or **www.fape4everychild.org**

Our children are in hospitals so much, what should we know from a legal standpoint? What are our rights for remedy?

Provided here are some legal words and definitions that may come in handy in schools, hospitals and any other place that cares for your child:

- **Due care:** Assumes that the doctor or medical professional possesses the qualifications and training to provide appropriate competent medical care for the patient, as well as competent office personnel.

- **Divisions of law:**
1) Criminal - cases against the state
2) Civil - cases against a person

A physician can be accused of assault and battery if he does not get consent to do a surgery. This can be criminal and civil.

- **Assault:** is an ATTEMPT or a THREAT to inflict bodily harm on another. As an example: *"I'm going to kill you."* then *"appears to try to run you over."*

- **Battery:** is unauthorized touching of any kind.

- **Negligence:** is the result of a doctor or nurse not using due care (failure to do something required, _normally_ performed for a patient for a specific problem).

- **Child abuse:** It is mandated by law that any medical personnel, teacher, etc. has to report any suspicion of mental or physical abuse. If they fail to do so, they can lose their job and/or go to jail for failure to report. This is why in many cases *"child protection"* agencies get called in by professionals to investigate.

- **Intentional tort:** happens when a doctor or nurse fails to use ordinary care; if a patient is restrained against their will; or the professional was negligent

- **Malpractice:** There must be _four_ things present to prove a malpractice case.
    1) Duty to care, was not done.
    2) How was the duty breached?
    3) The injury was a result of the breach of duty.
    4) Damages resulted from the breach, such as physical injury. There is a statute of limitations. The time starts on the day the injury was *discovered.*

- **Civil action:** A crime against a person. Cause for civil action? It is intentional infliction of emotional distress such as assault and battery on a person, negligence, sexual assault, defamation of character (harming or damaging ones character or someone's reputation that _causes some loss_. Defamation in writing is called libel. If it is verbal, it is slander) and invasion of privacy. These may be hard to prove. There must be an injury or loss of some sort, either physically or financially to prevail.

- **Contracts:** A doctor has the right to drop you if you do not pay your bill; but there is a sequence of events he must follow. He must notify you by certified letter, give you adequate time to pay, and give you a list of other doctors from whom to choose. Then all medical records must be made available to the new doctor. If these steps are not followed the doctor can be charged with *abandonment*.

Medical records

Medical records are legal documents; this is why they have to be protected. They are also property of the doctor and the hospital, even though they are your records and you paid for them.

A patient should have forms in place in: your doctors file, a hospital file, your personal file and a copy should also be in a file of someone you trust. *ADVANCED DIRECTIVE* forms can be obtained from a office supply store that sells forms. They can also be printed FREE under LEGAL on our site: **www.22qCentral.org** These forms are:
- Living will
- Durable power of attorney
- Health care proxy (someone to stand in)

Parents Speak Out

I wanted to share with you what we just went through for our son, who is almost 6 years old, to get him an aide this year in his kindergarten class. I fought for him to attend a typical kindergarten classroom with an aide. Of course, the school system said that they couldn't do it. "It's not budgeted that way." (Is that legal?) They stated to me that if a child needs a one-to-one they should be in a more "contained" classroom. (They meant special school.) Well, I took it to mediation through the State Department of Education and now he is in a typical classroom with an aide. He has made TREMENDOUS progress and is counting, writing his first/last name, reciting poems, etc. I feel that it was the right decision (for now); although in his IEP, they would not state one-to-one, because the aide can assist other students. But, she is there primarily for him. My advice to any of you who think your child needs special help, is to push for assistance in the classroom, if that is what you feel would best serve the child.

Years ago when we were involved with the school system someone explained to me that IF our 22q11 kids should ever be told that they had progressed to the point where they no longer qualified for special education services, that we should have them be designated OHI. This is _Other Health Impaired_ so that they can not ever lose their services at school. It's common for our kids to catch up and then lose skills and get behind again. Someone told me that their child had lost 2 years within 6 months, and then it took several more months to regain services or be reinstated. It was a huge fiasco! Because our kids have different issues that can cause them to fall behind and then catch up and then fall behind, and so on; they should always have services in place. If this should happen, you will most likely need a letter from your doctor. (A geneticist would probably be a really good one to write this or a developmental pediatrician.) Choose someone who really knows about this syndrome! Request that they inform the school system to change your child's designation/status and request that their services be reinstated based on the likelihood of the child having difficulty. This is especially important when the learning material gets more abstract and the child will be expected to begin to work more and more independently. The child is likely to have a hard time. (He or she may do just fine but my motto is, "Don't wait until they fail!").

I had my first parent teacher conference and was not able to find enough material to back up my claim that since my son has 22q, he may need some special education and tutoring. He is in 5th grade and I can see that it is becoming more apparent that he needs help. It is not just that he is not doing the work because he doesn't want to, he is having a tough time; and I know he is trying hard. So I was happy to hear there was going to be a book to give to teachers that will help them understand the things about this syndrome that we can't explain well.

Medical Trivia

If you were locked in a completely sealed room, you would die from carbon dioxide poisoning before you would die from oxygen deprivation

Blank page for personal data:

"Dreams are renewable. No matter what our age or condition, there are still untapped possibilities within us and new beauty waiting to be born." Dr. Dale Turner

HOUSING AND GROUP HOMES
What choices are there for my child if they do not stay at home?

Group homes and other alternative living

The objective is to discuss living arrangements for persons with disabilities before they become adults. This is a difficult topic for many parents to acknowledge, especially when your child is young and as parents, you are in a full time struggle for survival with other children and other kinds of issues. At this early time of their life you may be trying to find interventions to alleviate all the various symptoms of 22q11. You are also trying to deal with the school system to get the appropriate education; living with the alterations in the household and often impossible behaviors; plus trying to get funding for expensive interventions which are not covered by either government or insurance funding. Worrying about future living arrangements might be down at the bottom of your list of priorities; especially if it has never entered your mind that your child might not be able to live on his or her own. You might even assume your child will always live with you.

Without trying to scare parents, the reality is your problems don't end when your child is 18-22 years old. When the government says they are adults or mandated requirements of IDEA no longer are in play; there are often other struggles ahead. If you think you can simply place your child in a group home, you will be surprised to know that it is not as easy as one might think. It is every bit the battle you had with all the other issues you have already overcome. There are long waiting lists for the good group homes, even if you can get qualified for them medically or financially. You may think you don't want to go that route. Over the last few years, by government mandate, many large institutions that were run by the government and by private agencies had to be shut down and the people in them had to be placed within group home settings. With many institutions closed down, and the government having purchased and setup many of the group homes, the government has announced they will not develop any more group homes. So, the waiting lists are getting even longer. The popular method now is to encourage and help private agencies to develop group homes or other living arrangements. So now you have agencies that are very nervous about investing in real estate and running programs which require great infusions of government funding. You know how complicated this can get. When the government and private agencies print up brochures and give lectures on all the help they will give you, remember it's not an entitlement as it was in the education part. There is a lot of public relations hype involved. When you get to the point of looking for the right placement it may be a long search and then a long wait.

So what is the prognosis on the hunt for the right affordable housing of which you would approve? What can parents look forward to for living arrangements for their child? The first step is to develop a dream or ideal plan for what you want for your child and then work to make it happen. In this area there are no legal limits as to what you can do, it's usually limited by financial considerations.

What are the government group homes?

The first place most parents will look, is for government run group homes or privately-run group homes with mostly government funding. Each has its own requirements and levels of funding. There may be long waiting periods. Again, these are not entitlements like schooling was. You can't sue the government because they haven't placed your child in a great group home.

If we can't get into a group home, what are our options?

Another area, and one which is growing in popularity, is arrangements set up by several families with or without government help. The ideal arrangement is for the families to buy a home, set up a staff, programs, recreation, etc. and keep the government out of it. If you are able to do this it would be a good way to go. BUT, this could require a funding stream of upwards of $100,000 or more per year in addition to the house, depending on the level of attention the persons with 22q need. It gets better if three or four families pull together on it; not only for the financial reasons, but for safety and peace of mind. This arrangement would be a non-certified home, where you have to pay the property taxes, insurance and all other things of a home each year. For families who are pretty well established and creative, this is a good way to go. Another arrangement is for the families to take care of purchasing the house and have the staffing and programming done by a private agency, if running it themselves interferes with their schedule or employment. But, if there are 4 families in on the deal, each family can rotate a week for duties and care each month. This creates a sharing of the responsibilities, a co-op of sorts (more on that later). You might also think about setting up an LLC or non-profit corporation.

- **Co-op group home, example a:** Here is an example of how this can work: There are families X and Y. Say these two families each have a daughter with 22q. The parents grow tired of trying to get their daughters into a government-run group home, so they take matters into their own hands. They couldn't even get weekend respite arrangements made, the

---

- **IRA** is an *Individualized Living Arrangement*. **ICF** is an *Intermediate Care Facility*. Levels one and two are levels of certification by the government. The amount of SSI is dependent on the level of certification of the home, where the disabled person is living. If families get together and buy a house and establish an uncertified group home, none of this will apply.

system was so overloaded. So the parents hired a lawyer and incorporated themselves as a 501(c)3 corporation. Now they have created their own little non-profit entity. They set both their daughters up in an apartment or house of their own and hired a caregiver to live in the apartment/house with them. The caregiver is there in the morning to help the girls get through breakfast and clean the apartment. The girls go off to jobs or school as does the provider after the chores are done (has free time). The provider is there in the evening to make sure the girls are getting their supper, doing their chores like laundry, taking their medications, etc. and getting to bed.

Some other things the provider can do include: helping the girls make menus, and go shopping *with* them for food and other needs. The caregiver is paid through the Medicaid HCBS waiver program. The reason for establishing the agency is that the government will not fund families directly, it has to be done through a non-profit agency.

• **Group home, example b:** Now, there is a family Z who bought a 4 bedroom house for their son. They brought in the son's non-disabled uncle to live with him. They give this uncle free room and board at the house in exchange for looking after their son in the evenings. They also rented out two of the bedrooms to university students to help with the mortgage and taxes. The 22q boy goes off to do daily things in the morning with a provider who is paid for under the Medicaid HCBS program, and the uncle goes off to his job or school.

• **Trust home, example c:** There are many other arrangements that can be made. One couple deeded their home over to their 22q son's special needs trust, with this arrangement they can continue to live there with him; and he is guaranteed a home when the parents pass on. (See Special Needs Trust on page 326.) If the parents pay the special needs trust as if it were rent, they can funnel money to the trust without jeopardizing his eligibility for government aid/services.

• **Private home, example d:** In another arrangement a group of four families bought a house, moved their disabled children into the home, and then turned it over to a private agency for operation as a group home, with the parents having control.

There really is no limit as to what parents can do to set up living arrangements if they have some money, enough people that are compatible, imagination, and lots of determination. Each case is different. There is no magic formula, format to follow or set rules. To do it right and save money you would want to take the lead and do the coordination yourself; in your child's best interest. It also helps to consult with professionals.

Here are a few more ideas that are similar living arrangements yet a little different. Say, my goal is to remain my son's primary caregiver and for us to live together as a family for as many years as possible. While that objective may not be possible for everyone, the following ideas reflect how that can be accomplished.

- **Caregiver in exchange for rent, example e:** In order to stay together as a family, yet have affordable care for your child, you could move into a duplex and have a caregiver live in the other half of the home, in exchange for providing care for your child while you are at work; or a caregiver could move into the home to provide care. (That would mean less privacy and space for everyone.) Another option would to build a small casita or apartment on your lot.

- **Caregiver services, example f:** You could purchase a larger home, hire a caregiver and offer respite care to other 22q children or handicapped adults (during the day only). Or, find interested, responsible college students who could be recruited to help out; thereby providing a better caregiver to client ratio. Parents could pay a reasonable, minimal fee, which would offset the cost of the caregiver(s) and hopefully make it more affordable for everyone. The purpose of taking in other disabled adults would *NOT* be for profit but rather to bring down the cost for everyone. Activities could be provided during the day, and everyone would be encouraged to participate in daily living skills as well as other tasks, such as gardening, painting and crafts, according to their ability.

- **Group care co-op, example g:** A group could form a co-op with other parents of disabled adults. Everyone could contribute their time, as well as finances, for the care of disabled adults. Each parent's financial obligation would be adjusted according to the amount of time they spent caring for all the adults in the co-op. Parents volunteering the most time would pay the least, while those spending the least amount of time would pay more. College students studying psychology, child care, communication disorders or other related fields could be recruited, hopefully, as volunteers or as part-time caregivers. Other caregiver(s) could also be hired to fill in as needed.

These are just a few of the many possibilities for living situations. With creativity, many others are possible.

I want my child to see a specialist but the clinic is so far away, I can't afford airline tickets, what can I do?

Many families find themselves in this situation and feel desperate to see a doctor who might hold out hope and promise of answers for your child, if only you could get there. Well, there is an organization that helps families do just that. It is called *National Patient Travel Etc*. In some cases, with a doctor's request, this transport company will fly you and your family free to your medical destination. For more information go to: **www.patienttravel.org** or call: 1-800-296-1217.

Parents Speak Out

For those of you parents who know a group home is right for you, but you are afraid, I am a parent of a child with VCFS. I also have worked in many group homes as a house manager. Actually, group homes work very well; especially if parents will visit regularly. Group homes can provide stability, independence, lots of recreational activities, etc. The best place to start inquiring is the (MHMR) *Mental Health, Mental Retardation* offices. I'm not sure if it is called that in other states, but they are operated through Department of Health and Human Services which is federal. You can also check with the (NAMI) *National Alliance for Mental Illness* support groups. They are everywhere. They will know about group homes in your area and are a great resource, just make sure you do it before your child is 18 years old. You can look them up on the Internet.

Medical Trivia

One hundred years ago, the average life expectancy in the United States was 47.

Blank page for personal data:

"The country clubs, the cars, the boats, your assests may be ample, BUT the best inheritance you can leave your kids is a good example." Barry Spilchuk

CHAPTER 13: TRUSTS

FINANCES & TRUSTS

"Be financially responsible to those we love"

I recently read a book called **The Nine Steps to Financial Freedom** by Suze Orman, who is a renowned financial guru, when I found the instruction I was seeking for my personal financial planning needs. Then it dawned on me that this information should also be included in this book because many of us are in the same predicament. We want to fulfill what the title of her chapter indicates and *"Be financially responsible to those we love"* by making sure that we do all we can to prepare a smooth transition should something happen to us, the caregiver. (**www.suzeorman.com**)

My husband and I had spoken frequently about updating our *WILL*, the one we had made when our children were minors and now they are both adults. Of course at that time we didn't know that our son had 22q11 nor did we know anything about *TRUSTS* and thought a *WILL* would be all that we would need.

The more we spoke with savvy people and read on the subject, the more we began to realize that we needed to set up a trust. This meant sitting down and making a list of all our assets, gathering up all our paperwork and setting an appointment with a TRUST lawyer, to ensure financial care for our child(ren) sooner or later (hopefully later). By taking this step now, we knew it would ease our minds and guarantee that all transitions would go smoothly should we die unexpectantly.

We now know that a WILL is not the only thing we needed, especially because we have a child with special needs to take into consideration. What we personally did was set up a _Revocable Living Trust;_ one with a _Special Needs_ section in it.

What is a 'revocable living trust'?

This was explained to me as: a set of documents that states what a WILL would offer, but much more, it states who controls your assets while you are alive and what is to happen to the assets when you are gone.

The next thing I wanted to know was, what are the differences between a will and a revocable living trust, and do I need both? From what I have learned, I would say it would be good to have both, but not necessary in states that include it in their trusts. So, if you have a trust set up ask your lawyer if your state is one that includes it. For those who already have a will in place, you're doubly covered. A revocable living trust includes steps that are taken while you are alive to sign over your assets like property to the trust (as your representative) for your own use and benefit while you are alive, but when you are gone you have already told the trust where you want each asset to go. The beauty of a trust is, that it is now the _TRUST,_ not the probate courts or any other legal entity, that will handle all the financial arrangements for you. When you put a trust in place, it will not require a lawyer to fight for your estate. Another benefit about a trust is that it lives on as you set it up with no disruptions, even when you are no longer around to manage the assets. It will carry out the wishes you put into place. You also have the right to amend or change the trust at any time should you change your mind or should assets change, such as a sale of property or winning a huge lottery. (We can dream can't we?) Most importantly there is no probate and that is a real financial savings for your heirs. Does this mean there are _NO_ taxes what so ever? Ask your lawyer about your state.

There are many things that can go into this trust such as property, stocks and bonds, cash etc.

Do you have to have a lot of assets to need or want a trust?

The answer is no. It can be set up so that a disabled child is helped with how they spend the money, how much is delegated weekly or monthly, for special purchases, for direct payment of bills, for car expenses or purchase if they drive, and for handling other expenses, etc. Talk to a lawyer about these special arrangements.

A *Revocable Living Trust* can be set up by those who understand legal jargon by purchasing the paperwork that is available online in some programs; but, this is not recommended for those of us who have children with special needs. I personally opted to hire an attorney who is knowledgeable in this area. It is too important to have done it right when it is needed. I have learned that the charge is usually in the range of $300 to $1500 for this service. So it pays to shop around and ask questions. I paid $500 for ours.

I have been fortunate enough to have found some very knowledgeable and caring people who are specialists in this field of planning for the future of someone with special needs. One such source is at MetDESK, MetLife's Division of Estate Planning for Special Kids. You can reach MetDESK toll free at **877-638-3375** (Met-DESK) or at **www.metlife.com/desk**. MetDESK specialists are located throughout the U.S. and offer their services at *no charge*.

These estate planning agents who specialize in *Special Needs* also work in cooperation with lawyers who also specialize in this area. Companies such as MetLife and Edward Jones inform me that every state has this offered to those who seek it.

Estate planning for special needs, how can I plan without harming my child's rightful benefits?

Federal law states that anyone (whether disabled or not) who has assets in their name of $2000 dollars or more automatically is _NOT_ eligible for government benefits such as Social Security. (SSI) This includes cars or cash.

Estate planning comes in two categories:
- Legal estate planning
- Financial estate planning

How can we leave money for a special needs child within the law and not jeopardize their government benefits?

In other words, how do we legally shelter it? Yes, you can shelter the inheritance. There are many trusts out there but to get the right trust it must simply be called a <u>supplemental needs</u> or *special needs trust*. This should be done by an attorney to make sure it is done right and that it is legal. A *Trust* is similar to a *Will* but more in depth and needs special wording for *special needs* that usually only a lawyer can decipher.

There is a common denominator for *ALL* children with disorders and that is they will all need special assistance one way or another; some of course, more than others.

When planning estate provisions always plan for what is going on right now, not for the future. The future can always change and we do not have the ability to know in what direction life will turn. This is the beauty of a *revocable* or changeable trust. It can change as situations in your life do. The bottom line is having a safety net in place for anything that might happen at any time.

From a legal aspect: The *will* is the piece of paper that tells everyone who gets what of your worldly tangible assets, as well as proclaiming in whose care you wish to leave your minor children. Even if you don't have a written will; by way of the court system you do, to a certain extent. The court takes over and will decide where your assets go and to whom your minor children will go.

What is another difference between a will and a trust?

Wills are contestable; they will go through probate and are of public record. A trust is almost never contestable, anything that is in the name of the trust by-passes probate. A trust is private, not of public record.

What is probate?

It is an arm of the court system that validates your will or lack of one. If you don't have a trust and only a will OR no will at all, and if your estate goes over a certain amount of money it goes into probate court. So in many instances someone has to get an attorney anyway and the estate is out money to the lawyer for filling fees, etc. It is also tied up until the judge releases it. It is at this point that everyone is wondering *"Why didn't you set up a trust when you were alive?"*

In many states, assets can be tied up from 6 months to 2 years or longer, depending on the estate. During that time all assets are frozen. The problem with children with special needs is that their disorders/illnesses don't come to a stand-still, if something should happen to us as care givers, they need the money or assets right away, to continue to maintain their lifestyle, therapies, treatments etc.

Something to realize is that a special needs trust is totally separate from a regular revocable living trust or a family trust, and only goes into affect when you die! The *ONLY* need or function of this <u>special needs trust</u> is to protect government eligibility for the disabled person. It is a separate entity.

How does this work?

Example: If my husband and I have 4 children; one with special needs. We go to our estate planner and tell them that we want to split our estate 4 ways. One fourth to Dianne, one fourth to Cathy, one fourth to Alan and one fourth to <u>*Bob's special needs trust*</u>. Notice that technically we are not leaving it in Bob's name but in a special needs trust. Now this money is designated *ONLY* for the use of Bob's needs. So since Bob has no money or assets in his name he is allowed to go on collecting social security, or he can apply for social security and use other government services that Bob is entitled to because we legally have sheltered the assets in a special needs trust; this is its <u>*only*</u> function.

Now, if you feel you are ready to start your trust, you should first evaluate the condition and function of your disabled child. Are they high function? Do they need help with their daily living skills? When you go to the lawyer the first thing the lawyer will ask you is *"to whom do you wish to assign care for your child should you die tomorrow? Who is going to take over all your responsibilities?"* You may not think of this as a hard question to answer but you may be surprised. After you have made a decision you must go to that person and talk about it. Remember, it is not only who YOU want to care for your child(ren), but is that person willing to do the task you may be dropping in their lives?

I know from experience that one should not take a responsibility or request lightly as I ended up with 3 young teens myself, when 3 different families had 2 deaths and had asked us previously to care for their child should something happen to them. Did I ever dream that it would come to be? Not in a million years! And one of them was mentally handicapped. I was a young mother with 2 of my own (one with VCFS) and so it was a real struggle at times. So think before you ask someone this important question or before you say *"Of course"* to someone who asks this of you.

Then the lawyer will want to know who your trustee will be, who will be your successor? In my case it will no doubt be my adult daughter, Tishri.

In a *Family* or *Living Trust* <u>YOU</u> are the trustee while you are alive. Most of these trusts are revocable which means you can make changes to them. Who is going to distribute the assets and take over control when you are gone? In making the decision think of examples of questions to answer like: Is the person you are considering organized, honest, caring, shown to be dependable, not needing to go to "AA," "shoppers anonymous," "gamblers anonymous," or some other 12 step program, or will they expect an allowance/fee for duties they will be performing?

The Lawyer also will want to know who is going to be trustee of the special needs trust. Once there are assets put into the special needs trust, the government will say that Bob can <u>*NEVER*</u> manage his own money. There will have to be someone as a trustee delegating the money; deciding what is needed or necessary for Bob. It must be someone who will pay all out-of-pocket expenses that Medicare does not pay, etc. It must be used for things other than *"food, clothing and shelter"* because these are the items that the government says they are paying for with Bob's disability checks. (I have found that the checks never cover all of these needs.) Now, if you think about it or even look into your check registry each year, you will see that there are many more expenses that came up other than food, clothing and shelter. It adds up and it might surprise you. If there are enough funds in the special needs trust, one could even buy a condo or patio home where the older special needs person can live and supply assistance to help them like housekeepers, etc. The special needs trust would own it, but the special needs person could live there. There are many details and questions that only an attorney in your area can explain to you. *(See Housing and group homes on page 315.)*

The attorney is going to ask for at least 2 people (they always want a back up). If there is only one person you propose and that person dies or changes their mind, it gets into a tangled mess. The person may have agreed initially but might get scared and think they can't take on the added responsibility of care. Then what? The judge may have no choice but to make your child a ward of the court and they will more than likely be put with strangers. (The ultimate fear of parents!) Remember your choice does not have to be a relative. Many of us have close friends who would make wonderful parents with nurturing personalities. We would feel relieved to have them as our choice. However, just remember the list of questions to answer for yourself along with the consideration of their lifestyle. This way we don't ask people impulsively and then end up with hurt feelings if WE change our mind and the one we asked impulsively then feels insulted.

Remember, I know from experience what can happen, when *WE* were the ones impulsively saying yes to care for someone's child should something happen to them, the parents. It is not easy for either side to make decisions not based on emotion. All kids, special needs or not, will have issues to deal with. After all, if you are caring for them, that means that they just lost their parents, on top of the normal issues of growing up. Recently I was asked if I would become a child's guardian should something happen; this time I tactfully declined. I would rather stay friends and state why I felt I could not do so, now that I am older, than to cause any long-lasting problems down the road, should something happen. (With my track record, it just might.)

Make sure you don't become unreasonable yourself if those you ask say, "no". They might have too much on their plate already at present and can't see that far ahead and might not want to commit to something they are not sure they can handle. They are actually brave to say they can't. So don't get mad or hurt. We as parents see it as a privilege to be asked, because we love our child(ren), but others do not have the same investment as we do and may feel they would have to change or alter their life too much. Another realistic reason why more than one person should be on the list for guardianship is the fact that, it is possible that your first choice could decline when the real thing happens. They can stand before the judge at that time and refuse to take your child; and if there is no back up, well, you know the answer! If this scenario is a possibility you can get creative and have another back-up person to help the one who will be the "main" care giver, as it may be hard on one person or family who may have children of their own. There may even be someone willing to take the child(ren) for the summers to give the caregiver a break.

There is a Metlife agent who provided much of this information, and I thank her. I think the document that we are about to talk about is one of the best tools she told me about. The document would be very useful to the one(s) you ask to fill your shoes as guardian, it is a _LETTER OF INTENT OF CARE_. This can be written when you make out your list of questions to decide who will receive the the letter of intent. This will help them decide whether they are up to the challenge. It will either scare them off or help them be aware of the importance of their decision should they decide to accept the request.

What is a letter of intent?

A letter of intent is simply a step-by-step, play-by-play of a normal day or week of your disabled child's life. What should it include? List a 24 hour normal routine, covering *ALL* the bases. List what the child can do independently and what they need help with. Include the various appointments that are on-going. What diagnoses have they had? (There may be more than one, and there usually is.) List what medications they are on or have been on; and what if any, side affects or adverse reactions they have experienced. What seems to work best? List all the doctors your child has seen. List each doctor's name, addresses, phone numbers, and area of specialty. We may not think of this as being important, but think, if you were gone tomorrow and someone had to step into your shoes, would they or anyone know what to do, where to go, what to look for, what to ask? Does the one left in charge know your child(ren) as well as you do? Do they know your views on sex, marriage, birth control or religion? (not necessarily in that order). If we don't put it down on paper and/or video now, and these issues come up in the future, it may cause grief for everyone. So now is the time to let your wishes be known. The adults who may be appointed, will find their jobs as guardian or advocate easier knowing that decisions have been made ahead of time and that there are guidelines to follow. This also relieves some anxiety for us, as parents, just knowing that this step is done and in place. The guardian after all, would have to make what they felt were the best choices for your child either way. Since time changes many things including our views; this information should be updated *EVERY YEAR or two*. There may be new doctors to add, different medications, growths and digressions in areas, etc. Each time you update this information have it witnessed and notarized so that there can be no disputes of its origin. If there is more than one caregiver or family member that has an interest in the child, it is wise to have them look over what you have written and make sure they're in agreement so that it does not become an issue at a bad time. If there are minor children involved, make copies and give one to the two or more people you have chosen to be your replacement should you die. Also give it to those you have asked to keep tabs on the child(ren) to give emotional support should this happen; then go over the letter with all of them.

Since we are talking about special needs trusts, it would be wise to go to all your friends and relatives and let them know what you are doing. If any of them plan on leaving your child any inheritance money or assets, explain how this can harm the child with SSDI, SSI, services and benefits if it is not done properly as mentioned earlier. Any gift or inheritance needs to be funneled through the *special needs trust*. There is a formula that the government uses. If the child is cut-off from government funds due to a mistaken kind gesture, the disabled person <u>can go back on benefits</u> after the time period that the government allots for the value of the amount given.

Ask the contributor to simply leave the inheritance for the child in care of the parents and then the parents trust will automatically do the right thing for the child as directed by the lawyer who created it. If, by any chance, you are not able to explain how this special needs trust works to those who need to know, please have them call your lawyer/agent and most would be happy to explain it. It is, after all for their benefit as well.

There are those of us who have stocks, bonds, mutual funds etc. that we have put in our child's name, not knowing what we were doing. Fortunately these can have ownership reversed and they can go into the name of a special needs trust also.

There are a couple of ways that a child can collect Social Security: if they are a minor and mom and dad are low income; or when they turn 18, the child can apply for himself now that they are no longer under their parent's income. Then it legally doesn't matter if mom, dad or caretaker passes away. However, there still needs to be someone who will take charge to guide the adult child through life.

Many families have been devastated by all the out-of-pocket expenses that were not covered; that they *thought* were covered; or that they thought they had enough money to cover, such as: the expenses for the unforeseen illnesses or surgeries. It can quickly wipe out a person's nest egg of savings. In making this statement I am primarily speaking to those who may be prideful and think they will never need help from the government. Some people think that only a certain class of people do. Well, never say never, my friend.

Fortunately there are some states that now offer insurance to all children who are residents of their state. Of course that will depend on budgets.

Beneficiary

When naming a beneficiary (for something like life insurance), if you want the funds to go to the person with special needs, you need to name the special needs trust, and _NOT_ the child by name.

Should I put money in a special needs trust now?

The answer to that may vary depending on your circumstances but from what I have learned I would say no, not yet. You would set up your revocable trust with the wording by the lawyer to create or set up the special needs trust upon your demise. It will then be given a tax I.D. number and the assets will be dumped into it. Usually a person does not want to put anything into a special needs trust until it is necessary, because you want control over the assets as long as you are alive. If you are to dump any money or assets into a special needs trust you no longer have control over those assets and they can't be taken out, except for the child's care <u>after you die</u>.

Here again, we don't know what the future holds, so that if there is a marriage, a cure, a birth, death etc., this will mean we need to go back to the revocable trust and make some changes or adjustments periodically.

There may be a time, God willing, when our child will be able to live a full and independent life and not need the special needs trust. If that ends up being the case we never have to use it and we can have the "special needs" part removed.

What it boils down to is this: it's better to have a trust with a special needs clause in place and not need it, than to die and have your child need it and it not be there.

What kinds of assets can we leave our kids in a special needs trust?

Just about anything can be put into it. But it would be better to leave the children or other heirs any properties or things that need managed or are tangible; than to leave it to the child with special needs. Cash is what works best for them. You can put in a clause, if you wish, that states that certain properties are to be sold and liquidated as are all tangible items that are left to the child with special needs so that all that is left to them can be turned into cash for the child's expenses. It is much easier to deal with cash. Your attorney can guide you in this decision.

What if my child is 18 or older and can qualify for Social Security but can't handle money?

You may want to discuss a *Financial Power of Attorney* with the Social Security office. They can have the checks sent to you as

<u>payee</u> for dispersement. When you become the payee, you will have to keep records of expenses; for food, clothing, shelter and any extras. You will have to be re-qualified each year; and show where the money was spent each month. I thought the auditor was joking when he asked me if there were any savings at the end of the year. My son got $570 per month and it cost approximately $930+ per month to supply all his needs. *SAVINGS?*

Are your documents in place?

As our children head toward adulthood, they may still need help in deciding on the appropriate medical care. That is why we should have in place at the time of their 18th birthday a <u>*Durable Power of Attorney for Health Care*</u>. (This paperwork can be purchased in many office supply stores and only needs to be witnessed and notarized, unless your state requirements are different.) This is a legal document designed to make health care easier for the "loved one" when a person is not capable of decision making. The health care representative (usually the parents) is appointed by the young adult and is expected to act in the best interest of the person. However, this can be revoked at anytime if the young adult is determined to be competent and contests the decision.

It is wise to have another document in all doctors' and hospital files, and that is a <u>*Living Will*</u> for adult children. This is a legal document that allows the person to state preferences about the use of life-sustaining measures, what can and cannot be used. Many may feel this is not what they want and that is fine, but it has proven to be a good item to have in times of crisis and decision making. This is also a document that can be obtained at an office supply store. These can also be acquired free on our Web site at **www.22qcentral.org**.

> **DISCLAIMER:** Since laws are somewhat different from state to state, PLEASE have an attorney review ALL legal documents.

PARENTS SPEAK OUT

There is a program called *Children's Special Health Care Services* or *Medicaid Waiver Program*. It isn't income based, but there are guidelines, and some have to pay a small co-pay. It covers all of my son's specialists, hospital bills and prescriptions. This is supplemental insurance that requires you to have primary insurance first. On the SSA web site there is a reference to this insurance and it is available everywhere. It is usually run through the Medicaid Office.

State Children's Health Insurance Program (CHIP)
Legislation passed in 1997 created a new Title XXI of the Social Security Act, known as the State Children's Health Insurance Program (CHIP). This new program enable a state to insure children from working families with incomes too high to qualify for Medicaid, but too low to afford private health insurance. The program provides protection for prescription drugs, vision, hearing and mental health services and is available in all 50 states, and the District of Columbia. Your state Medicaid agency can provide more information about CHIP.

Other Health Care Services:
If it is decided that a child is disabled and eligible for SSI, they are referred for health care services under the *Children with Special Health Care Needs* *(CSHCN)* provision of the Social Security Act. These programs are generally administered through state health agencies. Although there are differences,

most CSHCN programs help provide specialized services through arrangements with clinics, private offices, hospital-based out- and in-patient treatment centers and community agencies. CSHCN programs are known in the states by a variety of names, including: Children's Special Health Services, Children's Medical Services and Handicapped Children's Program. Even if your child is not eligible for SSI, a CSHCN program may be able to help you. Local health departments, social services offices or hospitals should be able to help you contact your CSHCN program.

http://www.ssa.gov/pubs/10026.html

Medical Trivia:

A person will die from total lack of sleep sooner than from starvation. Death will occur about 10 days without sleep, while starvation takes a few weeks.

> "*Being considerate of others will take your children further in life than any college degree.*"
> Marian Wright Edelman

CHAPTER 14: SUPPORT GROUPS

22q11/VCFS SUPPORT GROUPS

By Maureen Anderson

Used with permission. The Velo-Cardio-Facial Syndrome Educational Foundation, Inc., January 2004 newsletter, Volume 9, Issue 01

It's March 1989 in Boston Massachusetts USA. Okay, now we're parents! Yeh! But wait...not only do we *not* have a Parenting 101 handbook, it seems our child is missing a slice of good ol' chromosome #22. Unexpected to say the least!

I remember when my husband and I first received *"THE DIAGNOSIS"*. We had tried for many years to have biological children, undergoing just about every indignity that doctors can throw at infertility parents. When we finally were fortunate enough to adopt a wonderful little boy, he brought with him some cardiac "issues", but, hey, we could deal with that. We did not hear the "V" word until Eamon was about 3 years old. VCFS came at us from out of the blue.

As time went on and our brave little boy's challenges were more evident; it also became evident that it was psychologically difficult for us to face these challenges alone. But it was clear that none of our parent peers could relate to our situation. They would nod, look uncomfortable, and sometimes say things like *"You'd never know—he doesn't look . . . (fill in the blank.)"* or *"Velo-what?"* And our own family members were in denial. I suppose they were trying to make us feel better. *"He'll be fine, just give him time."* *"You worry too much!"* (They continue to say these things to this day!)

We could deal with the more tangible issues, we thought. The physical challenges were mostly fixable. But our fears for the future were sometimes overwhelming: *"Will he ever be able to live on his own?"*, *"Are the real world sharks going to eat him alive?"*, *" Will WE ever be able to live on OUR own?"* and the big one (insert organ music here) *"What about the psychological issues?"* Words like *"normal," "typical,"* or *"average"* became like splinters that we couldn't dislodge from our psyches. Eventually, words like *"special ed"* and *"learning disability"* entered our everyday lexicon.

After allowing the diagnosis to percolate a bit, my husband and I decided that, since there were supposedly so many more of us out there in the same boat, swimming in the same gene pool, we wanted to find them. But the only name and phone number on any *"VCFS contact list"* was of a woman whose child was an adult and had just about every one of the 150+ manifestations (then) of VCFS. When we spoke, she proceeded to tell me everyone of those challenges in detail culminating with a daunting psychological profile. Now, the woman was sincere, well intentioned and most likely felt she was being supportive and informative. Of course, after I hung up from that phone call, I put my head down on the kitchen table and cried.

But, after I dried my eyes and wiped off the table, in a scene that closely resembled a scene *"Gone With the Wind"* I (in my own Scarlett O'Hara type voice) vowed (with my fist in the air) *"As God as my witness, I will never let anyone go through that again!"*

When my husband arrived home that day, I told him about my phone call and we talked about starting a local support group for VCFS families. (a.k.a. *"How to secretly help yourself under the guise of helping others."*) Now, we had no idea what this entailed but were sufficiently naïve enough to give it a try.

Not having a clue about support groups, I called upon someone who was integral in helping us get a diagnosis for our son: Eileen Marrinan, who was a SLP at Children's Hospital in Boston at that time. She worked closely with us in our new endeavor; calling patients (as she could not give out this information to us due to confidentiality constraints) and helping us gather a team of knowledgeable, caring Children's Hospital professionals to talk at our first meeting.

Six weeks later, the Northeast VCFS Support Group had its first meeting at Children's Hospital. We met in a room in the Harvard

Medical School Library and it was a standing room only gathering of 50 attendees! Now, this may sound like a small group but, when you think about the fact that this was about nine years ago when VCFS was recognized even less, this was an amazing feat.

I still remember the expressions on the many faces as they heard stories from other VCFS parents at that meeting. They were so encouraged to meet people (both parents and professionals) who could understand their unique challenges and joys. Although meeting parents and hearing stories was frightening to some, the overall tenor of the meeting was positive. We now have over 200 families and medical professionals on our mailing list.

The Northeast VCFS Support group has experienced ebbs and flows over the years. And the advent of the Internet has made the quest for information easier and makes the need for personal contact unnecessary for some families. But our group still tries to meet twice a year—sometimes socially so the children can interact. We also have a database by which we connect those looking to talk with others who have similar situations.

Starting our group has been a most rewarding endeavor. I have made lifelong friends and have learned so much about what the future might hold for our little guy. I have been privileged to share in the joys, and yes, the challenges of VCFS families. And, most importantly, I am constantly amazed at the strength of these families and the spirit of the children who are coping with VCFS. They are all inspirational!

Support Groups

*Compiled by: Maureen Anderson,
Northeast VCFS Support Group.*

Oftentimes, those who have received a VCFS diagnosis for their child (or themselves) feel they want to interact with others who have had similar experiences. This can be an important step in their ability to come to terms with the challenges of being a VCFS family.

Years ago, my son was diagnosed with VCFS. Whenever I mentioned to someone that my son had VCFS—they would simply gaze at me. It appeared as if they were trying to figure out whether I was telling them about a new type of savings bond or the latest, high-tech running shoes. This, of course, made us feel quite isolated with our concerns.

This prompted us to start the Northeast VCFS Support Group. We now have over 200 members. Following is a compilation of information gleaned from research and trial and error.

What is a support group?

Technically, a support group or self-help group includes two or more people who share the same health care problem or life situation. This can include both family and medical professionals. Support groups can be as informal as a telephone network or as close-knit as a group that meets monthly.

Unfortunately, it is sometimes difficult to find a VCFS support group. Until quite recently, there were relatively few people who were diagnosed with VCFS. There is a list of the VCFS support groups on the VCFS Educational Foundation website at: **www.vcfsef.org**

Those who are interested in starting or joining a support group must always remember the following.

Support groups can and should:
- Provide a safe, non-judgmental and confidential outlet
- Provide understanding
- Provide ideas
- Offer resources

Support groups cannot and should not:
- "Cure people"
- Attempt therapeutic interventions
- Diagnose, evaluate, or recommend medications
- Attempt to resolve conflicts
- Take an active role in unfolding a member's development

Starting a support group

Starting your own support group can be a rewarding yet daunting experience. Before you start down this road, there are things you must consider:
- Do you have the time and energy?
- Is your family ready?
- Are you prepared for a long term commitment?
- Can you put personal beliefs on hold?
- Can you share authority?
- Are you doing this for personal recognition?

AND . . .
- **Know what's out there:** This involves ensuring your group will not be redundant
- **Spread the wealth:** Share responsibility for the group with someone or a group of people who are energetic and organized. This can be another parent or a medical professional.
- **Be realistic:** It usually takes between 6-12 months to get a group going.

Develop a mission statement

One of the best ways to figure out where you're going with your group is to hone your ideas down to a mission statement. This includes asking the following questions:

- What will be the group's goals?
- How will the group accomplish those goals?

Since the process of developing a mission statement can be a really good way to get membership brainstorming at the first meeting, you or your core group may want to loosely define how you or they feel about the mission statement prior to the meeting.

There are four issues that can be considered when developing goals. For example, will your group target:

- Emotional Support
- Educational Support
- Social action or advocacy
- Combination of any of the above

Sample mission statement:

"We are a network created by families for families of children and adults with VCFS/22q. We connect people in similar situations with one another so they may share experiences, offer accurate and practical information and/or support. The material provided by our group is for information only and should not take the place of advice from your own health care providers."

A mission statement is a document that should be in the hands of all your members. It can be modified to meet the purpose and goals of the group as you grow in number. The mission statement is also a good thing to read at the commencement of each meeting.

When developing a mission statement, it is crucial that all group members adhere to the tenet and that no members ever try to lead anyone down a path, that is best left to the medical professional.

If you are going to try to have medical professionals refer patients to your group, they will want to be assured that no one is going to suggest something harmful to their patient. With VCFS/22q, there is so much variation in the manifestations, that this cannot be stressed enough.

Leadership and shared responsibility

A decision needs to be made as to who is going to lead your group. This is the person or people who make sure that things get done. This can be loosely structured until your group grows.

There are three general structures from which to choose:

- **Peer leadership:**
A member of the group is designated as leader.

- **Professional leadership:**
Someone who does not have a familial interest, either from the medical or psychiatric profession, a social worker, or someone who has had experience with support group leadership, is designated as the leader.

- **Discussion Group:**
You might want a group that is purely for discussion and all that needs to be considered are where and when to meet and a topic for each meeting.

Once you have a philosophical foundation, it is important to think about leadership structure. This can include one of the following:

- **Elected officers:**
You can choose a vertical leadership structure with formal, elected officers who serve for definite periods of time with specific duties.

- **Shared leadership:**
This model could have, for instance, an administrator, facilitator, and manager, all with designated tasks.

- **Flat structure:**
Everyone is equal and no one has ascribed rank over others.

Regardless of which model you choose, it is imperative that you deliberately cultivate a sense of shared responsibility. Tasks need to be shared. Nothing will sabotage a group more than one person doing it all. It is also important, if you choose a leadership model, to rotate leadership to avoid stagnancy. AND. . . delegate, delegate, delegate!

Group format

Group format can also follow different models:

- **Telephone**
You may only deal with people by telephone initially and then evolve.

- **Meetings**
Formal or informal gatherings

- **Combination**
This is most likely the model that will work best for a VCFS group

Telephone contact

Once you start promoting your group and even before your first meeting, you'll be talking with parents. Remember, the first call from someone is their first step in re-establishing some personal control.

Remember:
- Return calls promptly and from a quiet place
- Use caller's name
- Share YOUR story and info about the group – BRIEFLY
- Ask for name of child or family member
- *"How're ya doin'?"* How are the PARENTS doing?
- How can I help?
- Give info about next meeting
- Obtain info for mailing list

Crisis telephone call?
- Show concern, be calm
- Use active listening techniques
- Let caller know they are not alone
- Encourage caller to find solutions
- Suggest alternatives - but never tell a caller what to do!
- It's OK to say, *"I don't know."*
- Ask for name, address, etc.
- Call back

Meetings

Once you put your group's name out there, you'll start developing a membership list. Then, if your core group agrees, you may want to plan your first meeting.

• Where?
This can be logistically difficult for a VCFS/22q group as they most often cover a large geographic area. Look at your membership and decide upon a centrally located facility. There are many meeting places that offer free space, including: libraries and community centers. Some feel it is better not to have meetings in medical facilities because it makes some people uncomfortable. (However, medical facilities can also offer a built-in resource for networking and speakers.)

Private residences can be tricky as they tend to lend an air of extreme informality and this sometimes makes it difficult to stay focused.

- **When?**

This is up to the core group initially. Will you meet at a certain time of year? Evenings? Weekends?

- **How often?**

Weekly? Monthly? Twice per year? This, too, is up to the membership.

- **How long?**

The length of the meeting depends upon the agenda. If you have a speaker, the meeting will, of course, last longer than a purely social gathering.

Meeting structure

- **Set the agenda**

An agenda can be structured or casual. It can include formal speakers, an informational video, parent presentations, or just coffee and goodies.

- **Choose a facilitator**

Each meeting must have a facilitator. This person is usually familiar with the membership and sets ground rules. But they should also remember that their role is subsidiary and they should defer to all members.

- **Set ground rules**

Ground rules should be simple and based upon respect and concern for others. *These should be in writing*, and every member should receive a copy, and the rules should be reiterated at each meeting.

- **Determine a decision making process**

Eventually, every group must make choices about how to deal with particular situations. The methods used to make decisions will have an impact on the group's effectiveness. The book, *Robert's Rules of Order,* is a good source.

- **Contact speakers**
This can be an integral part of the meeting structure and success. Medical professionals are usually willing to speak to a group and will not charge a fee, but ensure that the speaker knows the issues and makes the talk relevant to VCFS.

Note: There are some housekeeping needs that must be accomplished prior to a meeting such as: sending out meeting notices four to six weeks prior with a RSVP requested. You need to ensure that A/V equipment will be adequate. Ask attendees to bring something for the snack table and ask new members to come a bit early to help set up.

Promoting your group

Following are suggestions for developing your promotional plan:
- Define your market
- Research contacts
- Announcements
- Press releases, brochure, newsletters, information packets
- National organizations
- Internet
- Professional assistance

Resources and services

- **Identify local and national resources and services**
You may receive calls from people wondering where they can go for general information. Have a list handy.

- **Recognize when referrals are necessary**
It's OK to admit that the group does not have all the answers. In fact, when in doubt, refer.

- **Develop a referral list**
This can be a list of VCFS clinics or it could include people like a local speech and language person who knows about VCFS. Encourage professional assistance.

Group maintenance

It is sometimes difficult to maintain an active group. Here are some suggestions:

Inactive membership
• Notify members of meetings by phone, by mail, or both
• Coordinate transportation
• Schedule interesting, varied programs and agendas
• Include a social component
• Enable members to take leadership roles
• Encourage members to share their talents
• Welcome new members and include them by giving them tasks to do

Not enough people:
• Utilize word of mouth
• Publicize meetings
• Schedule dynamic speakers
• Arrange for child care at meetings
• Include children in at least one meeting per year

Fundraising

Fundraising can not only raise needed funds, but it promotes camaraderie in the group. There are generally four methods for use in fundraising:

• **Personal contact:** in person or by telephone
• **Special events:** bake sales, walkathons, cookbooks, etc.
• **Direct mail**
• **Grants:** available through foundations, corporations, governmental entities. Application usually involves writing proposals, meeting with agency representatives and the ability to keep any records that may be required.
• **Dues:** Members pay an annual amount to the group

Non-profit status

When considering non-profit status, examine the following:
- What is the purpose of the group? Is fundraising going to enhance that purpose?
- How do our existing activities accomplish our goals? Is it necessary to add fundraising to the mix?
- Are there other projects in which the group would like to engage?

Advantages of non-profit status:
- Federal, and possibly, state income tax exempt
- No federal tax return if gross revenue per annum is less than $25,000
- Donations to and personal expenses for a non-profit corporation are deductible
- Group qualifies for lower postal rates
- Potential for fund raising
- Confirms you as legal and bone fide

Assessment

As with any ongoing endeavor, it is important to keep goals and attitudes fresh. Your membership should reassess the group periodically.

- What are the goals of the group? Have they remained intact from the original mission statement?
- Do these goals arise from member needs? Have these needs evolved?
- What methods are used to meet these goals and are they appropriate responses?
- Do the size and structure of the group promote or impede goals?
- If group needs are not met, were needs accurately assessed? Were goals appropriate? Did the group have an unrealistic idea of the resources available?

Conclusion

Starting a support group is an incredibly rewarding experience. The people who contact you will usually feel comforted just talking with someone who knows what they are experiencing or can find someone who does. As with any endeavor, things do not always run smoothly. Dealing with parents who are experiencing the stress of an ill child can be difficult, but the benefits far outweigh the challenges. You will learn much by talking with other families and will make lifelong friendships.

Acknowledgements

www.vcsfef.org VCFS Educational Foundation
Parent to Parent of Pennsylvania
National Spinal Cord Injury Association
SupportPath
Self-Help Network
4Non-profit Organization
www.vcsfef.org/support_group/us_groups.html

Maureen Anderson has a son with VCFS. She coordinates the *Northeast VCFS Support Group* and is affiliated with the *VCFS Educational Foundation.*

Let's learn to communicate with one another in a way that encourages

Most of us think we communicate well. That may be the case and it may not. When we are dealing with the most precious things to us (our children) we will need to keep living and learning about ways to make life better for them and for ourselves. We will need to try and get a universal language going; one that is based on courtesy and thoughtfulness.

E-mail, voice mail, and Internet use are revolutionizing the way we communicate and exchange information.

Communication is a process most of us take for granted. First there is an *"idea"* or *"information"* you want to share or impart.

Then there is the means by which you choose to send that information. When it is received, hopefully it is understood or interpreted the way in which you intended. Finally, there is feedback from the recipient. <u>Here</u> <u>is</u> <u>where</u> <u>we</u> <u>should</u> <u>pay</u> <u>close</u> <u>attention!</u> Ask yourself, when I communicate with someone am I adapting the message to that person? Am I predicting how it will be received by that person? Am I using tact? This is important for successful communication and in supporting one another. When sending a message, we must expect and want to receive feedback from the recipient. No one enjoys a one-sided friendship and neither does one enjoy a one-sided conversation. The sender should encourage feedback by including a question or two in their message. Language is important; so is listening. This is especially true for those of us who are dealing with disabilities. Emotions run high and can actually impede communication if we are not careful.

There are 4 stages of listening:

• **Perception:** This is the literal recognition of sounds and concentrating on them. It is the conscious act of listening and focusing on the sound being made.

• **Interpretation:** Once you have focused on the sound (this may not be literal sound, but tone, such as in an e-mail) of the message, this is when you start to decode or understand what is being said. This is where we can sometimes jumble up the intent of the sender's message; because of our bias or due to our expectations and prior life's experiences. We may interpret something differently than it was meant. This sometimes happens in e-mails and letters.

• **Evaluation:** After the interpretation of the meaning of the message, we analyze its merit and then draw a conclusion. To do this we have to separate fact from opinion and try not to be judgmental. Think about what a juror is asked to do. Consider ALL the information, don't be prejudiced, recognize your biases, and avoid jumping to conclusions.

• **Action:** This is the response to the message. This feedback is beneficial to the sender in different ways, the main one being whether or not the message was understood correctly.

Talk show host *Oprah Winfrey* stated, she owes all her success to the artful application of *"listening and then responding"*. Listening is the art of controlling your own talking and self-importance in order to allow others to express themselves.

Communication can either build up or tear down. Do we stop and analyze what we are saying and how it will affect the recipient? As we reach out to educate others about 22q as parents, family members and professionals, we must be cautious. Do we know the facts well enough to relay them to others? Do we know which are verifiable and which are opinions and hearsay? Don't be afraid to say, *"I don't know."* or *"I'm not sure."*

As the knowledge and exposure of VCFS/22q extends and grows there will be more and more need for us to communicate about what is needed for our children and for the community. This takes team work which means we work and communicate and interact closely with others. To be a successful team we have to treat each other with respect, minimizing misunderstandings and misinterpretations. We want to avoid cognitive conflicts and set rules that achieve harmony and strong bonds to achieve a worthy goal. One thing I have learned while writing this book is that there are competitive natures in many people. We must, if we are to be successful, unite, be humble and exchange information freely. We must collaborate rather then compete. The common goal is to reach out to educate and advocate for those living with VCFS/22q.

Make a good impression when presenting the educational material

Just as we want to make a good impression at a job interview; our first impression can make a difference when we meet with an individual whom we want to educate about 22q. This might be a teacher, or school board member, a non-profit organization board, a doctor or any other formal appointment. We want to dress appropriately and take note to avoid using some negative non-verbal signals that can also speak volumes, such as: posture, excessive hand gestures, gum chewing, lack of interest in what we are hearing as feedback, or types of questions, etc. *(See Negotiations on page 261.)*

We want to leave the person with whom we had the appointment, with a good impression, viewing us as a reliable source. So we dress clean and neat, we make eye contact, use good posture, relax and use a clear, pleasant voice, use appropriate hand gestures. (However, don't get dramatic like me, I swear my mouth couldn't move if my hands didn't.) Smile and have positive facial expressions.

Make sure you are on time for appointments and be conscious of the time spent in the meeting. Try to be friendly but don't go off task too long. Make sure you have a presentation or at least be familiar with your literature, not just your own case study so you will be confident. *(Hopefully this book will be one of the items.)* The person you're seeing is graciously giving you their time so you want to show appreciation for it and leave them wanting to see you again. One area that is important and rarely gets mentioned is the appearance of our literature that we present. It needs to be clean and in good condition. We don't want our literature to be crumpled or sloppy; which conveys a negative message. It conveys the nonverbal message that you don't care as much as your words state; and your credibility suffers.

Remember we are there as ambassadors of our cause to educate so *"express not impress"* and remember to KISS (Keep It Short and Simple).

Be a "critical thinker" when making decisions

I learned to be more of a critical thinker in nursing school; it was there that I learned that thinking is more than just memorizing facts. However, learning to memorize effectively enhances our ability to think critically. I believe it has been a valuable tool both professionally and personally. To think critically one must be able to recall facts to progress to higher levels of thinking. We have the facts, now we need to know how to apply and analyze the information. You may wonder why I would opt to put sections like this in a book of this kind. I guess it is because many of us have been so overwhelmed in life that there are times that it is hard for us to think clearly, and perhaps we could use a little direction. I hope that I will be able to instill these life's tools in my own son to help his path go smoother. First let's define what a *"critical thinker"* is. A critical thinker is one who is optimistic, careful yet doesn't give up easily, ready to improvise, deliberate, and goal directed. This applies in everyday life or for the goals of your child. A critical thinker bases all decisions on fact not guess work. We use it in decision making. <u>*This is simply the act of making an informed choice between a number of alternatives to either solve or maximize an opportunity*</u>. We need to have pro-active thinking; to be organized; systematic; flexible; realistic; and committed. We need to learn to <u>*look ahead*</u> to prevent a problem from occurring, to be a problem solver, and to learn logic by reasoning in a given situation. *(This includes the reason for the section on Finances and trusts, see page 323.)*

To do this it takes a type of brainstorming that we all do with other people we respect. This may require us to think *"outside the box"* at times when there may be no real answers to be found. We must do what we believe is right, even if the majority might not agree. We choose based on what the facts presented to us dictate.

Always base thinking on principle, constantly evaluating, self-correcting and striving to do our best with the knowledge we have obtained from our experiences. Remember we have to live with our decisions; and, in most cases, so does our family. Try to be confident and not use too much brain power worrying. I know, it's easier said than done, but it only reduces our energy level and is counter-productive to critical thinking.

When making decisions, we must be fair. There is a correlation between moral development and critical thinking ability; those with a clear sense of right and wrong are more likely to make better and smarter decisions. Always try to be humble and self-disciplined.

If another way or opinion presents itself don't resist a change; be flexible like a young willow tree that can bend with the wind and not snap.

To be a critical thinker, it takes a person with an open mind. One thing that was difficult for me and a hindrance to my learning critical thinking was making decisions based on emotions. This can be in forms of communication and interpersonal relationships. I have been told I am an *"old softy"* and that is not always good, especially when I am in a trying situation. Anxiety, stress and sheer fatigue make concentrating difficult. Sometimes I have made wrong decisions based on how I felt, or how I was guided by emotion.

I believe now that if I would have had resources that are available today and a strong support system in place, I would have made different decisions in the past for my son. This is another reason for writing this book. It is to help others know where to start; to make a life's road map, with a starting point, and a happy safe destination!

Active parent support groups, informational and resource organizations

These are organizations which are not VCFS/22q specific but can provide a wealth of information for all or us. They can also direct us to organizations which may be able to help us in the area in which we live.

National

National Parent to Parent Support and Information System
PO Box 907
Blue Ridge, GA 30513
1-800-651-1151 (voice/TTY)
706-374-2822
706-632-3826 (fax)
http://www.nppsis.org

Parents of Blind Children, Committee on the Blind,
Multiply-Handicapped Child
1912 Tracy Rd. Northwood, OH 43619
419-666-6212

Parent Support Group
27315 152nd Ave. E
Graham, WA 98338
360-893-1256
(206) 282-1334
E-mail: **info@parentsupportgroup.org**
Web Site: **www.parentsupportgroup.org**

National Dissemination Center for Children and Youth with Disabilities
NICHCY
1825 Connecticut Ave. NW Suite 700
Washington, DC 20009
1-800-695-0285 (V/TTY)
202-884-8200 (Voice/TTY)
(202) 884-8441 (Voice/TTY)
E-mail: **nichcy@aed.org**
Web Site: **http://www.nichcy.org**

National Organization on Disability
910 16th St. NW, Suite 600
Washington, DC 20006
1-800-248-2253
(202) 293-5960
Web Site: **www.nod.org**

Note: *There are many disability-specific national support groups listed each year in the January edition of* **Exceptional Parent Magazine** *This magazine also offers, in each edition, a section where parents can write-in to contact other parents of children with specific disability issues. Additionally, there are a number of Internet websites that offer support and information. Some that you may want to check out include:*

Velo-Cardio-Facial-Syndrome Education Foundation
http://www.vcfsef.org

The Family Village
http://www.familyvillage.wisc.edu/

Our Kids
http://www.our-kids.org

Sibshop
http://www.siblingsupport.org

The Beach Center of Families and Disability
http://www.beachcenter.org

The ARC of the United States
http://TheArc.org

MUMS National Parent to Parent Network
http://www.netnet.net/mums

Disability Resource Organization
http://www.disabilityresources.org

Focus Adolescent services
http://www.focusas.com

National Mental Health Information Center
http://www.mentalhealth.org

NIDCD
http://www.nidcd.nih.gov

DMOZ
http://dmoz.org/Health/Conditions_and_Diseases/Genetic_Disorders/Velo-Cardio-Facial_Syndrome

Meddie.com
http://www.meddie.com (Go to: "conditions and diseases", "Genetic disorders", "Velo-cardio-facial syndrome" for a list of articles.)

http://www.kumc.edu (Type "VCFS" into Search box.)

PARENTS SPEAK OUT

Many parents have stated *"I find it very often depressing to read the VCFS support groups. It breaks my heart to hear about all the suffering of the kids."* I can understand, this is called fellow feeling, my sensitive friends. It can be a good thing, but it can also blind you to the fact that there are many success stories that we don't hear. The VCFS support groups don't mean to be negative, but that is a safe place to vent, to share pain and to ask questions in the areas of concern about what we are living with at the time. When I say success stories I mean in a VCFS/22q way of thinking. I believe our thinking will need to change about our view of *"normal"*. Our 22q children may take longer to get through school and they may not get a degree (some do). But that doesn't mean that they can't take one class at a time in a community college if they are able, or get into apprenticeship programs of some sort. Working with your child is important. I do have to say that if your child has impulsive behaviors, don't jump at each thing they say they want to have or want to do in school or as a job. You may find yourself in my old shoes. I was so bent on helping my son build self-esteem that without realizing it; I would indulge him, and spend money I didn't have, to try and fulfill each dream and desire. Just when I would get everything set up and had jumped through hoops to get him at the point that he swore he wanted and would promise to do his best; he would say he didn't want it now, and that he didn't remember saying certain things. Then off we would go on another tangent. So be cautious and don't beat yourself up if you don't open up all the doors of *opportunities* that you think your children may need. Get to know your child and do what is realistic and makes sense long term for both you and the child.

Medical Trivia

A sneeze can exceed a 100 mph.

Blank page for personal data:

"Experience is not what happens to a man. It is what a man does with what happens to him." Aldous Huxley

CHAPTER 15: DIARIES

THE DIARIES

This chapter contains actual stories from families who have opened up their lives to share and teach from their experiences and their hearts. We thank them.

The story of T.G.

My name is Sherry Gomez and I am for the most part, the author and collector of information for this compendium/book. Since my son is 32 years old, most of what I am about to tell you is a condensed version of his life. We had 18 of those years dealing with VCFS/22q without knowing it. I want to share our experiences, in hope that it will help, encourage, or save others some unnecessary grief from lack of knowledge and support, thus minimizing wrong decisions. It is said, that *"Knowledge is POWER"* and I believe that it's true. For if you know what you are fighting you can collect the information that can empower you to defeat the enemy called *"FEAR of the unknown."* (Aren't I the philosopher?) Actually the statement is true.

Let me first set the stage with a little family history. My husband and I have been married since 1975. We met while I was on vacation in Cottonwood, Arizona. After 4 *DAYS* of intense talking on the hood of a car (well, of *me* talking . . . ahhh, to be young again!!!) We decided to get married. YES, I know, *"Are you nuts?"* Probably, but that is another story, for another book.

The reason I mention this, is to confess I knew *NOTHING* about my husband Gil's family, and I was to learn that he didn't know much either. His sister died of leukemia at age 17; his mother died of cancer at age 41; his natural father died of Parkinson's disease at age 57 (he didn't know him well); and his brother was killed in a car accident. So there was not much to go on, right? Information on my side of the family was not much better. The only things that I am aware of are that my maternal grandfather had mental problems (I met him once and I concur with the diagnosis) and my mother showed some symptoms of heart problems, lung and bronchial weakness when I was growing up. I didn't know my father or his side of the family, so there was not much to go on there either. We were both 20 when we got married, and being young we had no reason to be concerned about family genetics and therefore were oblivious about their importance. It is believed that T.G.'s condition may be a *mutation*.

T.G. was born with long fingers. His face was that of the average cute baby. His features have changed as he matured or course, now taking on some definite characteristic of VCFS; the longer face and bulbous nose. The kids in middle school used to call him "Gonzo" (from *Sesame Street*®) because of the shape and length of his nose. (We will touch on that later.) Now, lets start from the beginning . . . .

Birth

T.G. was born in 1979 in Salem, Oregon, at home on a bean-bag chair that was positioned on a waterbed. My daughter was born in the same fashion, 19 months earlier and it was a marvelous experience, so I decided to do it again. Once again it was a wonderful experience, no problems were noted. He was on time, no labor problems, no delivery problems, just a normal 8 pounds of bouncing baby boy, with nothing out of the ordinary, except excruciating pain on my part. (We mothers need credit.)

Then the FIRST red flag appeared.

I breast-fed my daughter and knew what to expect, at least I thought I knew. T.G. had trouble nursing and wasn't able to latch onto the nipple (no suction). The milk would drain out his nose, causing him to flail his arms, gasp for air and struggle. This, of course, scared the daylights out of me. Because of this he was not getting enough to eat. So I needed to try an alternative feeding method, but *WHAT?* I ended up in the hospital myself within a couple of days with mastitis (infection of the breasts) because he was not able to nurse and I wasn't able to pump milk out as fast as it was coming into my breasts. I was full, full, full;

and needed to be relieved. I was nervous for my baby, so I enlisted the help of the *Le Leche League* and requested a wet nurse to come in to nurse my baby as much as she could, or supply milk since I could not nurse him at the time. I was buying time so that I could figure out what I was going to do. That helped at first. The volunteer could pump milk to feed T.G., but we could not keep that up for long, and so the doctor suggested formulas. At the time, the doctor thought that T.G. was allergic to something in the breast milk and it made him regurgitate the milk, sometimes violently. That of course was not the case and he continued to have feeding problems. He was soon labeled *"failure to thrive."*

In the meantime, I started to read more books on nutrition for infants and my mother became an avid herbalist. When I look back now, I wonder how that child ever survived at all; with all the concoctions and gross tasting teas and things. We tried everything we could think of, but of course, nothing corrected the problem to stop the liquid from coming out of his nose. He was finally sent to an ENT when he started his long siege of persistant ear infections. I asked the ENT if it was possible that something was wrong with my son's palate. He looked and felt and said, *"No, I don't think so."* Well, we found out *MUCH* later he was wrong. We continued to deal with the feeding problem.

By reading, I learned that goat's milk is the closest thing to mother's milk and that this was important for digestion in infants. So, I found a lady who raised Nubian dairy goats, not too far from where we lived. We paid her extra, to feed a special diet to the one goat we used. This insured us of a consistent diet for the goat; with less chances of intestinal disturbances for our baby. While the whale-like *"blow hole"* problem in his nose still existed, I learned to hold him in an upright position to feed him using *"Newton's law of gravity"* to our advantage (30 degrees or more works best.) Not only did he start to thrive, but he became a butter ball. Goat's milk has a very high butter fat content so with less food he got more nutrition. I also added supplements to the bottle which we also believed were helpful. This, however, only caused another problem; that of constipation from the rich butter fat. Now I realize he probably was having problems eliminating due to slow peristalsis (the snake like movement of the colon to move solids down and out) but I didn't know that at the time. (I only learned that after becoming a nurse). He would wake at night crying and crying, bring up his legs in a colic pose and would try to grunt. I just knew he was trying to fill his diaper, so, I tried a lot of things, like rubbing his tummy, etc., but to no avail. What I ended up doing is something that might sound bizarre to most, but it worked and that was the main thing. I would lay him on his tummy on the changing table and take a rectal thermometer, grease it with Vaseline® and insert it into the rectum a little ways (mostly the tip) and move it in and out, and side to side to stimulate the bowel. (I don't recommend this, but I was desperate and didn't feel I had any alternatives.) This also helped to relieve a lot of built up gasses. I actually got this notion from living on a farm and knowing how animals stimulate their young to go potty. I love my son, but I drew the line at that act. If any of you have seen a litter being born, you know what I mean! Not long after, I learned about molasses and its properties.

Ear infections!

That was to be a term in our vocabulary that was used often. They always came, it seemed like every month. (I now know he had immune deficiency as well.) It was the secondary infections which followed that seemed to do the most damage, not only to the child, but to the check book. Pneumonia was the most frequent end result, over and over again, until his chart was labeled with "lung disease".

When I think about it now, I cringe. With high fevers T.G. would go into convulsions. Instead of the doctor wanting us to bring the child into the ER when I would call in the middle of the night, we were told, that one of us should get into the bathtub with the baby, fill it with cool water and stay in it with him till his fever went down. Brrrrr... I was not brave enough, so my husband had the honor. After so many bouts with pneumonia, T.G. had an accumulation of scar tissue in his lungs, especially the right lower lobe. He was sent to the nearest medical teaching hospital in Portland, Oregon to have test after test done. Cystic Fibrosis was their main concern and when that was ruled out it was back to the drawing board.

Although he was sick a lot, his development (except speech) still seemed in about the same range as others in his peer group. He walked at 13 months and was potty trained by 2 years of age. What I noticed most and had a real concern about, was that he never spoke. He would point and grunt at times, but, no words. He was just there. He smiled when smiled at, played by himself quietly, and was generally a perfect baby.

I mentioned this fact many times to the pediatrician but he would say things like, *"Boys are just slower learners."* or *"It is because he has an older sister who talks for him."* He always said it was something other than what I suspected; which was a hearing problem. Well, one day when he was almost 3 years old and not talking, I got a little more forceful. Okay, I got *MAD!* We were once again sent to a *specialist.* Without even looking at our son, the doctor asked if T.G.'s father and I would come into his office to talk. The first thing out of his mouth was, "You mustn't deny the fact that your son is retarded." *WHAT?!* Oh boy! I went into "mother-bear-protecting-her-cub mode." That poor doctor didn't know what hit him. I think back now, and both men had their mouths hanging open, the doctor and my husband. Finally after telling him he had a lot of nerve making that kind of assumption and assessment when he hadn't even examined my child (and wasn't ever going to do so!). I put my foot down and

said, *"I want him referred to the deaf school for testing!"* I was tired of listening to all of those doctors with their opinions, and learning nothing; while my son was getting older and going nowhere fast. From now on, I thought, I am going with my gut. Do I need to tell you that mom was right? He was clinically deaf. I was told it was from all the ear infections he had. The 3 bones in his ears had calcified and they had to be scraped. Tubes were put in his ears. This was the beginning of one speech therapy program after another. Now that he could hear enough to make sounds and form words, we noticed that he sounded kind of like a munchkin from the *Wizard of Oz*. Hmmm . . . I thought, something still is not right. What is making him sound so high pitched and nasal sounding? We would not know the answer to that question until he was 18 years of age. Even after all the years of speech therapy, the therapy work could only do so much, because the real problem was VPI. (See Glossary.)

Right after his ear surgery T.G. ended up in the hospital again with double pneumonia and things were not going well. The doctor told Gil and I that if we stayed in Oregon with the damp cold weather, we might not see our son through another winter. The doctor's suggestion? *"Go west young man!"* Go to Arizona or New Mexico, he needs the dry heat. Since we were familiar with

Arizona, we came back. In doing so, we practically gave away our home and everything else, and left Oregon to do what a parent must do, with love and with *NO* regrets.

Arizona 1982

We landed in Wickenburg, Arizona at first. My folks had lived there as snowbirds. It was while we were there that T.G. demonstrated he could stay in good health longer. He learned to ride a tricycle, as well as climb the fence next to the cactus garden, into which he managed to fall. Oh my, what a pain that was. (No pun intended!) It was not easy to extract different kinds of cactus needles from his body, which was, literally, covered from head to toe. We were starting to see some real progress in him; he had energy and an interest in things. It made our hearts rejoice!

We soon realized that the commute for my husband to his new job in Mesa (2 hours one way) was too hard on him and was taking its toll. Phoenix was our next move. It proved to be a good move. At this writing T.G. has only been in the hospital 3 or 4 other times for pneumonia since our move 20+ years ago. Oh yeah, he did, however, develop asthma and had to have breathing treatments 2 to 3 times a week. That went on for about 5 years. Fortunately he outgrew it. No more Nebulizer! And we celebrated.

Now we were in a city and we needed a new pediatrician. I asked around and found one who was close to our new home. He was an older gentleman, and on T.G.'s first visit, the doctor heard right away that there was an unusual sound in T.G.'s chest. Because we had no insurance we qualified for medical testing through what was then Crippled Children's Hospital in Tempe. (It is now CRS-Children's Rehabilitation Services in association with St. Joseph's Hospital, as of this writing). That was the beginning of a list of *"odd"* illnesses and occurrences to come with symptoms I wanted to understand. I was not to get my answers for many years. The exam at the hospital revealed that he had a heart murmur. The cardiologist believed it was only a crimp in an artery and that he would outgrow the murmur as his chest grew. It was not until he was in the hospital at St. Joseph's, on one of the pneumonia visits, that we found out that his heart is reversed and he has a displaced artery. At first, I was told that he had a tumor in his chest. Fortunately, that was not the case. The bulk from the displaced artery made it appear, in an x-ray, as though there was a thickness or mass where there should not have been one. To date, it's *"so far so good."*. I told them at the time, *"If it's not broke, don't try to fix it."* While at Children's Hospital (CCH) T.G. was put though a wide range of developmental tests, due to the ear surgery and delay in speech. This proved to be really beneficial because other delays were just beginning to surface. He qualified for a special education intervention preschool. It was a wonderful school and I believe it gave him the needed edge to catch up, and start school on time. It was the special training they gave him that made the next few grade school years more successful.

School years

In T.G.'s kindergarten school days I heard, "It hurts, mom". That proved to be an inguinal hernia. One side of his scrotum was double in size and red. So off to a surgeon we went. He fixed the one side and re-enforced the other side. T.G. hasn't complained, so I suspect all is well.

During the years of falling baby teeth and permanent replacements, we had a real test. It turned out that T.G. had *TWO* sets of baby teeth and a few extra permanent ones. So he visited the dentist regularly to have teeth pulled, until X-rays finally showed that he had the right count. Then came the braces. Now remember I not only had T.G., but other kids to care for who also needed braces, so I became friendly with the orthodontist. We supported him well!

By the time T.G. was about 8, he was regularly complaining of leg pains and his knee kept popping out of joint. I was told it was growing pains. By this age he had spent a lot of time in bed at home, sick with viruses and bacterial infections etc., or in the hospital. I sure wish I *"knew then what I know now,"* I could have eliminated a traumatic surgery for him by doing simple ROM (range of motion) exercises for him while he was bed ridden and not able to walk for extended periods of time. From not walking or being stretched, the tendons in the heels and toes did not lengthen with the growth of his foot and toe bones. This caused T.G. to walk on the balls of his feet and toes and for his toes to draw up and hammer. Needless to say, he had to have the tendons lengthened and the toes broken and reset/pinned to straighten them. He also had to wear shoe inserts for a while, due to an arch problem.

In 1995 our daughter got married and moved to Mineral Wells, Texas. It was then that we decided to move to Flagstaff, where we were advised it would be healthier for T.G. Ahhhh . . . a new start . . . or so I thought! This move would get mixed reviews from me, if you asked me today.

While we were there T.G. had several serious nose bleeds. Of course, everyone thought it was the altitude of 7000 feet. One night in 1997 he had an unusually bad nose bleed, and we took him to the ER. He had lost so much blood that he was pale. The ENT on call that night was a young female doctor from Colorado and she did all she could to cauterize his nose and get all the bleeders. She had never seen anything like this before and was certain something was wrong. When she heard T.G.'s history, heard him talk with the sound quality of his voice she said, "Something is not right here. I think I might know what it could be, but I will send you to Phoenix to a specialist." That was what I had been waiting to hear from someone for a very long time. So back down the mountain to Phoenix we went.

Answers

When the new doctor was told T.G.'s history, he was certain that he was one of those kids who fell through the cracks; and did not get the diagnosis of VCFS early in his life, for proper intervention. T.G. was referred to a geneticist, at CRS (St. Joe's) who in turn sent him to have the FISH test done. T.G. was given the FISH test and it was confirmed he had 22q11 deletion. This *FINALLY* explained all the odd illnesses, as well as the learning disabilities throughout his life.

This step started the ball rolling. T.G. was referred to the ASU speech and language department for the velopharyngealoscopy and it revealed VPI. So the velopharyngeal flap surgery was performed and T.G. has done well since then. We have since moved back to Phoenix, a move which brought on the next adventure.

Oh no . . . now what? Is this a nightmare?

It was about the time we moved to Flagstaff that I also started seeing a different kind of anxiety in T.G., about odd things, but I fluffed it off because he was a teenager.

He started wanting to prove himself and would think he could do things that he couldn't possibly do. Everyone could see that but him (grandiose delusions). He also started wanting to participate in activities that were not safe. Knowing he was slow on the draw, I thought perhaps it was just that he could not see the consequences of certain actions. Now I realize that this kind of thinking, with the added impulsiveness, can be part of VCFS and the start of my worst fears, those of mental illness.

It wasn't until he was 18 that it all hit the fan, so to speak. We had moved back to Phoenix after 3 years in Flagstaff, Arizona and were trying to settle in our new home; when the first solid sign appeared. Although we didn't know that it was a *sign* or *symptom* at the time. He joined a kick-boxing club and was starting to fantasize about becoming Jean Claude Van-Damm. About that time he began to lie, he refused to eat, he would sneak off to kick-boxing bouts (that we did not know about) and he would not sleep all night. Then he worked all day with his dad. At the time, I thought this was all due to the fact we had made another move and we were now closer to his old childhood neighborhood, thinking that now he had more to do and was going out with "the guys". How naive I was.

One week T.G. got a flu bug, he was almost over it and decided to go out with a friend. About 9 p.m. we got a phone call from the friend who was with him, saying he thought T.G. was on drugs and was acting weird. We knew T.G. wasn't on drugs and we asked the friend to bring him home ASAP. When T.G. got home he was burning up with fever and I asked him to go to bed. This was not to be. He started to say strange things to me like, *"He knew what I was trying to do to him."* and then he proceeded to say that, *"I was trying to poison him but he was too clever, so he wasn't eating."* (Hmmm . . . OK that was why.) He was sure I was trying to keep him from God and he started to cling to his Bible. Everywhere he went in the house, he would see Jesus standing behind me, in door ways, etc. There was more, but it is too much to relive.

The next morning he snuck out of his bedroom window and he headed for the church we used to attend. He was on foot and the church was at least 10 miles away. He hadn't dressed, or even combed his hair. His dad found him on the road. We told him he had to go to a hospital and he refused. (In his paranoia, we were his enemy and he had to escape.) Since he was 19 we were not permitted to force him to get treatment, so we had to call the police. We met them outside to tell them what was going on; they said we could follow them to the hospital. Fortunately, I had worked as a volunteer at St. Joseph's Hospital and T.G. had been a patient there in many prior years. I called one of the nurses who I knew that had experience in this area. She worked at Barrows Neurological Center at St. Joseph's. When I told her what was going on she told me to tell the police to take him to St Joe's and she would meet us there. So off we went. We got there all right, but he was 19 and refused to allow anyone to touch him. We had to get the hospital social worker to advise us. She got some paperwork together; and with the help of the ER doctor on call, we petitioned a judge to give us emergency medical authority so that we could get T.G. some help. The help was given and it turned out he had a viral pneumonia, and malnutrition. Something had caused *"white spots"* in the basal ganglia area of his brain that weren't there before. These showed up on the CAT scan.

I thought the worst was over, but it was not. His pneumonia got better while he was in the hospital but the mental illness didn't. We had to let the court take over, and they took him by police escort again, to the Maricopa County Mental Facility. Oh my God! I hated to leave my baby in a place like that, but I had no choice. We were fortunate actually; the doctor who saw him that day knew right away that he had bipolar disorder (manic-depressant) with paranoid psychosis (schizoid-affective disorder). It had been triggered, perhaps by the move, puberty, the virus, or who knows what; but it was going to take time to get it under control. The doctor tried to reassure us that T.G. was going to be fine once they got his brain chemistry stabilized and found the right drug combination. This was not as easy as it sounded. He was hallucinating and at times non-compliant. I will never forget the days he spent in that hospital and how devastated I felt. When he did finally stabilize there were side affects from the drugs.

When he was released to go home we had a long row to hoe (so to speak), so we rolled up our sleeves and got to work learning about mental illnesses as they relate to VCFS and so on. At one point, I had to request that the medication be changed because he was getting such horrible side affects. He was shaking and shuffling like an old man with Parkinson's disease; and his face, eyes and tongue would twitch and dart. He was sleeping 24 hours a day and had to be awakened to eat. So, it was on to the next drug combination. The routine of no exercise and the drugs increasing his appetite, caused him to gain 80 pounds. (His closet looks like a second hand store with all the sizes we had to buy in such a short period of time.) He is now diabetic, on insulin, and has liver problems. The weight is not coming off easily. I wish I could say his friends stuck by him, but no, I think he scared them.

One of the manic relapse episodes was a real eye opener. When a person is in the manic phase of the illness they can exhibit hyper-sexual tendencies. At this time, the person can't reason or think clearly; they just know how they feel and need some relief. So it was a shock to my system to have my own son expose himself. I have since heard this is very common. It shook me up and I felt nervous, that I could not trust him or be alone with him for several months. There were only 2 times that this happened. My son feels very awkward that it ever happened at all and states that it seemed unreal to him; although he knows that it happened. He says it was like another person was doing it and he was just an observer. At those times of mania he would lay and moan for hours. It was very disturbing.

Needless to say it was emotionally draining for all of us. T.G. is, at this writing, doing well and has been stable with no bipolar episode relapses of significance in over 2 years. I think we have finally found the right combination for him. Because of the hypersexual part of this illness, and knowing that VCFS can be inherited; T.G. opted to have a vasectomy when he was 21 yrs old. This was successfully done and paid for by insurance. By doing this, it safeguards against the possibly of creating another child with all the health problems of VCFS and saves my sanity!

This leads to what we are dealing with now
T.G. has now been diagnosed with medication-induced diabetes and hepatitis with the possibility of the first stages of cirrhosis from all the medications he has had to take throughout the years. We do not yet know the extent of the damage. If there is no necrosis (death of the tissue) the liver has a miraculous ability to rejuvenate itself. In the meantime he is trying to learn about his diet and has joined a gym for exercise. So with supplements, a workout and better control of his diet, we are determined to win this current war. Wish us luck!

Biggest challenges

If I had to choose the one area that gave us the longest and most difficult challenges up through the age of 18 and beyond; it would be that of cognitive/learning problems. I now know that T.G. has ADD and was never diagnosed. Therefore, he never had the appropriate information added to his IEP. Since no one knew he had VCFS, he was not always given the right IEP in my estimation. I am not talking about just in the content of goals, but the way his learning environment was conducted. He really needed to be put in more of a one-on-one setting with a tutor or TA; or a small structured group setting of children who also needed more help. Instead, in grade school and middle school, many of his classes consisted of 30-40 students and only one teacher and a TA. He was in special education through middle school, and had a very good special education teacher. I can't thank her enough for what she contributed in his 2 years of learning at Cholla Middle school.

After 8th grade graduation, as with MOST things in T.G.'s life; he fell through many of the cracks again. Instead of recognizing that T.G. had problems, when he did not do well, I was told he was *"a lazy thinker"* and a *"kid with no ambition"*. I have learned that in the school system, unless the child has extreme health problems that are noticeable to the eye or *"tangible"* handicaps, they can be over-looked, and don't get the added attention they need. VCFS kids are, for the most part, fully functioning but slower with diverted thinking, so if they were just given a little more personal help, they could be productive 100 fold. I believe the lack of understanding about VCFS has been the major culprit for our children being left behind in society (socially and educationally); in programs for services. This is true, not only in school but in life in general, as they get older. After 18 you have to fight even more vigorously for them. Society expects more from them than they can possibly give. Instead of trying to understand, most people just push them to the side and go on with the next individual. I have seen this done socially, in employment, and in recreational areas.

What is needed is for certain types of companies to step up to the plate and hire the disabled first in areas of assembly lines and the like, to work at things they can do. We need recreation programs in places like *Parks and Recreation* that do not lock all adults together by age and that will allow young adults who are slower to join in the older teen group (where they are emotionally) or to form a team of their own. We need a GED program that can be taken by the special education person with the emphasis on _special education_, a program/test that is not as difficult. The VCFS kids/young adults surely do want to give their all, but they do not understand the deeper mechanics of life or have the *"know-how."* They cannot confidently defend themselves in some situations in order to make people understand and give them a chance.

Working and housing

Remember when I mentioned that when the child hits 18 years of age, you have to advocate for them even more? It reminds me of the time when T.G. was about 21. I was told by a state vocational rehabilitation counselor that the state would not approve of any funds for any training schools for him, because after testing, they did not consider that T.G. was worth the educational expense. He could not perform sufficiently enough in a class room or on a job placement to warrant having the state spend money on him. These are not the exact words, but it all meant the same to me. Needless to say I was not really impressed with the system then. We have since changed counselors and can see that not all are created equal. She has been marvelous. The first counselor made the statement after results of a test they had T.G. take when he had just started on new medication and his mind was cloudy and his health was struggling. I suggested that they reschedule but I was told, he is over 18 let him decide. How could the test results be accurate? This also backs up what I said about lack of understanding.

There is the _ticket to work_ program that _social security_ offers to give all people who are on social security a chance for gainful employment. Back then, there were about 3 or 4 options on the list for offices in our area that we could contact to have them help with job placement. They told us that they either were not taking any more clients or that they could not help us at the present time, but we could contact them in the future. As you can see this can be very frustrating and time consuming. Yet, I must admit it can be a good tool in some instances. I think it just needs to be fine tuned. Perhaps getting some employers on board, as I suggested before, could be part of the solution.

One thing I have learned from all this, is that if your child wants a job and can hold down a part time job, it is up to *YOU*, the parent, to give them the needed help. This is what I ended up doing many times. First, I sat T.G. down and had him make a list of things that interested him and a list of what he thought he wanted to do. Then we put a line though the stuff that was absurd (like lion tamer) and only kept the ones that seemed realistic. Knowing your child is the key to success. What are their strengths and weaknesses? (Write them down.) Are they a morning person or night owl? Do they require a lot of sleep? Do they get tired easily? Do they get distracted easily, get nervous in a crowd, etc.? Then, with that information in mind, I narrowed down the choices of jobs that were possibilities. Then I got out the yellow pages section of the phone book and went from A to Z, making a general list of jobs and places in our area that might have something that T.G. could do. I made up a fact sheet for him to use. It included things like names of people to write down for references, phone numbers, who to put down for emergencies, what work he had done in the past, what he was good at doing, etc. Then I called and asked if the different companies were doing any hiring. I did this with temporary services also. If they were hiring for positions I knew T.G. could do, I would take him there or go with him. (He now does this successfully on his own and he drives himself.) I let him start learning how the process works; filling out applications, dressing for success and conversing at interviews. (You may want to work with them on that one.) If you can't or don't know how to help, see if there is a person or program through your local employment division, etc. Your child may or may not get hired, but it is a good life's lesson. I have learned that these experiences are as important as taking a college class because the child is going through a "life" training program. It is wonderful for them to experience job searches and experience different jobs, as well.

When a child has 22q, it is not the same as it is for other people who expect to stay at one job. If that happens and your child stays on one job for a long period of time, you are lucky. I used to get so angry and impatient with T.G. when he would job hop, but now I can see that it has all been for the best. T.G. has grown and matured with each experience. We also have found out he is a natural keyboarder and can do data entry. He also is a very good driver and does little courier jobs now and then. He now has enough experiences to have his own résumé.

Did he ever get to go back to school?
Yes, he did homeschooling and earned his high school diploma (after trying 3 times to get his GED). He graduated in October of 2002. I had a cap and gown ready with plans to celebrate a long and hard, but rewarding goal achieved. I am very proud of him. He has talked about taking one class at a time at one of our local community colleges.

One of the other lessons we have all learned in our household, is that T.G. does better not living by himself. We would like to try a group home, but because his diagnosis was not made before he was 18 (what difference should that make, right?) he is not eligible for group home placement through any agency that we can find, as of this writing. So we will try one of the options given in the *Housing and group home* section *(See page 315.)* to create a group home. Too much togetherness is not always good, and all young people like a little freedom from being under their parent's thumb.

We tried four times to mainstream T.G. into a life on his own, as recommended by his doctor, but it has not worked well. He did not eat the way he should, he spent too much money impulsively and his personality is so easy-going that people (strangers and acquaintances) took advantage of him. We, as his parents, end up picking up the cost—literally. We are not rich by any means, just the average "working Joes" and so when we had to come up with sums like $1800 to pay off someone else's debt because T.G. was conned into getting involved, it was a problem. T.G. has admitted that he feels better living with someone, anyway. It is safer, less lonely and he no longer yearns to be on his own, at least for NOW, that might change any hour. He has stated that he would like to try a group home in the future, but as I mentioned before, unless we try to start one up ourselves, I don't know how that can happen.

T.G. often says he wants to find someone to love and get married. I have mixed emotions about that desire, at times, not that he wouldn't make a good mate, but that he will not find someone who can truly understand him. If not, it would make his life even more confusing. He is a great kid and if he finds a gal who understands, it just might work. He is very patient, a good friend and willing to share all things! Some people may ask, "What could he offer a wife?" Yes, I can see where that might be asked, after reading this diary, and I agree it would be hard for them; but I believe he has as much to offer in a relationship as many of the men I have known who didn't have VCFS. So if he finds his kindred spirit, I will be happy that he is happy and has found someone to love and be loved in return.

Update on TG 2005-2010

Since I wrote the book ***Missing Genetic Pieces*** in 2004. My son TG, now 32 yrs old, has been through some trying times. He has been hospitalized twice for mental illness issues and continues to have extreme impulsive and indecisive difficulties.

At one point, he had applied for Section 8 housing 6 yrs previously and wouldn't you know it, the very first month he received his voucher and moved into his new apartment he decided to take himself off his medications. Since he had been doing well, he reasoned he didn't need them any more. (This is so common, they don't realize they are doing well BECAUSE of the medication). TG then had a severe episode of mania with Psychosis and he believed there was someone trying to use VooDoo on him at the apartment complex and he took a baseball bat to the walls to silence the noise he believed he heard. Someone called the police, the police found him with the bat and tried to talk him into surrendering it, only to find he believed they were soldiers from an enemy country and fought with them (so much for encouraging him to watch the news). The police, not understanding the situation, tackled him down and tazzed him twice. Cuffed him and took him to jail. This led to one of his hospitalizations to stabilize him, and legal problems.

We went to court and fought the 4 felony charges (resisting arrest). They reduced it to a misdemeanor and he was put on probation. He is recently off the probation and so I hope he has learned a valuable lesson. Prior to this, I had asked many times for his physician to request court ordered medication but they didn't believe it necessary. But mother knew that when an authority demands it of him, TG will comply . . . finally with this episode, we got the court order we needed.

However, I had to fight the local State mental health organization to give him services for the mental illness, as 22q/VCFS was not recognized as a qualifying disorder and they tried to deny him on that basis. We won the case and so he is now getting services such as SMI, SSDI, ADA transportation, monitored medications and so forth. But he lost his sec. 8 voucher and he ended up back with us. (Remember, unless you get your child labeled and in the system BEFORE 18, you may miss out on many services such as group home housing, etc.) So this leads to the next adventure . . . .

A couple of years ago. TG was taking a medication called Zyprexa. Even though the medication worked well, he developed diabetes from it and he was in a class action suit. The case was won and he was awarded some money, which was put into a trust that was set up before the suit had settled. This type of trust is called an IR TRUST. (This is different from a "special needs trust" the IR TRUST can be used when you are still alive. The Special Needs Trust is for the time you finally get to rest--permanently.) It is good to remember that if a person has anything of value over $2000 the government can disqualify them for benefits. So the IR TRUST is necessary if you are to use it while you are still alive to disperse for what the child needs NOW (other than food, shelter and clothes). With that money we were able to make a substantial down payment on a small home not far from us, so we can have him on his own with a roommate and still monitor him.

This kind of arrangement makes it MUCH easier for TG's frequent visits (now that he doesn't have a car) and for us to do the maintenance and chores he does not know how to do, can't do or refuses to do, for himself.

In the mental health clinic he goes to, they found him an older roommate who was compatible in personality and they lived together for several months. So far so good, until TG cycled again and it scared the roommate and he moved back with his parents. TG took too many pills one night and almost OD'd on them. (He is now on monthly injections.) He now has a different older roommate.

It was nice to have the roommate, except that he was not as high functioning as TG and TG found he was the one doing the cooking and cleaning (what cooking and cleaning got done). The roommate's family literally dropped him off and had not called or showed from that day forward, until he moved out. Because of this, we had the added responsibility of the roommate as well. Before the roommate TG would barely go out of the house (like a little hermit) and didn't want to do ANY thing on his own. So having a companion really did help to get TG out, willing to go grocery shopping on an assigned weekly shopping day with his dad, etc.

TG has become antisocial over the years and lacks motivation to do most anything. I don't know if it's from the medications, depression, his diabetes that he refuses to keep in check, all of the above, or what (I suspect all of the above). All I know is, it can be very trying. My guess is some of the problem is severe ADD without Hyperactivity. One only need read the symptoms and it is like reading his chart.

TG will get into a cycle for a few days each month where he wants to get a job and go back to work. He will reason clearly, be so full of energy, to the point that he doesn't know what to do with himself. But then the mood will change and he won't remember that excited feeling and the day he is to start a new job he will have a bad day and not show up. I try to explain to him what happens each month and that there is a pattern, but he assures me it is going to be "different THIS time" . . . I used to encourage him to work, but now I tell him that he doesn't have to work and in fact could mess up his services which I worked years to get for him. I tell him that he could do volunteer work or something he enjoys to fill his days. He will understand for that moment, but by the end of the day, it is like the movie *Groundhog Day* with Bill Murray. If you have never seen the movie, please do, then you will understand what you may be facing in the future . . . over and over and over again!

For several months TG had been corresponding with a gal from the Philippines on the Internet. (She has now become my PenPal.) I wanted her to know and understand the situation with TG. So that neither of them would get their hopes up if she didn't want to care for a person with this type of illness, or if she can not get a visa to come here to the USA in the future. She seemed to understand and wanted to be able to be there for him, she even started school to learn to be a caregiver. But CAN she really understand? I think not. But I don't have to worry about that any longer. While in one of his odd moods (my husband and I were on a RARE vacation) TG came into our house to use the computer. He got angry with the girl for some reason (I suspect out of jealousy), he called her vulgar names and sent her dirty pictures. I had NEVER known TG to be so vile and rude. I was shocked to say the least. He has since apologized and they are once again friends. I now know that the computer is toxic to him. And he seems to realize it himself, so no more computer for him, unsupervised. He said he didn't feel that way until he had been on a couple of chat rooms and girlie sites. These had people with filthy minds and mouths and he started to imitate the behavior and become agitated, which made him cycle. The environment and associations can play a big part.

TG was married for 2 years, from 2005-2007. I am happy to say, he was a good husband. She was a gal he met on the Internet from Mexico and was not the right complement. She was a user and ended up an abuser. She took his car in the divorce, forged his name to get credit cards and ran up enormous bills. When I bought them large gifts like furniture, a new sewing machine, dining set, etc. . . . she sold them cheap to her family or friends in Mexico for cash to spend on silly non-essentials, then asked me for money for necessities. When I put a stop to it, and found out what she was doing, she decided I was the enemy and tried to keep TG away from me, trying to turn him against his whole family, but especially me as I held the power to tell her no.

When I found out that she was mentally abusing TG, it was time to step in. I asked him if he was happy. I reasoned if he was happy, I wouldn't interfere. But then he said he was miserable and told me what was going on. He asked if he could divorce her and get away from her. I said "by all means" and so this was done, post haste. He has done well after that ordeal and we have once again become close. But I can still see the influence of the time spent with her and he now seems to be afraid of commitment. He is suspicious and unsure about women. But still seems to enjoy the ego head games and nudie photos with girls on the Internet . . . to my dismay!

Since the last writing in 2004, I have developed many ailments myself. I believe much of it has been due in part, to continuous long term stress. My doctors believe this as well. (This is why TG does not live with us.) I have been battling cancer for the past 2 yrs. Have had 2 surgeries, I have an aortic root aneurysm, leaky valves, fibromyalgia, arthritis and diverticulitis, with a couple of minor things thrown in . . . so needless to say, I find it more difficult now to handle all the ups and downs in TG's life, especially in the mental health dept. I have had to turn over many of the tasks I used to do for him to my daughter (like the accounting) and his father does all the other stuff. While his sister Tish is a responsible and reliable person, she is not a push over like mom and we find the kids can butt heads on occasion, usually over money. He has no concept of what things cost, what he spends or that there is not an endless supply of it.

I'm happy to report, that since Tish has taken over her brother, my stress level has declined 50% or more. So please watch your health and give yourself some much needed preservation time on a regular basis. (It is not selfish to do this.) If you know of someone who needs a break, please be there for them. It's hard for someone to ask. Even if we think we can't afford to now, we pay dearly for it later. Know it is alright to delegate to others who are willing and able.

I have been concerned for a long time about TG spending so much time alone. Even with a roommate, this seemed to be the case. The roommate watched TV all night and slept all day. So TG would be up all day by himself. He had even developed a habit of talking and laughing to himself. It was very bizarre and unnerving, we tried to bring his attention to it. He would just laugh and think it was funny. UNTIL . . . the old roommate started doing the same thing.

One day TG asked me if he looked and sounded that weird too and I said, "Yes, that is what we have been trying to tell you." Needless to say he has stopped the habit. It has been good for TG to have had to care for someone who was less functioning than himself. Getting a "dose of his own medicine," so to speak. TG really does want to have a friend and companion. I believe this really is the key to helping our kids be happy and find a way to eliminate loneliness. Hearing this scenario play out in so many of our kids lives, is one of the reasons our Organization at **www.22qCentral.org** has a long term goal of trying to help bring parents of older 22q young people together, to set up group housing to support them and in turn will support us as parents. But like all organizations, we have to take baby steps and it takes money.

A few months ago we were informed of a city program that has dances and refreshments for the disabled on Friday nights every other month. TG and his dad went to their first one about 3 months ago. (TG would never go by himself anywhere and he almost backed out, as he usually does on most things when he doesn't want to leave the safety and security of his house). But he went and I was happy to hear him say, he had a good time and MAY go again. Now he's going to a new international organization called "Best Buddies." I keep being optimistic!

In the past 12 yrs. We have tried to get TG qualified for many services that he would have qualified for if he would have been diagnosed with DDD (Developmental Delay Disorder) or some other qualifying disorder like seizures before age 18. we have tried to reason with the authorities of the programs, showing that he was born with 22q11 and so he would have HAD it before 18. It is not his fault that the physicians did not understand he had a disorder. So we are still at it, chipping away at barriers. But you know, little by little we are seeing great things happening for 22q and this makes my heart joyful!

Sherry Gomez, Mother, AZ

Lauren's story

Lauren was born on a beautiful white December morning in 1992. It was the perfect day for a baby to make an entrance into the world. With the help of modern medicine I was able to stay awake and watch the birth of my first and only child.

She was a healthy looking 8 lb. 2 oz. little angel, so there was no hint that anything was wrong. In the first year she seemed to be developing just fine; eating, sleeping, growing, etc. She was always on the chart. It was in the second year of her life that I started to notice that something might be wrong. Although never by a large amount, she started to slow in her growth pattern. Now, as I look back, I can see that there were many pieces of the puzzle on the list of VCFS symptoms that were happening in front of us; but we did not know what we were seeing. We were clueless as new parents. I remember now seeing little oddities; but I dismissed them, thinking I was being too critical as a parent. These were things like small eyes, and thick, unruly hair, just to name a few. Was I being overanxious as a new mommy? Or were the symptoms becoming less subtle? I decided it was time to see the doctor about her slowing growth and I was assured that she was fine since neither her father nor I are very tall. But as time passed the growth charts became more important and with the increase of infections I really started to worry. On one of the doctor's appointments I just knew the doctor was going to see a big difference from the last visit and there was going to be something wrong. (I felt it in my gut.) I wanted to cancel the appointment; I wasn't ready for any more bad news. I had written down a list of the things that my husband, his mother and I had noticed. We just knew there must be an answer, one I felt I didn't really want to hear. Lauren was now almost 4 years old and just about to enter preschool. The normal tests were run, which all children have to enter school, BUT along with those, there was a blood test taken and it came back that she had chromosome 22q11.2 deletion.

I have a hard time putting into words how I felt or what happened next. It was as if someone had sentenced us to life in prison for a crime we hadn't committed. In a way I guess they had, as life has never been the same since and never will be again. After visiting several specialists and geneticists, we still walked away with only a few photo copied papers of information on VCFS and the comment that Lauren had a chromosome disorder that is fairly common. COMMON? Then why wasn't there any more information on it, or a support group in our area set up to help with the shock to us parents? We felt in desperate need for more information on the subject. We now know that we were grieving a type of loss. We didn't really know where to turn, on whom we could rely or even what questions to ask. It was my sister who happened to come across an article on genetics. I contacted the writer of the article and asked for some referrals to others who were dealing with the same thing, that of 22q. I was given the VCFS Education Foundation. Unfortunately they were aware of no one in our area who could offer any enlightenment on the subject or encourage us. We felt we had been catapulted into the *Twilight Zone*®, a strange world where we were unprepared, with a disabled child.

Lauren seemed to be susceptible to every germ and virus that came around; and therefore, many doctors and specialists would soon be in our lives continuously. When she was ill so much as a baby, I just thought it was natural, now it was scary. Yet it turned out to be a positive experience in some ways; we learned more about health issues, or was it just more acceptance?

Throughout all of this (It has been 6 years since we learned of her VCFS.) we have always tried to be truthful with Lauren as much as we could, with our limited knowledge. As she gets older her frustration is getting more and more evident. I hope as time goes by that she will be able to have a better understanding of VCFS and the limitations she will have in some areas, such as in her small stature and difficulties in school. She is still small even with hormone therapy. At 10 years of age, she is only half the size she should be. I am most fearful of the teenage years.

Our days of thinking of having more children have come and gone because I feel the gamble is not worth the chance of bringing another child into this world who may have the same deletion. As parents we are *ALWAYS* wanting to know; are we saying or doing the right things? I am saddened to say that after the newness of the diagnosis of Lauren wore off, many acquaintances forgot about us. Those who seemed interested, and many who were friends, started to keep their distance. I guess it is true, that only a very few people can really understand what we parents of disabled children go through; and those are usually the ones who have walked in our shoes. I have also learned that many professionals can appear indifferent; I believe it is because they do not want to be exposed to the fact they know little if anything about the disorder themselves.

We are very proud of our daughter and always tell her so. I now feel lucky to have her. She has taught me and her father a lot, including patience and love

Noah's story

Noah is doing very well for the most part, considering his start. He had major heart defects (Tetrology of Fallot and only one right coronary artery). He had feeding issues (reflux coming out the nose and textural issues when he reached table food age) but all feeding issues seem to be resolved. Although, occasionally if he throws up it will still come out his nose. We now know that this is related to VPI (velopharyngeal insufficiency) and this also affects his speech. We have learned it is possible to have corrective surgery on it, or Noah can learn to compensate through speech therapy. The speech therapy is slow and he is not progressing as we would like in that area, so surgery may still be the option we will take.

Noah has speech therapy 3 times a week, instead of the 2 times he used to have and even though he may be a candidate for a pharyngeal sphincter revision surgery, his palate is good otherwise—no cleft or other real issues that can come with VCFS.

His growth is slow but steady (5 yrs on July 4th; 42" tall; 37 lbs —about the 20% percentile).

His immune system is good. He is on his second set of ear tubes but with the last set they removed his adenoids. This has made a huge difference; only two infections in the past year. Even though I am discovering that they interchange VCFS, 22q11 deletion and DiGeorge; Noah has none of the immune/calcium issues that go along with DiGeorge only.

We also had Noah evaluated for ADHD and he ranked in the severe category for *Attention Deficit Disorder*. (Unfortunately, the therapist was not familiar with VCFS or 22q11 deletion.) We have found a vitamin/mineral drink that helps Noah tremendously but we also started him on a low dose of medication. With the two combined, Noah is now doing even better in pre-kindergarten and his teachers are starting to see all of the potential in him. He is a smart boy and can learn, if we can just hold his attention long enough for him to do so.

Noah sleeps well during the night and almost always has, except for foot pain that seems to come in cycles. I know this is common for this deletion, but sometimes I think this must also be linked to growth spurts as well. These kids get the double whammy. We have had arthritis and other orthopedic issues ruled out.

Noah has had surgeries for a hernia; strabismus (crossed eyes); urinary reflux, and at two weeks old he had a brain hemorrhage, developed hydrocephalus and had to have a shunt placement. He has since had two shunt revisions. I know this sounds like a lot; but really considering, he is doing much better than many.

Developmentally he is about 18 mos delayed. Noah is in pre-K and so far doing well with resource help. He also gets OT help as well.

We live near Atlanta, Georgia and actually lived in Atlanta when Noah was born and have been very fortunate to have Children's Hospital near by. They are 2nd in the country in pediatric cardiology and are following CHOP closely in genetics and keeping up with new developments.

Tracy, Mother

Alex's story

Alex was only diagnosed in August 2002 at the age of 15. He had a difficult birth and was in intensive care for 5 days afterwards because of an apnea attack. However, he was the largest and healthiest-looking baby in the ITU having weighed in at 9lbs 10 oz. These are the things that Alex has had to cope with over the years in rough chronological order:

What?	**When?**
apnea attack	new born
small hole at base of spine	neonatal
noisy breathing problems	1 month to 18 months
only a few hyper-nasal words	2 years old
speech therapy	2.5 –9 years old
grommets in ears	3.5 years old

long-lasting mucous in the chest, colds that made him vomit	when 5 years old, that one notable episode went on for weeks-on-end
occult sub-mucous cleft palate & VPI	tentative diagnosis at 3.5 years, finally confirmed at 5 years old; operated on twice aged 6 & 9 years Heinz pharyngeoplasty and muscle replacing; able to make all English sounds in isolation; connected speech still sometimes difficult for those who are unfamiliar to follow
mildly obsessive	4 – 6 played only with model cars in reception class; talked frequently about different sorts of windshield wipers on cars, lorries & buses
lack of stamina	always (5–7 would sit down on football field when ball was at other end of pitch, still loses all his energy suddenly after exertion)
aneuristic bone cyst on fibula	leg broke when kicking a football aged 7; cyst removed, non-weight bearing bone has not rejoined but no real problem now
frequent ear infections and deafness	ongoing
began to fail at school	aged 8; put it down to continual deafness; frequent ear infections and perforated eardrums

changed school (3 tier system)	13 – more caring school boosted his confidence, but still not performing as we expected
difficulty swallowing	prompted FISH blood test for VCFS. Pediatrician expected negative result but the result was positive; explaining all the health issues that Alex had and his learning difference
possible arthritis	The painful finger joints are a serious problem because Alex is an accomplished guitarist

As you can see from the above, Alex had problems from birth onwards but these were generally not sufficiently severe to prompt a diagnosis of VCFS. Consequently we took each thing as it came as another example of *"bad luck"* and dealt with it in isolation. The fact that we were at one time traveling to hospitals in different cities for consultations on cleft palate, the bone cyst on his leg as well as ENT treatment, never added up to a coherent diagnosis in anyone's mind until Alex started having problems swallowing. Initially, we put this down to the pharyngeoplasty; but the surgeon concerned said it was more likely to be a genetic problem. This eventually led to the blood test that confirmed his diagnosis of chromosome 22q11 deletion.

Alex has been a talented musician since the age of five, and an interesting, extroverted character. From aged four he always started planning his next birthday party from the day after his previous one and always had lots of ideas about what to do. Friendships were not a problem when he was young but became a bit thin between eight and thirteen. He now has a small group of good friends who appreciate his humor, his skill as a musician and the fact that he is a reliable friend. He also has strong bonds with adult family friends.

Alex recently made the decision to ignore the ear problems and go swimming regularly in order to catch up on a skill that he largely missed out on earlier. He has just achieved an adult swimming certificate level 2. His confidence in himself and particularly in his musicianship is growing; but he is not achieving well academically. It would be great if he could earn a living as a musician but the recent appearance of swollen finger joints, which might be arthritic, has put this in doubt.

Nevertheless, I expect Alex to continue with his education for another couple of years, and by the time he leaves school, to be able to earn a living for himself in one sphere or another.

Physically, Alex is 5'10" tall and still growing. He used to be a bit overweight but this has fallen away with puberty and he is a slim, athletic-looking figure. He is very flat-footed and this is reflected in his walk, although inserts in his shoes have helped a little.

His speech is far clearer than it was before the pharyngeoplasty and sounds particularly clear on the phone. His heart was said to be fine after the initial diagnosis, but I do notice the very blue extremities indicative of poor circulation. I realize that Alex has been very mildly affected by the deletion compared to some children who have the same overall diagnosis. Compared to a healthy child, he has had a lot of problems with illness and some problems with learning. However, he is independent and fun to be with. He has strong feelings about moral and political issues and has a very creative streak. We are lucky people.

Phillipa Partington, Mother, UK

My son Sam, living with VCFS

Our son Sam—born in January 1999—has VCFS. At birth we had no idea that there was anything wrong with Sam and I suppose it was just by "luck" that we found out he had any problems.

We had gone on holiday in September 1999, and within 30 minutes of arriving at our hotel Sam had a fit (seizure), something he had never had before. It really freaked us out. We spent the whole week of our holiday at the hospital in Intensive Care (ICU), where they diagnosed a heart problem, but could not be specific as to what it was. We came back from holiday and went to Great Ormond Street Hospital. Sam was diagnosed with Tetralogy of Fallot and he was operated on—successfully—in November 1999. Sam had other problems including occasional fits, development delays, sleep problems, etc. After further investigations by a number of specialists, our cardiologist said that Sam had VCFS/22q. This was in April 2001. No one seemed to know much about VCFS and our biggest frustration was the lack of cohesion between the specialists to give us an overview of Sam's condition and future prospects. Sam now attends a VCFS clinic along with the other specialist clinics he has to visit to eliminate or treat other conditions associated with VCFS.

Sam was 5 in January 2004 and he is a kind, lovable boy. He is delayed in his development; particularly his speech and his toilet training. He displays many of the symptoms associated with 22q; low muscle tone, obsessive personality etc. Sam attends a *Special Needs Nursery* 2 days per week and a mainstream nursery 3 mornings per week. There does not seem to be much knowledge about VCFS/22q and this is reflected in the level of treatment available in the UK.

As parents to Sam, it is still difficult to come to terms with what he has been through and more poignantly, with what he faces in the future. No one really understands the feelings you have unless they are in the same position. Even our parents seem to have the idea that *"everything will be all right"*; when we know that it is going to be a struggle to even get Sam into mainstream secondary school. As an immediate family—there is also Katie aged 7—we are very close knit and enjoy doing things together. Sam really looks up to Katie, most of the time! They play well together.

We hope that Sam can enjoy a full and rewarding life and we will love and support him all the way.

By Anthony & Marian (parents), London, England

Daylan's story

Daylan Anthony was born on March 11, 1980. I was a young single Native American mother and had more than common issues to deal with and understand than most people I've known. My culture, my situation and the reality of my child not being _normal_ were very difficult to handle.

He was a beautiful little boy when he came to be with me. He was one week premature and weighed 6-lbs. At birth the doctor informed me that he had what was called an imperforate anus and would have to have surgery to create an opening for him. That was hard to deal with and understand all by itself. I was also told that the birth had caused him to have an umbilical hernia. This could not be repaired until he was older, because his development needed to progress before they could repair it. At two weeks old he began experiencing an unusual high fever. The fever was 103 degrees and there was no explanation for it. He was placed in the hospital for observation.

My father was a _medicine man_ and began to see these issues as something other than what it was in reality. The situation was blamed on superstitious beliefs; with no thought given to the fact that it could be something other than someone else's fault or a spiritual doing.

While Daylan was in the hospital he began to have difficulty eating and keeping his food down. He was diagnosed with gastroesophageal reflux, and an allergy to milk. He exhibited symptoms such as: nasal regurgitation, nasal vomiting, feeding difficulties, irritability, and chronic constipation. He was labeled as experiencing _"failure to thrive"_. He could only drink soy milk, and had to be placed on a reflux board, a board at an almost upright angle to sleep on; or he needed to be held for at least a half an hour upright after every meal so that his food would digest and not be brought back up, causing him to vomit or choke.

He was a true fighter. He strived with difficulty and overcame many of his early feeding symptoms by the age of 8 months old. He fought hard in his early days only to face more difficult issues as he went along.

My family situation resulted in Daylan staying with my mother and father for a time when he was young. I saw him on an almost daily basis and I could see that there was something wrong with him. I had another son not long after and he began developing at an enormously faster rate than Daylan. They almost began going through their developmental stages at the same time although they were a little over a year apart. My second son soon passed Daylan developmentally and new struggles began for all of us. Daylan's early development was at least 8 months behind. He didn't walk until he was 18 months old. He never did learn to crawl. As a result of the delays and over indulgence on my father's part, at the age of 5 years old, Daylan was unable to dress himself, feed himself and did not speak clearly. His comprehension was very limited and this made it difficult for him to communicate with us. Daylan returned to live with me and his brother and sister. At this point he received the diagnosis of another umbilical hernia. They repaired both of them at the same time.

Daylan has had dental problems since he began getting teeth. He had very small teeth and was later found to be missing his back molars. The dental issues did not result only from the VCFS but also because of lack of dental hygiene. Daylan refused to take care of things that should have concerned him. Through ages 5 to 13 he refused to even acknowledge the need or the necessity of personal hygiene. He would have difficult behavior issues frequently. It would be a major endeavor to get him to bathe, dress and pick up after himself. I know that most children have difficulties, but Daylan's were much more severe than my other children. He would act out when he was in the safety of his own home but not anywhere else so no one would see it. There were times when he would get upset and isolate himself for days without eating or even coming out of his room. The only thing I could do was just be available to him. I did not know what to do.

The symptoms of *velo-cardio-facial-syndrome/22q* have the same physical features as *fetal alcohol syndrome*. Although I had not used drugs or alcohol during my pregnancy I had used before my knowledge of the pregnancy. I was made to believe that I had caused Daylan's condition and that compounded the emotional weight I carried.

As he began growing he was a wonderful little boy. He was cute, funny and had a wonderful personality. He got along well with his younger brother. Actually his younger brother would take care of him when they were at school or out playing. No one could actually tell there was a disability right away.

He showed great intelligence but lacked in comprehension. It is a very difficult situation to get others to understand that there is a difference in the mind functions of both. His learning difficulties began at birth but it would take me a lifetime to convince others that they were there.

He began to go to school when the *special education* and the *Individual Education Program* systems were just beginning. I could probably write a book on the experiences we had all by themselves about the situations we dealt with through his early school years. No one understood what an *INDIVIDUAL EDUCATION PROGRAM* even meant let alone how to implement one.

One young teacher who had convinced me to allow Daylan to go to public school, rather than homeschool; got fired because he went out of his way to teach me all about *special education*, the laws, and my rights as a parent. There were other teachers who would become very demanding about the fact that they were *NOT* going to give my son any special considerations in their classrooms.

Daylan struggled with his learning disabilities, incomprehension, low IQ, and a mental retardation label throughout all of his school years. Although there are thousands of people who deal with having a disabled child, I have spent the last 22 years trying to convince the world, and my family, that there really is a disability in Daylan.

If you were to see Daylan on the street he would appear *normal*. You would not be able to tell there is anything wrong with him. His physical symptoms are only apparent to those who are trained to see them.

The *fetal alcohol syndrome* label stayed with us until he turned 13 years old and I was finally able to convince the *Department of Developmental Disabilities* in the state of Washington that we were dealing with something undiagnosed. We were sent to the University of Washington's Fetal Alcohol Syndrome Clinic in

Seattle, Washington. He was evaluated at that time and fit the symptoms of VCFS/22q beyond a doubt.

When Daylan finally received his diagnosis of VCFS, assistance from the state disabilities offices became easier to obtain. Daylan's behaviors began to worsen. He was not overly aggressive or physical with me. He would just refuse to do anything. His *"shut down"* episodes began to become more frequent and it was harder to get him out of them. I would have to physically take Daylan and put him into the car to go to school. He would become very angry and not want to go. I would have to take him out of the car and force him to walk to the door of the school. When the principal met us at the front door, Daylan's entire demeanor would change, and he would act like a perfect angel. He would be very good until he got off of the bus and then his behavior would return or he would isolate himself and appear very depressed. I finally learned that he didn't like to be in a place where he couldn't understand what was happening.

The special education programs always said that Daylan did an excellent job in school. When I would talk with him at home his understanding was totally confused. He was able to answer any question they wanted but could not transfer the information into reality. They had attempted to teach him to count money and tell time. He could answer their questions on the little worksheets they gave him with the pictures; but he had no idea that it was the same thing as the  money we spent at the store or the clock on our wall. He was getting straight A's but not learning anything. Transferring his information to reality was just not possible. He had to be taught with the real thing and in the real place. He is still that way now. I never experienced a special education program that succeeded in meeting Daylan's individual needs.

In 1993 we were allowed to have a personal care attendant come into our home to observe and help me deal with Daylan's behavior. This was not only for Daylan's benefit but also for my own mental health. We were linked up with a behavioral therapist to come into the home and set up a behavior modification program to help Daylan deal with his surroundings in a more effective manner. It was the first time any other person had experienced Daylan's severe behavioral issues and someone finally believed me when I said that we needed help. We were then allowed to have an in-home behavioral therapist.

Daylan has had a sexual fixation and obsession since he was around 4 years old. I believe part of it is a result of some early abuse, but not entirely. I would even venture to say that the abuse might have just opened the door to his obsession. Whenever Daylan tries something new and he likes it, it can evolve into an obsession instantly. As he has gotten older his sexual obsession has gotten worse and more difficult for him to control.

We tried everything we could think of to help him with it. We could not find a caregiver who would deal with this issue or help us understand it.

The behavior modification program was put into place to help Daylan deal with daily living issues. I initially requested the program for the sexual obsession, but the therapist began dealing with the daily living skills areas first. The personal care attendant implemented the program for the first 30 days so I could see it working and learn to implement it myself.

My mental and emotional condition was not good. I had been in counseling for my own issues with a dysfunctional family and dealing with Daylan, since 1990. Together Daylan and I began to modify our behaviors. The program was put into place and Daylan and I were learning new ways to live together more effectively. It changed our lives.

The in-home behavior therapy program was very successful in providing Daylan with the very structured environment that he needed to be able to learn and understand how to function within our family. He worked very hard and made tremendous progress.

With his daily functioning improving, after 8 months we attempted to address the sexual obsessions with no success. The therapist basically said he couldn't help and gave up. We never found anything that would work. We had another behavior therapist a year later who was supposed to be an expert in that field, but there was still no success.

Getting the diagnosis and finally finding out that there was an explanation for the difficulties we were having, was a tremendous relief and I began to take a whole new angle of dealing with it. I began to become educated on *velo-cardio-facial-syndrome (22q)* and the reality of advocating for an individual who will never have the capability of living an independent life.

Daylan's dental problems began to worsen and he would not even allow the dentist to examine him, he was so scared. All of Daylan's dental examinations and repairs had to be done in the hospital as an operation.

Daylan began middle school. He was unable to be mainstreamed into regular education classes because of his low comprehension. The school did allow him to play baseball. We took him to have a physical and an inguinal hernia was discovered. He had his fourth surgery. After surgery, he went on to hold the record for the most stolen bases for the team.

I thank God every day that Daylan has not had to deal with the physical difficulties that VCFS/22q can produce, but we deal with the totally different reality of having to convince people that Daylan lives with a disability and needs to have assistance in a wide range of living skills in order to live a reasonably normal life.

Daylan graduated from high school in May of 2001. He was 21 years old and still does not understand enough to function completely on his own. I spent the last three years of his school years trying to convince the school that they should be getting him ready to live on his own not asking him where China is.

Daylan was, until now, in The Redfield Developmental Center in Redfield South Dakota. His obsessive compulsive interest in sexuality got him into a situation. It did not look like I would be able to advocate him out of it. I needed help from someone who was knowledgeable about VCFS. If we could not succeed in proving Daylan's condition he could have been put in prison for an Internet violation. I mention this because so many of our children might be held accountable for their actions; yet they are actions that they don't even understand. VCFS *MUST* be recognized in our courts for our children to get help and justice.

I am concerned about Daylan's ability to deal with these situations. He has always struggled with the obsessive compulsive disorder, impulsiveness and depression. He functions at the age of around 11 or 12 years old psychologically and emotionally. We have worked so hard to get where we are today but when I stand in the courtroom and hear them say my son belongs in a penitentiary for something he doesn't even understand, I wonder if I have gotten anywhere through the years.

I do know one thing. I would not be the person I am today if I did not have Daylan. I dedicated Bette Midler's song *"Wind Beneath My Wings"* to Daylan when he was 12 years old and I still believe it today.

Fran, Mother

Brynlee's story

Brynlee is my sweet baby girl and we have dealt with feeding issues with her from day one.

She had severe reflux and would choke and turn blue after most feedings for her first four months, even though she was being strictly NG (nasal gastric) tube fed. We had MANY LONG days and nights of no sleep, major stress and anxiety as first time parents. We finally found a great doctor (in our opinion) and trusted him when he told us that surgery could correct the reflux and would greatly benefit our child. Out of pure desperation we chose for her to undergo the surgery at four months old. We are not sorry we went that route. The choking and blue episodes stopped immediately and we were so relieved. (She also had the G-tube placed in the abdomen during the surgery).

The things we were told by GI doctors and dietitians at the time were that Brynlee would not eat a _normal_ meal by mouth until she reached kindergarten. This either motivated me or really ticked me off because once I heard that I was determined to try for her second birthday. The first several swallow studies showed aspiration and food pooling in the back of her throat and it also took many swallows to get the food down. At her 9 month swallow study it no longer showed aspiration but still some pooling with a weak swallow. We were also told that she had VPI and this is why she suffered from slight nasal regurgitation.

As soon as I heard that she was no longer aspirating and that we could start offering her tastes of food, I really went for it! I was scared and nervous at first, but thought, *"If I could just get her to swallowing, giving her the practice, she will improve."* So everyday several times a day, I offered her drinks from a bottle and bites of food. She did choke and liquids would come out of her nose, but within a week this really improved. Within a few weeks she was hardly choking and was not as reluctant to try bites or sips. She actually seemed to enjoy eating. I am proud to say that by 11 months old she was eating! I mean chewing, swallowing and hardly ever choking. By 1 year old she ate _normal_ meals and would drink like it was never an issue. The nasal regurgitation slowly disappeared and we soon forgot that she had ever struggled with an eating problem.

I have to say we are still thrilled today to watch her eat; it still makes us smile and shed tears of joy too! We dealt with 24 hour continuous feeds for six months and then bolus feeds during the day with continuous feeds at night until she was 10 months old. At 10 months old we slowly cut back on the amount of bolus feeds in hopes that she would feel hunger. It seemed to work and by 12 or 13 months old we stopped the tube feedings. She is 18 months old now and we have not supplemented her at all for the last five months. She gains weight slowly and weighs 18 lbs 4 oz's. (About the same as I was at that age). She enjoys so many foods now and drinks from a regular cup, from a straw and a sipping cup. It was kind of hard weaning her from the bottle because she struggled with the faster uncontrolled flow of a sipping cup. Now she eats three meals a day with some snacks in-between.

Trust me. It was not easy. We worked with her and did not quit. We knew that the sooner she ate by mouth the easier it would be for her to learn. Being diligent is what I feel got Brynlee where she is today. It was obvious that Brynlee really had to work hard to learn to chew and swallow. Just as it is obvious today that anything to do with her muscle tone is a more difficult task for her than most kids. (But, she did learn to eat at 10 months and to eat normal meals far before kindergarten.) We have set expectations for her far beyond what most doctors have set for her. We have learned that some MD's tend to give worst case scenarios and do not know the child like the parents do.

I realize they are just doing their job and trying to inform us about our child's medical issues. Yet, we have been very discouraged and frustrated with many of them. Many don't understand VCFS and they have a tendency to put fear in us as parents. We have relied on faith and our instincts, doing what we feel would benefit Brynlee in all areas. By the way, we did not know that Brynlee had VCFS until she was 10 months old. Regardless, we wouldn't have changed a thing we did. Brynlee's doing great and we have God to thank! He has been our strength when we needed it the most!

Denae Thompson, Mother

Jeffrey's story

Jeff was born November 23, 1992.

He wasn't tested for VCFS until he was 6 years old. He is now 9. Here are some of the things that Jeff has that are on the list of symptoms that are *typical* of VCFS:

Cleft of the soft palate: The cleft was discovered at birth, when we were referred to a plastic surgeon. Jeff had the cleft repair at one year old, a fistula repair done at 2, a pharyngeal flap at 3 and another fistula repair at 8. We sure hope this is the last for the palate.

Retrognathia (retruded lower jaw): Jeff was originally diagnosed as having *Pierre Robin Sequence* and *Sticklers Syndrome*. He had feeding problems at birth and the small lower jaw contributed to this.

Tortuous retinal vessels: When Jeff had his first eye exam this was noticed, but so far it has not caused any problems.

Sub-orbital congestion (allergic shiners): This is redness under he eyes. Jeff has had these forever it seems and since he has allergies we consider that they are a contributing cause.

Attached lobules (soft lobe or tissue): These were noticed by our plastic surgeon when he was about a year old. The only comment made was that they could be fixed. We don't have any plans to do anything about it at this time.

Frequent otitis media (ear infections): Jeff has had 2 sets of tubes and I am happy to say by 4 years of age, the infections had somewhat subsided. Now we mostly have sinus infections.

Narrow sinus passages: We have comments from every doctor that looks at him that he has narrow nasal passages, but then he also has narrow ear canals as well, which makes it difficult for them to look.

Right sided aortic arch *AND* vascular ring: When Jeff was 6 years old (after being treated for 6 years for asthma) our asthma/allergist specialist referred us to a pulmonologist in Omaha who discovered the right-sided aortic arch and vascular ring. Jeff underwent surgery in January of 1999 to separate this ring. It took another year following the surgery to get rid of the croupy cough. He still coughs a lot whenever he has a cold or sinus infection, but it is much better. Finding this heart defect is what prompted the 22q11 FISH test. Jeff was always very thin until after this surgery. We had never noticed that he also had swallowing problems until after the surgery and then he began to gain weight rapidly. Now I am happy to say, he is doing well.

Seizures: Jeff had a febrile (fever) seizure at about a year old from a high fever. He had no more seizures again until he was 8 years old. Then he had a seizure in the middle of the night and they did not discover what caused it. He was on antihistamines and we were told that medication could lower his threshold for having seizures so we try to avoid antihistamines at this point.

Mild developmental delay: Jeff was just a little slow at sitting, walking and things like that. We were just starting to get worried about something and then he would do it. He walked at about 14 months.

Upper airway obstruction in infants: This was caused by the small lower jaw *(Pierre Robins Sequence)* and therefore it improved as he grew.

Reactive airway disease: Jeff seems to cough *VERY* easily with the least bit of irritation but it's getting better every year.

Feeding difficulties and failure to thrive: Jeff was admitted to Children's Hospital in February of 1993 for *failure to thrive.* Originally they thought it was because of the *Pierre Robin Sequence*. Later when we discovered the vascular ring we realized that it, too, contributed to his feeding and breathing difficulties.

Nasal vomiting: Since we knew at birth he had a cleft palate, we expected this. When that was repaired, the nasal vomiting stopped.

Gastroesphageal reflux: Jeff had this until he was about 18 months old. It seems to have repaired itself.

Language impairment: Jeff did not talk until he was 3 years old. Once he started talking he never stopped. He did then and still does have a problem with following any complex directions. Sometimes you have to tell him over and over again before he gets it. I noticed it sometimes helps if we explain it in a different way than we did the first time.

High pitched voice: This seems to have been corrected with the palate surgery.

Learning disabilities: Jeff reads well and can tell you what he read out loud, but he can't seem to answer questions on paper. He aces all his spelling tests. Since Jeff is only in the 4th grade he is still in basic math facts and he seems to be doing fine. But we can already tell story problems are much harder for him.

Concrete thinking and difficulty with abstraction: This is hard to explain, but we have to tell Jeff *EXACTLY* what we mean. For instance, if the cat scratches the couch and I say, *"I'm going to kill that cat."*, he gets very upset until I explain that I am not going to actually kill the cat, I am just mad at him for scratching the couch. We have to be careful about what we say so that he does not take it wrong; and worry about it until we can straighten it out. Sometimes we find that he worries about something someone has said and we have to clear it up or he won't be able to sleep at night.

Borderline normal intellect: Jeff's IQ test scores fall in the lower part of normal.

Spontaneous oxygen desaturation without apnea: This condition was noted on his first admission to Children's Hospital in Omaha when he was 3 months old. It happened when he slept and ate, therefore for several months he would have to wear oxygen when he slept and ate. Originally it was thought to be the product of the lower jaw, but now we believe it was from the vascular ring.

Frequent upper respiratory infections/lower airways disease/reduced T-cell populations: Jeff was constantly sick from birth and still gets sick frequently. After the diagnosis of VCFS, they checked his T-cell count and found it to be low. He also had a low IgA (another antibody). At one year old he was on I.V. antibiotics for six weeks. We have hope that this will improve on its own.

Hypocalcemia: Jeff was found at birth to have low calcium. He was jittery. The problem was taken care of by giving him formula right away. Since then he has been tested and it seems to be normal.

Abnormal scalp hair: Jeff does have a lot of hair but so does most of my family so we don't know if that is an anomaly or just a trait that he inherited.

Thin appearing skin: This is a condition where you can clearly see though the skin. In Jeff you can clearly see the veins in his face and chest.

Robin sequence: As I stated earlier Jeff was originally diagnosed with *Pierre Robin Sequence* and *Stickler Syndrome*. He had a small lower jaw and a U-shaped cleft palate.

I would like to talk a little about how our family has been affected. I believe that things could have been different had he been diagnosed at birth. We spent the first 6 years or Jeff's life being told we were over reacting to Jeff's problems and that nothing was wrong with him but a cleft palate. Jeff was sick all the time. We were accused of not feeding him properly, not giving him antibiotics, etc. No one ever had an explanation for us. Our medical bills were staggering at times. Once he was diagnosed, our life changed. Finally I had the medical profession believing that I wasn't totally crazy. Though after six years of trying to figure out what was wrong with Jeff I probably was; crazy that is. Our medical bills are still large, but at least we are making improvements in our life and most of all in Jeff's life.

Our daughter who is 2 years older than Jeff had it hard. She was bounced around to different relatives while Jeff was in the hospital. There were things we couldn't do with her because Jeff was sick or because money was tight because of the medical bills.

In some ways our family is still affected by it. Jeff does not accept change well. We had a very hard summer this past year because I went back to work after being home with Jeff since he was born. He had a hard time sleeping and his behavior became a problem. He is adjusting nicely to it now, but it has been a long row to hoe.

I could never work until now because of Jeff's constant illnesses. I was glad to be home with my children, but missed working; especially because of the fact that we could use the money, due to the high costs of Jeff's medical care. I am very lucky, in that I work with people who understand Jeff's problems, and work with me. I only work part-time right now and when Jeff gets older I hope to be able to work full-time again. That may not happen for sometime, but I am happy doing what I can.

As far as healthcare providers go, in Lincoln, Nebraska, our new pediatrician is wonderful. We went through two other doctors until Jeff was diagnosed. I am sure the pediatricians who saw Jeff when he was younger would have done fine had they had the proper diagnostic information available. Children have just recently (in the last 5 years, to my understanding) been diagnosed regularly with this syndrome in our area, so, the only thing I would like to recommend to new parents is to interview a healthcare provider before you use them. See how knowledgeable they are about the syndrome or how willing they are to learn. Our doctor sometimes tells us he doesn't know and that is OK. He then refers us to someone else and we appreciate that.

Update on Jeff, 2005-2010

I never thought that I would need to write much of an update on Jeff's journey through the VCFS maze. We thought that our worries were just about over after Jeff was diagnosed with VCFS when he was six. We figured that by the time we had all his medical problems diagnosed and treated that our life would go back to normal. Boy was we wrong!

Jeff's 4th grade year started out pretty much like every other year. Up to this point, Jeff loved going to school and loved every one of his teachers. Going to school was never an issue. Jeff missed a lot of school because of his immune system deficiency, but when he was at school, he was happy.

When I heard other families talk about their school woes, I really didn't understand what they were talking about. Like I said before, Jeff was happy in school and he would always come home full of things to share with us. I knew something was up when he started not wanting to go to school in the morning. My first instinct was that he was probably getting sick and just couldn't find the words to tell us what was going on. It didn't take long to learn that his special education Math teacher was making his life miserable. She was keeping him in at recess and making him last in line for lunch as a punishment for not getting his math problems right.

I thought, no problem. I will talk to the teacher. Maybe she hasn't read any of the materials that I took to school. I made copies of his neuropsychological exam and copies of as much information about VCFS and the learning disabilities associated with the syndrome as I could find. However, I soon learned that she thought she knew more about Jeff than anyone. She told us that he was just lazy and that she was going to "fix" him.

I then went to the principal and she said she would help us. Prior to this, teachers always commented about what a nice young man Jeff was. They would also talk about how hard he tried and how much he wanted to please everyone. So, we were puzzled as to why things were so different. However, nothing changed. Jeff still was absolutely refusing to go to school and was very upset all evening. Doing homework with him was impossible. He was so afraid that he was going to do something wrong and get in trouble the next day.

Jeff was due for his three-year evaluation that year and I even remember her arguing with the school psychologist who said that Jeff presented just like the neuropsychological exam that had been done the year before in Omaha. This teacher said no, that's not what I see in the classroom. She then proceeded to say that Jeff "looked normal" and that she could "fix" him.

The final straw came when a report that was supposed to go to Jeff's psychologist came to us instead. This teacher had talked about how Jeff had a "learned helplessness" and all kinds of digs about what horrible parents Mike and I were. We met with the principal and asked that she no longer be Jeff's IEP manager or have any other contact with him.

However, by this time, the damage was done. We knew children with VCFS had mental health issues, but nothing prepared us for the storm that followed. Jeff would never trust another teacher at school again. He tried, but he had his 4th grade teacher in the back of his mind. Had we known what would happen, we never would have waited so long to give the school a chance to work things out.

The next four years consisted of major ups and downs in his treatment and I felt we would never have anything close to a normal life. In fact, at that point I didn't want a normal life; I just wanted a manageable life.

Jeff's fear of school blossomed and he began having fears of all kinds. We live in Nebraska, so Tornado season is a complete nightmare. We finally handled this by making Jeff a "fort" under the stairs in the basement. When the weather looks threatening, he gathers his favorite belongings and goes under the stairs until all is clear.

Being home alone really scared Jeff. I just started back to work a couple years prior to this period and I loved my job. He would call me at work multiple times a day. Thank goodness that I had such wonderful and understanding people to work with. However, it finally became necessary that I quit my job and stay with Jeff until we could manage his fears better.

For almost two years Jeff would not eat any solid food because he was afraid he would choke. This was complicated by the fact that Jeff was born with a right-sided aortic arch and vascular ring and had major feeding issues as an infant. We had no idea if this issue was anxiety and fear or if there was a real medical reason why he felt he was choking. We had to go through a lot of medical tests to make sure that he was able to swallow before we knew whether we could push him in therapy to eat solid food again. With help from his pediatrician, psychologist and karate instructor he is now eating solid food again. In fact, a little too much solid food!

Jeff worries about his health all the time. It is so hard to sort out what is a true medical problem and what is brought on by anxiety. Last summer Jeff complained of stomach pain and heartburn constantly. We chalked it up to anxiety for a couple of months and finally decided we could no longer ignore it. He saw a gastroenterologist and had an endoscopy. They could find nothing wrong. Again, it was anxiety. He still complains of heartburn from time to time and we treat the symptoms, but it isn't a constant thing anymore.

With the help of a wonderful team of doctors (pediatrician, psychologist and psychiatrist) and psychiatric medications, we have Jeff's anxiety under control. We know that can change any day, but at least our life feels manageable again. Jeff is now 17 and has never been able to return to school. He is being educated at home at this point. Just talking about school makes Jeff's anxiety rise to unmanageable levels. Our first indicator that Jeff is too stressed is when he starts picking his skin. He will have holes the size of quarters all over his arms and legs (and sometimes his face) by the time we start to get things calmed down again. Jeff also has had some OCD issues. They are a lot easier to manage than anxiety, but can really wear you down. When he is "stuck" on something, there is no stopping him. We have gone through a Hot Wheels phase, a Dinosaur phase, a World War II phase, a Star Wars phase and now a Resident Evil phase. His whole life revolves around whatever phase he is in. He wants everyone to know everything he knows about whatever he is interested in at the time.

He also has a need to check and recheck what our plans are for every day. We try to keep our schedules vague because he can't handle too many changes from "the plan". To make sure we are not going to change our plans, he will call Mike and me at work several times to check and recheck what we are having for supper, if we are doing anything after supper, etc. Routines will always be important to Jeff. We are trying to teach him to be more flexible, but I don't see that happening any time soon!

I feel like I've mostly talked about the problems we've faced raising Jeff up to now. However, there are so many wonderful things about Jeff. He is a very loving and caring person. He always makes me smile and I know I can count on him no matter what. I had someone ask me once "I'm sure you wish that Jeff didn't have this syndrome". I thought about it for a minute and I said "No, I wish he didn't have to go through all the bad things associated with the syndrome, but if he didn't have VCFS, he wouldn't be our Jeff". I couldn't imagine my life any different. I am better person because Jeff is my son.

Contributed by Anne, Mother, Nebraska

Richard's story

We were forced to deal with the physical consequences early on, because Richard was born with multiple serious heart defects, no immune system, and low calcium problems, etc. He didn't make it home from the hospital for the first two months of his life. My husband absolutely refused to look at or deal with any potential brain/learning/emotional problems. I had read a lot of the available information on VCFS, but 10 years ago there wasn't much around that was accessible. I soon realized there were significant speech issues. We were able to get Richard into a program here where we live, that I'm sure is offered in many cities. It's an early intervention program run on a volunteer basis, and provides speech as well as both gross and fine motor skills; a kind of physical therapy. I believe this has made all the difference for him.

When Richard was a year old, and still not trying to speak, I finally convinced them to help me get Richard into a special program; he wasn't making *normal* sounds even by the age of two. He may have known half dozen words, and his new therapist taught him to say the word *"more."* More was like a break through for him. It became *"mo d'ink."* His speech, although not clear, slowly started to develop. By the time he was 5 he was maybe just a year delayed, although this was with ongoing speech therapy; at least twice a week until he turned 7.

Now at 10, his speech is pretty good. There are pronunciation problems; and of course just complete gaps in what words go together, *IF* he even understands what the words mean. I don't know if we were just lucky, or if the early aggressive intervention we got him helped. I like to think so. Richard also had significant hearing problems, which contributed to his speech problems. The head of pediatric ENT at the Children's Hospital here were I live, is an absolutely phenomenal man, who I believe saved Richard's hearing. Since he had the sort of classic small, poorly formed tubes, I think this was a fairly critical factor in his speech development.

One final factor that may interest some people with very young babies, or if you encounter someone with a new baby; is that there is some support for the idea that breast feeding encourages speech formation. I haven't looked at this closely myself, but believe it has something to do with the mouth motion required. I know it is offered as one of the secondary support reasons for breastfeeding after providing a stronger immune system, and since Richard did nurse, I guess it could have potentially played a role.

Connor's story

I had a great pregnancy and delivery with our son Connor. He was our second child, and Connor's brother was only 20 months old when he was born. About an hour after Connor was born the nurse announced they were going to give him oxygen. I didn't think anything about it since they had done the same thing with his brother. Another hour went by and the nurse wanted to call in the pediatrician because he still didn't have good color after the oxygen. I still didn't panic because I had such a good pregnancy and delivery. After the pediatrician examined Connor he thought he detected a heart defect called Tetrology of Fallot, due to the boot shaped heart. That is when we got scared. The transport team took Connor to children's hospital and my husband and I followed soon after. He was taken to the NICU. He had trouble feeding from the beginning. I would pump my breasts and try to feed him with a special bottle. We would try nursing sometimes. On his 4th day Connor unexpectedly coded. There seemed to be no reason for his coding. They ran a bunch of tests and determined that it was due to reflux. We were careful to sit him up after feeding and we didn't have any other problems after that. He was released 9 days later.

Connor's first year was a mix of doctors' appointments and trying to keep him healthy. His oxygen saturation stayed low, and if he cried they went even lower, so we tried not to let him cry. I would pull over in the car, and we held him all of the time. Needless to say, he became quite used to that. At 9 months his oxygen saturation was in the 70's (one hopes for 100) and the doctors decided it was time to repair the heart defect. After the repair Connor did really well. He walked at 16 months; he was a good eater, and a pleasant baby. He had frequent ear infections, but nothing major. He was slower to speak, and was evaluated for speech therapy when he was 4. Again, it wasn't bad, but there was a small articulation problem. He was evaluated by the school's system, and didn't even qualify to receive speech. When he was 5 we went to our annual check up for cardiology. We had a new chief of cardiology who examined Connor. He asked questions, and I mentioned speech. He asked about his feeding history as a baby, and his frequent ear infections. The doctor wanted to test Connor for a syndrome and we gave our consent. A week later we received a phone call telling us that Connor had VCFS/22q. We were stunned. We knew nothing about the syndrome and all that we read about it sounded so horrible. It was hard to believe that our little boy had this terrible thing.

Connor is now 8 years old and in private school. He is doing pretty well. He receives extra help, and math seems to be the subject with which he needs the most help. He does get frustrated, but he does very well socially. He is a sweet kid, with a very caring demeanor. He plays baseball and loves sports. He is always riding a bike or on wheels of some sort. It is hard to say what the future will bring.

We know there will be problems with school and learning as the years go by. Hopefully, we will be able to keep him at the same school where he has so many friends and where his brother goes as well. When asked what my goals are for Connor, I just try to take it one day at a time. He has terrible leg pains at night and it is so hard to listen to his screaming from the pain. Other than that, he is pretty healthy. We go to the cardiologist yearly and are checking in with our pediatrician as needed. Connor is a happy sweet little boy and we work with him all of the time to help him keep up with his school work. For now we are just taking it one day, one step at a time.

Jeannie, Mother

Connor's teacher copied this and handed it out to all of the kids, I think it is worth sharing. (See next page.)

Don't laugh at me

I'm a little boy with glasses, the one they call a geek,
A little girl who never smiles 'cause I've got braces on my teeth.
And I know how it feels to cry myself to sleep.

I'm that kid on every playground who's always chosen last
A single teenage-mother tryn' to overcome my past
You don't have to be my friend, but IS IT too much to ask?

Don't laugh at me

I'm fat, I'm thin, I'm short, I'm tall,
I'm deaf, I'm blind, Hey aren't we all

Don't laugh at me
Don't call me names
Don't get your pleasure from my pain

In God's eyes we're all the same
Someday we'll all have
    perfect wings.
Don't laugh at me.

(Author unknown)

Kaitlyn's story

Kaitlyn Redding
Age at testing: 6 years old; Age now 8 years old

When reading this please remember it took me six years to figure out what was going on with my daughter. It seems that she has a much milder version of VCFS/22q than some others. At the end I am going to give a timeline because I had no idea what to look for until after the diagnosis. Also, there could be many things on the list of symptoms that she might have and I just don't know it yet.

Craniofacial/oral findings:
She has a crooked mouth, but it's only when she smiles or cries. The explanation I got when she was two is that the muscle just didn't form and it wasn't a big deal because it would not affect her speech.

Her palate also does not seem to work properly according to her ENT. The first time I heard this he said it would work itself out. Low and behold she is now eight and a candidate for pharyngeal flap surgery.

Kaitlyn has terrible teeth. When she was five she had every molar capped, two bottom teeth pulled and the others had cavities fixed. We work hard on dental hygiene but it seems many children with VCFS have bad oral problems.

Eye findings:
Everything here to date is wonderful. She just has the eyelids of other VCF kids and tends to get very red under the eyes at times.

Ear/hearing findings:
Her ears are cupped, very small and she has a sensory-neural hearing loss in both ears. She has very small and narrow ear canals which constantly trap wax and debris. This leaves us to fight ongoing ear infections. In fact, we usually have one ear infection a month. She has hearing aides in both ears.

Nasal findings:
She has the wide nasal bridge like many VCF children and very narrow nasal passages.

Cardiac and thoracic vascular findings:
I am fortunate enough to say that I am one of the lucky parents. My daughter to date has no known heart problems.

Vascular abnormalities:
She has small veins; they collapse easily.

Neurologic, brain and MR findings:
She has always had a developmental delay; mostly, in part, due to the hearing loss.

Pharyngeal/laryngeal/airway findings:
Her flap does not work properly and she is a candidate for flap surgery. She has recently, been diagnosed with asthma.

Limb findings:
She has tapered digits, and short hard nails. I can remember people commenting on how adult-like her hands looked when she was born and how people are always commenting how she will probably play the piano or basketball with her long fingers.

Problems in infancy:
Kaitlyn ate very little until she reached the age of 5 and now I can't stop her from wanting to eat. Under the age of 5 she was severely underweight and the school considered this a problem so we put her on an eating regimen of 5 little meals a day. In infancy she always had nasal vomiting with every bottle.

She suffered from severe constipation in infancy and battles with it today. She was born with a crooked rectum.

Speech and language:
She is very delayed in all aspects of speech and language. She has severe hypernasality, language delay, hearing impairment, and severe velo-pharyngeal insufficiency (VPI).

Cognitive/learning:
Her performance has been showing learning disabilities and difficulties in school rapidly, since she hit first grade.

Immunologic:
She has had many lower airway problems and is often sick. Colds seem to linger and not go away.

Skeletal/muscle/orthopedic:
She has been diagnosed with mild scoliosis. Also, we recently found out that her spine is fused in certain areas. She has always suffered from chronic leg pain and complains every morning about putting her shoes on.

Skin/integument:
She has an abundance of scalp hair.

Timeline

December 19, 1994. I had a beautiful baby girl. We were released from the hospital with a clean bill of health.

Soon after she started to spit up everything she ate through her nose and would not have a bowel movement for days. I contacted the doctor about this and they just blew me off like it was nothing.

Basically, I just accepted the doctor's word until it got worse. I mean she stayed constipated. The vomiting I could deal with but having to literally dig poop out of her hiny wasn't fun. I had to actually lie and tell them she had not used the bathroom in 2 weeks. Well it had only been about 5 days but when they took an x-ray of her, they found she had a wad of poop in her the size of a large grapefruit. Oh, so now they claim she has a problem and referred me to a pediatric gastro doctor.

In between all this, at 6 months she was hospitalized with pneumonia. She spent a week there.

The gastrointestinal doctor said she has a crooked rectum and there is nothing we can do about it but to regulate it with MOM (milk of magnesia) and diet. Well this is a disaster either way I go because I couldn't regulate it. I don't want to make you sick, but I need to explain what would actually happen. I was either being shot at with poopy or digging poopy out because it was so hard. Sad to say but there are many different types of poopy and I know them all. This has worked its way out (no pun intended) over time and she can go on her own, but I still have to give her MOM every once in awhile and watch what foods she consumes in quantities.

Shortly after finding out about Kaitlyn's rectum (1-1/2 yrs old) I started to question the doctor about her hearing. It didn't seem like she was responding and didn't speak at all.

Again the pediatrician said, *"Oh, its nothing and she is progressing just fine."* Well this time I took it upon myself to take her to an ENT on my own. After putting her through some tests they said they didn't think there was any hearing loss, but they wanted to put tubes in and remove her adenoids because she had a bunch of built up fluid. So they went in and did this. After healing we went back to be retested and found out the tubes helped a little, but Kaitlyn was still not hearing well in either ear and it was determined she had a moderate to severe neuro-sensory hearing loss. Wow what a shock! From here I had to fit her with hearing aids, learn some sign language and enroll her in speech therapy.

Once we found out about the impairment they sent us to the intervention program at Children's Hospital because they had no space for us in the local intervention program in our county. They gave us a full evaluation there. Let me see, they gave me a list of things wrong:

Hearing impaired	Developmentally delayed
Speech impaired	Crooked rectum
Crooked mouth	Low muscle tone
Her feet didn't work quite right	

When I left there I was overwhelmed. At this time she was 2 years old. They told me at 3 that she could get services through the county and start school. From 2-3 she went to speech therapy twice a week. Sadly enough, somehow I fell through the cracks and no one contacted me about school enrollment. Luckily, I remembered and called. They admitted I was treated badly and got me an immediate appointment with the county. Not knowing, I later found out she could have been enrolled at two. This is why it is good to have information from those who have been through it. I wish I would have had someone who could have shared this information with me.

The IEP and eligibility process was hard, in itself, to get through. They were asking me questions I never thought to remember, like "does your child alternate feet when going up stairs". In the end Kaitlyn was eligible for services in a self-contained hearing impaired class with speech and developmental delays.

Wow, let me tell you how hard it was to put my 3 year old on a school bus. Things went well in school; she really was learning the basics and progressing well. Although, she was in preschool it was still fun. It wasn't academic at all. Her only problem in preschool besides recurrent ear infections and a new set of tubes was her eating problems. She would go all day with no food. So the doctor put her on a special diet of eating several small meals a day. She began to eat!!!

Oh, in between all this her teeth became decayed and she had to go under anesthesia to pull two teeth, crown all the molars and fill cavities in the rest.

It was her kindergarten IEP when things got better and worse. The speech teacher noticed how nasally she was and suggested I have it checked out. So I did. Our doctor said, *"Yes, her palate isn't moving like it should, and I thought it would work itself out."* He proceeded to tell me she would probably need flap surgery to repair this. They referred us to the MCV cleft team.

So here we are visiting another group of doctors. First we saw the genetics department and as soon as we walked through the door, she said we are going to do a blood test after reading your files and seeing her. I can most likely assume that she has something called 22q11 deletion or VCFS. Well, she basically just left that in my lap and let me go home. Two weeks later it was confirmed she had this thing called 22q11 deletion, but I had no idea what it was.

Then they sent us on to the cleft team where she was looked at again. Here they told me she would need this palate surgery but would need her tonsils removed before that. We had that done in August of 2001. Shortly, after having her tonsils removed and researching 22q11, I found this mailing list or web site; it had so much stuff on it that I was just taken aback by it. I started reading about the flap surgery and how we should make sure the arteries are right and there are no heart problems. No one told me about these issues. What if I had done the flap surgery without checking this stuff out? Thankfully she has no heart problems but she still needs the MRI of her neck. I let the cleft team kind of fall off to the side line. If they weren't aware of these things how could they take good care of my child?

After reading the list more, I watched the _Discovery Health_ show that was taped at CHOP. All I could think of was how she resembled them. Is all this stuff going to happen to me and to her? Where do we fit in the whole 22q11 deletion scenario? So I decided to get more information and set up a week to visit CHOP *(Children's Hospital of Philadelphia)* in April of 2002.

We went to CHOP, and got a lot of information, but I still lacked some information that I needed as well. I learned her spine was fused, and that she was a candidate for flap surgery. Everything else I pretty much knew because I figured it out by myself along the way.

At the present, these are the new things with which I am having trouble:

• Falling behind in school (math, reasoning, comprehension)
• Flap surgery – I'm scared of this and don't know why.
• Lack of reasoning at home
• She falls apart over everything done, even a simple thing like getting a spoon out of the drawer
• She asks questions over and over, although she knows the answers
• EAR INFECTIONS!!!!!

Kaitlyn is very loving and protective over me and her brother. She is also very clingy. I know she might be different and not succeed the way I had wished, but I wouldn't change who she is. I love her as is.

There are still so many things I want to learn, like: things about hearing impairments and ear infections. I want to be able to understand an easier explanation of this flap surgery and the side effects; how to handle math and what works. I know repetition works, but what kind of repetition? Why do they fall apart so easily and ask questions over and over again? I also think it would be nice to have a list of parents and doctor contacts for people in the future by subject or living area. Maybe we could even learn how we can *ALL* raise the awareness of VCFS/22q in our world-wide community.

Family dynamics

I'm so happy I was allowed to say anything I wanted to say in this book, and I want to get this out! I feel I have been nothing but alone from day one. My ex-husband has never been involved in any of Kaitlyn's *medical* problems. For goodness sakes, he doesn't even know what VCFS is. And as far as my other family members go, either they are in denial or they just don't care. If ever there is a time when a person could use some help it is when they are dealing with a crisis with their children. I can now see the wisdom in having people get educated in different areas, as we do about AIDS, cancer, etc., even if we personally don't have it. People seem to be able to accept those problems which are ones they cannot see visibly. Why can't they accept disabilities like VCFS that cannot be seen?

I have to say the only people supporting me right now are my dad and step-mother. My dad stayed by my side as I visited CHOP. My birth mother seems to handle it nonchalantly or indifferently, as if there's nothing wrong. There *IS* something wrong! Why can't people see that? My current husband is quite supportive but doesn't truly understand what it is. He can't understand her falling apart or being so clingy but all in all he is interested and will do whatever he can to help out.

I guess I just feel alone in this journey. I wish there were people and groups, who have 22q in common, locally accessible to me, with whom I could be friends. I have been told this is a common syndrome, so why do we not hear about it? Why is there no marathon, or *"Jerry's Kids"*, or any other help for them or for us as parents?

My special child-Kaitlyn

As I watch while you sleep, the tears are in my eyes.
The love I hold for you is more than I realized.

Never would I have dreamed of the special care you need.
The doctors, tests, and time in life that I now lead.

To hear "no cure" is scary, but the strength has found its place.
To have you here with me puts the smile on my face.

Slowly I am learning how to cope with what I feel.
But some days are spent dreaming that none of this is real.

I see you look to me when the doctors poke and prod.
Your tears burn through my heart as I put my faith in God.

I've tried to wish it all away, more than you could know.
But God has other plans as to how our lives should go.

I treasure every moment as if it was our last.
With health our blooming flower, and pain our dingy past.

Each parent should be able to know this kind of love.
Our children are the greatest gifts given from above.

It should never take an illness to have a "special" child.
Just lots of love within your heart and hugs and kisses piled.

Michelle, Mother

God bless all the parents who love and support their special children with unshed tears and quiet strength. How lucky we are to have such miraculous children. It's not always easy, but a smile and hug from our children let us know that we're not doing too badly of a job.

Andrea's story

Andrea was born on February 16, 1990 at Phoenix Baptist Hospital.

My husband and I were married May 22 1989. Shortly after we were married we realized that I was pregnant.

During this time my husband was working at McDonalds® and I was working at Dillard's® Department Store as a retail clerk. Because of making minimum wage we were not able to afford any type of health insurance program nor did we qualify for any state insurance program because we owned a vehicle; back then it was an old Ford® truck that wasn't worth more than two hundred dollars and this was our only means of transportation but it was enough for the state to claim it was a valuable asset. So we did not qualify for their medical program.

My husband and I ended up taking out a bank loan so we could have our daughter at Phoenix Baptist Hospital. By this time I was about 5 months into my pregnancy and was finally able to start prenatal care.

Being new parents, one has certain expectations and needs. Because of our situation, our needs were for both financial and emotional support. This would have been most valuable for me during the entire duration of my pregnancy. Unfortunately because this did not occur, I believe I had a harder time accepting the pregnancy and labor. The delivery was extremely tough both physically and emotionally. It should have been a joyous occasion. Since our loan only covered the prenatal and birth experience we ended up back at square one with trying to support this wonderful new baby of ours.

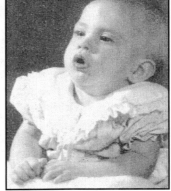

Our baby girl was beautiful she had all her fingers and toes attached although she did have club feet and swollen eyes and head. She weighed 9 lbs 8-1/2 ozs. To us she was beautiful even if she did have some deformities that raised a number of concerns for both of us.

We had been concerned about our daughter's delivery with the use of the suction cup extraction; but we had no where to go to inquire about why this swelling of her head did not go down; except to take her back to the hospital where she was delivered. We tried to take her in for an appointment, only to be turned down because we owed them a hundred dollars. No one had any idea where else we could take her to follow up on our various concerns, and ask the questions about our daughter's situation.

During Andrea's first pediatrician's appointment it was determined that Andrea's skull did not have a fontanel (soft spot). They said this was due to the suction cup that was used (This is an extraction method sometimes used at birth.)

During the first two years our daughter was extremely slow at developing. She was a fussy baby and screamed the entire duration of her first year of life. We could not afford anything more than our one bedroom apartment so we all felt trapped. Our daughter ended up most of the time out in the front room just so we could get some sleep at night. We tried everything we could think of to calm her down but nothing seemed to sooth her. I often begged my family for a rocking chair to sooth her and rock her to sleep but I never received one. Once we used a sitter to watch our child to receive some much needed respite, but the gal ended up feeding our daughter a whole can of formula in one day thinking that the screaming our child made was due to being hungry.

I can recall going to my mother-in-law's house; and having her say that I should not hold my baby, so that she wouldn't become dependent and spoiled. She had no idea what my husband and I were going through at home when making such a comment. This upset me. I can recall that it seemed that everyone had all the answers. We were blamed for our child's slow development and people would say it was due to our _social_ _economic_ situation and that I was an inept mother. People can be fast and free to make comments and give opinions, but where are they when you need them? There were lots of opinions for reasons why we were going through this with Andrea, but no one seemed to show understanding or compassion; not even later when our child was diagnosed with a genetic deficiency. I too was diagnosed with 22q and you would think that this would open some eyes. But no . . .

When Andrea was eight months old, we did sign her up for treatment of her club feet at the Children's Rehabilitative Services (CRS) associated with St. Joseph's Hospital. The center did repair her club feet. She was cast for 8 months.

Our daughter slowly started to walk at age three. I had no idea during this time that she had trouble with muscle atrophy. I had no idea that her muscles ached or that she would have been labeled *failure to thrive* during these developmental years. Looking back I wish that we would have had early intervention, acute care and a social worker assigned to us. This would have been extremely helpful and I believe would have made a remarkable difference.

Having all these medical problems with our daughter made our first year of marriage a rough experience. We had in-law problems and family members who thought they knew it all, yet, gave us no real support, only words. We needed emotional and financial support; as well as ideas about where we could go for medical treatment for our child. Believe me; this would have saved us a lot of anguish and heartache.

Because of her slow development and my limitations, I was labeled as an *inept* mother. I was accused of not providing the proper environment. Family members thought I was not able to handle raising a child. This attitude went on and on until Andrea was diagnosed with the genetic condition called VCFS/22q in 1995.

Andrea was age four when she entered the school system. We had her tested for many of the concerns my husband and I had and she was diagnosed as *failure to thrive* and entered into a developmental preschool program. It was during these school years that recommendations came, about what we could do to help her along. It was recommended that we research the Arizona Health Care Containment Cost System DDD program, to get the medical care she needed.

We also had our daughter evaluated by developmental psychologists. During the reading of the results of the developmental report our concerns were validated. It was a relief to know that there was nothing that we could have done that resulted in Andrea's delays. The developmental psychologists did recommend that we look into help based on the findings and that we should be eligible to receive all the medical and state support necessary for our daughter, up until she is 22 years of age.

It was recommended that Andrea undergo genetic testing called FISH. The developmental psychology report also indicated a possible cleft palate. He recommended that our daughter receive extended medical and it was necessary to order physical therapy (PT) and occupational therapy (OT) outside the school district and to see a family physician on a regular basis.

After my daughter had tested positive for the VCFS/22q DiGeorge syndrome the geneticist gave us a FLIER on the syndrome but it really gave us no real information. We searched everywhere we could for more information on VCFS but there seemed to be little to no information on this condition. My husband built our first home computer shortly after we discovered that there was no information locally on this disorder. We ended up becoming glued to the Internet to find more information on the syndrome with which we were dealing.

I ended up doing research on the syndrome through the net and came across many personal home web pages of parents who had undergone similar paths. It was interesting to learn that 22q/VCFS/DiGeorge syndrome patients usually end up having a heart condition and further health impairments as well as psychiatric complications.

After learning that Andrea could quite possibly be afflicted with a heart condition I took her back to CRS to undergo further testing. The physicians had a difficult time accepting the Internet information which indicated that there could be more related to this syndrome. It took three physicians to shuffle through her life to finally come up with a diagnosis of aortic arch dilation. This was at age seven.

There were further acute medical tests run that also proved that the developmental psychologist's report was correct. Andrea ended up having a nasoendoscopy test performed which proved that she needed surgery to repair a submucous cleft palate.

When she had surgery to repair the submucous cleft, it was recommended that Andrea undergo intense speech therapy.

During the course of her early years, Andrea oftentimes endured constant ear infections. We had a number of tubes placed in her ears during her early school years. It was recommended that she have her adenoids and tonsils removed. Andrea was ill so much that we were accused, by the school, of helping her make up illnesses.

Through my research on the Internet I ran across a psychiatrist who was in the states on a grant by the University of California. He was here doing studies for *JAMA* with regard to brain activity. I contacted him to find out more about his research on brain activity and development. Through the course of our interview, the psychiatrist inquired if I had similar symptoms or problems in education or any medical conditions, similar to Andrea's? I told him that I was raised having to go through special education programs and that I was born without a uvula and had scoliosis and Hashimotos (autoimmune thyroid disorder) and Endometriosis. He suspected that I too could have the 22q11 chromosome deletion of VCFS/DiGeorge.

The university sent vials so we could be tested locally for the disorder. My mom, and dad, as well as my other two girls were tested for the deletion. The results were that neither of my parents nor my other two daughters carries the deletion. But *I* did test positive for the 22q/VCFS/DiGeorge syndrome just as my daughter Andrea did.

Andrea's educational experience has been rough on the whole family, as it has been tough to acquire a proper academic atmosphere for her. Through most of Andrea's school years (she is now 13), we have had numerous confrontations with school districts to get them to understand Andrea's various educational requirements or special needs. Oftentimes we have had to utilize the assistance of the IDEA law and various individuals involved in the field of special education and Federal requirements.

Many times I have had my own condition thrown back at me in a critical fashion, while trying desperately to advocate for my child's needs. This gets extremely frustrating and it also leaves us emotionally drained.

When I was growing up, my parents did not understand that I had a disability with an actual diagnosis. Much of my life was sheltered and mom only ventured around in certain groups that would accept me. Back then, these groups were few and far between. It was unheard of (or not spoken about) to have a child with a disability, everyone wished for a perfect *Gerber*® baby. Society did not want to accept that many households had one or more children with disabilities, because these children were viewed as flawed!

It is also important to understand that even though there is a diagnosis within an individual or adult, we must not stop trying to communicate with them.

We must accept them as individuals and accept their disability. As a disabled person, we need this acceptance to live a completely whole life. Everyone must understand that we cannot change the fact we have this disability; but must learn to live with it. We ask that everyone learn about this disability. We hope that by people learning about us, it will help to change views of the disabled as a whole.

Today, as more and more people are being diagnosed with genetic deficiencies, it is important to accept the things we can not change. It is also crucial that we accept an individual as just that, an individual; instead of looking upon them as a diagnosis.

Tanya S. (Mother) 22q/VCFS/DiGeorge Syndrome and Andrea age 13 22q/VCFS/DiGeorge Syndrome

Tanya and Andrea Update 2005-2010

When Andrea was in the fifth grade we decided to pull her out to home school her. By this time, Andrea had officially been diagnosed with 22q11/VCFS/DiGeorge syndrome. We felt that the staff at the educational level did not want to help our daughter. It was the one-on-one experience that our daughter required and demanded, if she was to learn, with the diagnosis we had received.

With an exhaustive amount of prayer and sheer exhaustion in general, to get her needs met at the brick-and-mortar level, we decided that homeschooling was the next best thing for Andrea. Homeschooling proved to be a godsend.

I have found I can give her the one-on-one attention she needs, without having to struggle to fight to get her needs met.

Every day we started out with prayer and Bible reading. I have seen tremendous growth with all subjects. She has grown into a beautiful young woman.

Over time Andrea, who has two other sisters, wanted to try out public High School. So we decided that we would give her this opportunity. Exposure to public High School and that of being involved once again, in Special Education turned out, this time to be a positive experience.

Andrea made numerous friends and had a wonderful time. Though because of Andrea's psychosis she was not able to deal with a lot of social content. Because of her inability to handle social issues all at once we ended up dropping out of the program.

Since she has dropped out of the program she has tried numerous day programs. The day programs seemed to provide too much stimuli and it was difficult for Andrea to keep up.

As time has gone by we have found that life sometimes throws us more than we can handle. With this in mind we both have decided to take one day at a time . . .

She is now enjoying attending church and enjoying texting her friends.

Our story can be found via the internet. If you wish to write to us please feel free to email us at: **tms65@cox.net**.

Andrea is now a freshman in home school again and is involved in her church. She loves going to youth group every Sunday evening and participates on mission trips.

Here at our home, we truly do believe in miracles. Our daughter, was never expected to move past the fifth grade, she has overcome all the negativity [and] truly is a miracle.

Our expectations for Andrea are that she continues to remain happy and healthy. We hope that she continues to love life.

Andrea and Tanya Scarpitto

22q11 deletion – Richard W.'s story

I first came across this syndrome when I had my son, Richard who was born with complex problems.

I was tested only because my husband and I wanted to have further children and to ascertain Richard's diagnosis. Another factor was that my sister-in-law was going through a difficult pregnancy and there were worries that the child had the same condition.

At the time of testing, I had no idea what test we were being given. Had I known about the condition prior to the testing, I may have thought differently about it as I am minimally affected by it.

Unfortunately for me, it proved positive for the 22q11.2 deletion. Coming to terms with the diagnosis initially was difficult. I was more concerned about my son's problems than getting treatment for me. However, at times, I felt as though I had given this to him and it was my fault. It had an effect on our sex life and the guilt of having another child.

Both of my parents tested themselves for the deletion. So too did both my brothers, and the result was that the deletion was negative in all of them. My deletion is therefore a de novo deletion in me.

After the positive deletion test, I vowed never to have another child with the condition as morally I did not believe that I should pass this deletion on to another person – it should stop with me and Richard.

Having two children with the condition would have been very difficult to cope with. I was finding it difficult to cope with Richard's condition, never mind starting again with another.

After Richard's diagnosis, I felt that I needed to research the information.

I was astounded to find that there were close to 200 clinical findings of the condition. Richard fit into a large proportion of all of them! I, on the other hand, did not; there were minimal problems associated with my condition.

I, for instance, did not have learning difficulties or speech problems. I have never had any special needs and did not receive any extra help at school in any way. I was an average student in a private school; the other students on the whole were well educated and clever.

My siblings too were incredibly intelligent, as were both my parents and grandparents. Historically, they came from a family of physicians.

People often say to me that I had the advantage of coming from a medical family. In fact, nobody had known about the condition apart from one brother-in-law, at the time, a paediatric registrar in anaethestics. He mentioned DiGeorge to the cardiologist years ago regarding Richard and casually dismissed it purely on the basis of, *"Richard has T-Cells."* In fact, Richard had all the clinical signs of 22q/DiGeorge syndrome right from the start!

I, on the other hand, did not; but was clearly different from my other intelligent siblings. They were all good at sports, which I wasn't. They were all good at the things that I wasn't, including: chemistry, physics, and pure maths. They were good at horse riding, rugby, and running. However, I believe I was average at everything, although I hated sports. I had the disadvantage of coming from a family of people who were incredibly bright; and therefore in comparison to them, I did have learning difficulties. (My niece read a Harry Potter book in 2 hours at the age of 12!). I learned to ride, to swim, and had friends etc. I just had to get on with life but often wondered why I was slightly different than my siblings. I just thought, *"That's life."*

My minor problems associated with the condition include:
- Small hands and feet (size 5, narrow B fitting and different sizes)
- Small ears!
- Delayed puberty (aged 19)
- Small stature (smallest girl in the class for years)
- Sudden growth spurt at aged 16
- Astigmatism (I wear glasses)
- Raynaud's phenomenom (I suffered this as a child and under stress. My Dad always wondered why I couldn't cope with the cold.)
- Candidasis (fungal infections).
- Depression (due to learning that my son had to have yet another operation, on one occasion)
- Thrombocytopenia (my normal levels are about 100; normal range is 150-300). I bruise easily.
- Constant ear infections as a child

During school, I happily played the trumpet and piano (Grade 5 Piano and Grade 3 trumpet). I rode and achieved Grade C Pony Club although I was very good at falling off and breaking bones! (I have broken an arm three times and I broke my leg once.)

I have been investigated for heart problems and kidney problems, which are apparently normal.

Academically, whilst I struggled at school; I came out with six o'levels. I then went on to do a medical secretarial qualification which included medical shorthand at 100 wpm and typing at 70 wpm. Amongst the studies undertaken, I did anatomy and physiology, medical terminology and clinical procedures. I then did an HND in business and finance and topped it up with a **BA (Hons) Business Studies Degree in Marketing and Finance** and have a perfectly acceptable 2 (ii).

My experience has included dealing with paediatrics. One of my first medical secretarial jobs was in the paediatric nephrology department of a major London hospital, another job was working for a consultant who specialized in cystic fibrosis and asthma. As a result of these jobs, I had the experience of dealing with seriously sick children and their parents. I vowed that I wouldn't have a child with all sorts of problems. How wrong I was!

To say that I have learning difficulties would be incorrect, the same can be said for communication problems. Okay, I may go on but throughout my experience with the condition, I have educated people about the condition and written an article about Richard. People have learned about the condition, and a number of children have been diagnosed as a result. I work for a senior director in a government department and have worked for other directors in other companies. Good communication plays a vital part of the job! Had I had learning difficulties, I would not have been in my current role for more than a week nor would I be able to manage a budget of £7 m. Therefore, I do not consider myself to have communication problems!

I was delayed in puberty and did not start my periods until I was 19. I always felt inferior to my school pals as I was flat-chested and they weren't. There is always a stigma regarding this and I was bullied as a result. Okay, I have mood swings but then who doesn't have (PMS) Pre-Menstrual Tension!!! I was always considerably smaller than my classmates. Sometimes children, who were younger than me by 4 years, were taller than me!

In 1997, I found out that I was pregnant again. However, as I said earlier, I always felt that it was morally wrong to have another child with the condition. I certainly could not face having an abortion at 18 weeks, I knew I couldn't go through with it. I opted for CVS at 11 weeks. Even though the consultant said, *"You have a healthy foetus at 11 weeks!"*, the results of the FISH test proved that the foetus had the same condition. After a great deal of anguish and argument between my husband and myself, I had a termination of pregnancy at 14 weeks – the latest time it can be done by anesthesia.

In 2000, my son Richard sadly died from complications of sub-aortic stenosis and endocarditis.

My husband and I decided that we would try and embark on having another child. We were referred to Guy's Hospital for *Pre-Implantation Genetic Diagnosis* by IVF. Funding was obtained for two cycles by IVF PGD. After many months of further tests and diagnosis of my genetic phenotype etc. I then went ahead with one cycle of IVF. The chances of becoming pregnant at my age of 40 were somewhat minimal; something like 10%. Bear in mind that 50% of all pregnancies would carry the deletion. However, this does not mean that there couldn't be a chance that 8 follicles would produce 8 eggs without the deletion or that 4 would have the deletion and 4 wouldn't, it is a pure matter of chance, rather like tossing a coin, heads and tails. Unfortunately, for me I was only able to produce 5 follicles which was insufficient for pre-implantation genetic diagnosis and the cycle had to be stopped.

As a result my husband and I have decided not to go for further children. We may decide to adopt.

Over the years, I have supported people in various support groups. Sometimes, I have found parents to be judgmental assuming that those affected with the deletion have learning difficulties and psychiatric problems, etc. and they do not want to listen to those affected by the condition. I have made many friends throughout the world, but also a few enemies! Though obviously there will always be some parents who will just assume that you have problems.

Comments such as, *"You gave me hope that my child may be like you."* have spurred me to write this look into my life. It is possible that you can be perfectly *normal*, live a happily married life, drive, have a job and even own a home. I have owned my own house since the age of 21. Nobody looks after me. I am totally independent; I always have been and I don't expect anybody to care for me.

However, one thing is for sure, to have a label because you happen to have a 22q11.2 deletion is wrong. You either are or you aren't affected. If I had my time again in support groups, I certainly wouldn't have mentioned anything about having it.

Despite the loss of my son; outside of work, my time is spent now on our new Yahoo **VCFSandyou** group and for those in the UK, the **vcfsgroup**. I advocate for people who genuinely want the help for Munchausen Syndrome by Proxy (falsely accused) for exaggerating or fabricating their child's illness. I do work for **http://www.parents-protecting-children.org.uk**, Heartline and fight for treatment for those affected with sub-aortic stenosis. I have an interest in rare syndromes as a result of spending time on PDHeart years ago. I have also spent time supporting those who have been bereaved and contribute occasionally to Broken-Hearts-Infant-Loss group.

Fiona W., Mother of Richard, UK

My Angel's story

My Granddaughter, Angel was born on August 23, 1999 to a developmentally disabled mother and father. I am the maternal grandmother. Right from the beginning, it wasn't good. Angel's mother had no prenatal care. I found out she was pregnant the day Angel was born.

Looking at Angel at birth, she was a beautiful little baby all pink and winkled. When she was going to be discharged, the nurse noticed a heart murmur just hours before we were going to take Angel home. A specialist was called and it was recommended that Angel be transported to Children's Hospital in Oakland, California. More specialists were involved now. An x-ray showed that Angel had an interrupted aortic arch, just one of the characteristics of 22q/VCFS/DiGeorge. Angel had open heart surgery at 2 days old to re-construct her Aortic arch. The surgery went well. Seeing that little girl hooked up to machines and tubes was heart breaking.

While in the hospital genetic testing was done and it was discovered that Angel had 22q11 Deletion Syndrome. Some of the other characteristics of the syndrome that she has are the deep set eyes (almost a *Down's Syndrome* look), long slim fingers and feet, reflux, and stridor (wheezing while breathing, which she doesn't have anymore). Angel does not have a thymus; the thymus helps in fighting an infection. This causes her to be immune deficient. Angel had an ASD and still has a VSD of the heart. She also has a seizure disorder involving, grand mal seizures as well as apnea seizures, during which she stops breathing.

After 3 months, Angel was ready to leave the hospital. She had a gastrostomy tube and 10 different medications. Angel and her mother went to stay with Angel's aunt, uncle and cousins. After a week or so Angel developed pneumonia and went back to the hospital. It was becoming apparent that Angel needed more care than we could provide. In a family discussion we chose to place

Angel in a medical care home. Unfortunately it was a nightmare. For some reason Child Protective Services had gotten involved, and it became as I said, a nightmare. We could only visit Angel for 15 minutes at a time and at Christmas we could only take her for two hours. After going to family court 3 times, the judge gave us back our Angel.

Angel's development was slow. She was getting physical therapy twice a week which was a big help. From the time Angel was 6 months old until she was 18 months old she appeared to be perfectly healthy. Angel started George Miller East, Developmental Center in Concord at 18 months. She was eating regular food but not gaining weight. In December of 2000 Angel began to get ill. From December 2000, Angel was in and out of Children's Hospital constantly. Most of her admissions were due to aspiration pneumonia. At one point a *"swallow test"* was done and it was noted that Angel's epiglottis was opening and closing involuntarily which meant the food was also going into her lungs. In December of 2001 another gastrostomy tube was placed. Angel had continuous feeds during the night and one bolus feed during the day. She began gaining weight and was doing very well.

In the later part of August 2002, Angel's apnea seizures worsened. She was hospitalized for almost a week. With those seizures, she would be just sitting down or playing and turn completely blue. Talk about scary!

Angel had a couple of more bouts of pneumonia and that's when Angel's cardiologist suggested another EEG to see if they needed to patch the VSD. It was recommended that it should be done.

Just a couple of hours before the surgery was to be done, Angel's cardiologist came into the waiting room with the surgeon and said they wanted to do another procedure. What they had noticed was when they did the reconstruction of the Aortic arch; while pulling it up it was then pressing on Angel's airway and compressing it which meant that it was difficult for air to pass through the airway. The surgeon wanted to go in through Angel's left shoulder blade, pull the heart off the airway and stitch it to the rib cage. They said it would help with the pneumonias. That procedure was done in November of 2002. The recovery time in the hospital was supposed to be only one week to 10 days. There were problems. Because the airway was compressed for so long it continued to be compressed. Angel's saturation was still low. Every four hours, Angel was given breathing treatments to try and open the air way, it didn't work very well. The next step was a C-pap breathing machine. She had this monstrous machine on with a mask which just about covered her entire face. It helped open the airway and Angel was doing well. A month later, Angel came home.

Angel has progressed enormously since that surgery. When she's playing she doesn't get out of breath. She has began to run and do things that were virtually impossible for her to do before the surgery. She has gotten taller and has put on weight. She's thriving! She now knows at least 25 signs. (She isn't deaf but she can't talk, yet.) Angel is a special little girl, who when she meets you, crawls into your heart and you fall in love with her. Her little ways will make you smile and sometimes laugh. You can't get her face out of your head. Every time Angel went into Children's Hospital in Oakland, not only did the nurses fall in love with her, but the doctors did too, and so did the housekeeping staff. They all brought Angel bubbles when they knew she was there.

Angel has fought for her life many times; which I believe is because she knows how much she is loved, and she knows that she is on this earth for a reason and a purpose.

The doctors who have been involved with Angel's care have been a cardiologist, a neurologist, a gastroentronologist, and a pediatrician at the Sunshine Clinic in Concord, California. All of them are extraordinary doctors and extraordinary people as well.

Angel loves her Uncle Stephen and her Uncle Adam her Auntie Amber and her Auntie stinky (Jennifer). Not only do her cousins, Alexis and Kaitlyn adore her, Angel adores them as well. We feel Angel was truly sent from Heaven!

My son Christian

April 1st of 2004 was the date my son, Christian, had his 4th heart surgery. We thought we might have the luxury of waiting until school was out in May, but Christian's blood pressure started going up again, he started having really bad headaches, and was very dizzy. We tried waiting to see if this was just transitional, but it wasn't, and we had to adjust his medication. He had a really tough week that last week, and thankfully within a few days with the higher dosage of medicine he finally was feeling _normal_ again. We went to the cardiologist and had to change his medication to a different drug to prepare him for surgery. Christian is doing as well as can be expected, but he has a pretty high level of anxiety, and this is quite normal within the realm of his bipolar disorder. I spoke with his psychiatrist and we are adjusting his medication to also help level out his anxiety. We don't want to compromise his immune system, which is a very real possibility if this type and level of anxiety goes on too long. (You learn so much about the physiological aspects of a brain disorder that can impact physical disorders in VCFS, in this case, his heart.) Christian recently expressed to me that he had a fear of dying in this surgery. As anyone can imagine, hearing this was very difficult for me, yet it was something I couldn't escape and had to deal with, and help him as best as I could to comfort him. I am so very thankful I am a Christian person and that my son Christian is as well. I can't even begin to imagine how we would

face these times if we weren't. Christian asked me on the morning of his surgery, if he might die during the surgery (you know kids like Christian just ask things, flat out, without warning). Within the few seconds that it took for what he had asked me to register, and for me to ask God for the words to say; I breathed deeply, and just told him that, "Yes, it could happen, but we would pray that it wouldn't." He then told me he was too young to die, and he wasn't ready yet. He went on to ask me to pray and ask God to let him live so that he could have longer to be with his friends, because he missed playing with them and being with them. (He has missed the entire school year due to illnesses of one kind or another.) That part just about broke my heart. I told him I would pray for that, and I hoped that God would answer my prayer; and that he would live to be just as old as his Nana and Papa. I went on to tell him that really, only God knew how long each of us would live. I hugged Christian really tight then, and reassured him that he had the best doctors, heart surgeon, nurses and hospital and we would always pray for the best. Christian seemed content with that. I am happy to report that although the surgery was more difficult than originally anticipated, it was successful and Christian is doing well.

Since I learned of my son's VCFS; I have met many families dealing with this same plight. For the first time in my life, I have experienced a love that has touched my soul with a comfort and reassurance beyond measure. It has left me without adequate words of gratitude and thanksgiving to those who have been there for me and for my son.

Christian and I left March 27th of 2004 for Kansas City, Missouri and Children's Mercy Hospital. This is a wonderful hospital. Christian had a 3-D diagnostic C.T. scan to help doctors to reconstruct his aorta in preparation for his surgery. They needed to address the congential narrowing in his aorta. We always knew at some point we would have to surgically address this problem. One of the most challenging aspects of this surgery was the scar tissue that had formed from his previous three surgeries. (This is why they try to limit surgeries to as few as possible. Scar tissue can be as problematic as the defect they are going in to repair.)

I am very thankful that at times like these families have somewhere to go, like staying at the Ronald McDonald® house. This was my harbor in the days that followed.

Pam, Mother, SC

Blank page for personal data:

Medical Trivia

The average human body has enough fat to produce 7 bars of soap.

"Just because someone doesn't love you the way you want, doesn't mean they don't love you the best that they can." Author unknown

CHAPTER 16: MARRIAGE & 22q/VCFS

MARRIAGE, RELATIONSHIPS AND VCFS/22Q

Great relationships of any kind require constant work and patience on both sides. This is especially true when dealing with relationships that have the added burden of living with health issues such as those seen in VCFS/22q. Only after understanding the differences in men and women, will there be any allowances given when one of the parties seems not to be holding up their end of the marital bargain. Once we understand and accept that we (men and women) are not the same in our thinking and emotional make-up, we can start to explore the many ways we are different. Then we can learn to have a meeting of the minds, for a happier relationship. A couple working together in love will guarantee more support in the family arrangement. This will lead to success in that relationship no matter what trials it may have to overcome. There will undoubtedly be obstacles with children and especially children with special needs; if not with your mate, there will be trials to overcome with other people.

To learn these differences that will lead to success in our relationships we will need to learn how BOTH parties can show respect and appreciation. In the 36 years of my marriage I have read many books, spoken with many successful couples and learned from trial and error. I was even a marriage counselor for a while. In this chapter, I will try to share some of the things I have learned and read throughout the years, and I hope to give you some ideas that will lead to marriage longevity. Of course if you want more in-depth information, I would recommend you read books by relationship experts. There are some knowledgeable relationship gurus out there, and there are dozens of books from which to choose in your libraries and bookstores. However, not many pinpoint the issues involved in dealing with an ill child in the home.

It goes without saying that not all men and women are the same, we are not clones; and contrary to the thinking of some men (who don't talk much) women are not mind readers. There are basic studied differences and stereotypes that do set a basic standard for all men and women. A woman intuitively knows that life in a relationship is not easy and that work will be needed; no matter what crosses her path she will try to deal with it. She will put forth even more effort after learning that her child has several different health and learning issues on the horizon. She is already planning, researching and hunting for answers and will begin to handle the problems when they arrive. Men, on the other hand, usually keep to a routine, go to work, come home (mind still at work), have a drink to relax, seemingly to go with the flow, nothing seems to have changed after *HE* learns of the life challenges ahead for his family with 22q. (Some men have fooled themselves into thinking they have been supportive when actually they wasted precious energy, thinking of ways *OUT* of responsibilities. Notice I said *SOME* men not *ALL* of you fellas.)

Sometimes a man may even appear to be self-consumed, to the woman. At least that is how it appears to her. On the outside, he doesn't seem to want to talk about "it" nor think about the challenges ahead and what steps need to be taken. Nor do they think of how their lives have changed thus far and will continue to change more over the years (in many cases). This is the point where spousal misunderstandings can start to arise. If there has not been any deep confidential talk between them, the man will not understand why the woman may appear to be overwhelmed when he walks through the door. If the man is not kind and gentle in his handling of her, the woman may feel the man is cold and detached, not feeling anything and self-centered. Some men might subconsciously think, *"She thinks SHE'S had stress? At work I..."* or feel, *"She knows more about this medical stuff than I do."* or, *"She's better at this medical stuff then I am, so let her do what is needed to find out what to do for the kid, I would just botch it up."* Therefore he doesn't appear to give the support that they both need from the woman's viewpoint. In many cases, I believe that men just don't know how to support someone or know what to say. So they believe that keeping life as *"normal"* as possible is doing their part, even if it means (in their mind) not complicating it with too much talk. After all, don't most of us feel that the more we know about something the more we feel a responsibility to it? Think of all the causes that are out there that are supported, even by strangers, things like rescue organizations of one type or another. After learning of the need, supporters felt obligated to do something.

A man may believe that keeping the pay check coming in regularly and taking out the garbage is what will keep his family life orderly. This is their way of supporting. Of course, it is, to some degree, on the normal side of the coin. On the flip side of life's coin, having a child with 22q is seldom *normal* and requires a unique approach.

In many cases the whole family will experience the roller coaster ups and downs in their emotions, and in their lives. You notice this most (at least I did) when it comes to being able to do things and go places like other families, families who do not have disabled children in their household, who are able to do and go as they please. Parents with disabled children always have to plan, think about, care for the child's needs and comfort and make special arrangements if necessary. This is not always easy. There may not be enough funds (medical bills eat up any excess) for appropriate special accommodations. Often, no one is willing to sit with the child so that the rest of the family can do things and go places. If the VCFS child is on oxygen, feeding tubes, etc., there are few, if any, dinner invitations or other small, but important functions, that the general public takes for granted.

One mistake that many of us women make, is believing that our mates, friends, parents and others around us feel the same as we do about things. When they do not say, offer, or react the way we expect them to; we are confused, hurt and disappointed. Because of this, there may be tension in the household and the man hasn't a clue why the woman is acting distant. So women are labeled as moody, when just a little communication would do wonders to remedy the misunderstanding. This is usually a time when a woman answers, *"Nothing"* when a man asks, *"What's wrong?"*. Here again, she expects him to know, so the resentment builds. This is ammo for the blow up that is waiting to erupt someday from stress or just from being too tired and overwhelmed.

The secret to a good relationship is supporting the emotions that go along with normal everyday life; along with those that are heaped upon us dealing with VCFS/22q. We also have to learn how our mates think on certain subjects and not *ASSUME* we know their feelings. Realization could be at a critical moment in decision making, like in a hospital when a doctor has to have a decision made right away to do a procedure. You might feel that anything a doctor says must be right; where the child's father might want to ponder on it a moment, ask questions and then surprise you by deciding it is not really necessary at this time. I had a hard time learning these *"thinking"* differences myself. As an example of how some women might think; I always thought I would marry a romantic sensitive guy who just *KNEW* the right things to do when I was blue or lonely, so, for the first year of our marriage, every time Gil was a little late coming home from work I would visualize him stopping at a florist and getting me flowers or a sweet sentimental card. But of course when he would come home, there was nothing for me but an empty lunch pail. I would get pouty, resentful that he didn't think of me 24/7. (After all, aren't I special?) He would think, *"Now what did I do?"* Well, 36 years later, I have finally learned that we do not think the same way and we have to *"talk"* regularly to learn how the other feels and what they believe. Sometimes it is a real eye opener when you learn how your partner really thinks. Often, you get insight when you discover information about how they were raised or past experiences that were unknown to you. I mention this because when women are alone at home many times with sick children, we *THINK* too much! We have active minds along with those warm hearts. A lot of the time our thoughts are on our mates and how they *should* see our sacrifices for the family and how they *should* show their love and respect for us. When this does not happen and, in fact, they are thinking the same thing about themselves; it may cause us to become emotionally needy, even lonely. We *NEED* to feel appreciated and not taken for granted. (So do they.) While this is normal to have such feelings, it can cause a riff in the marriage or relationship if the guy hasn't been privy to how you feel. If we don't tell him and we continue to act moody, he will dread coming home and that is not a good situation either. This may lead to his wanting to spend more time with his friends out somewhere, or starting a trend of coming home late or it can even lead to infidelity. More communication is needed; you have to *"share to show you care"*. This lack of communication can lead us into a path of clinical depression on both sides. If you feel this way, talk to your family doctor or therapist.

While we are on this subject of *"sharing"*, I might as well tell you what I did that seemed to draw my husband and I closer together. We all know that marriage usually is not a 50/50. But to get close to this percentage, it takes finding a common interest you can share to encourage your loved one to talk and relate. If they can relate to you in other areas of interest, they are more prone to talk and relate to you in those areas where you need to come together such as the area of the health of your child. Here comes one of my personal examples: When Gil and I got married he was an auto body man and he worked at a dealership in a smaller city. This meant that it was commission work, with the amount of his paychecks always being a surprise; a feast or famine affair. This was not ideal when we were on a budget. My grandmother had a heart condition so we bought a mobile home. On a hardship permit, we moved it next to her home on some acreage. In exchange for caring for her and her grounds etc., she allowed us to build a shop so that Gil could build a business of his own. He specialized in classic cars from car clubs. I soon learned he was prone to be a workaholic (not a bad trait) and if I didn't show any interest or wasn't in the shop, I didn't see him much. So I learned how to use a wrecking hot line, wet-sand cars, and mask-off for paint, etc. I got him a coffee pot for the shop, turned the oldies on the radio, took the kids and spent many fun happy hours in that shop working, relating, teasing and bonding. Then when T.G.'s lungs looked like they could not make another winter in Oregon and the doctor said to move to Arizona, we did as Neil Diamond sang in his song *Cracklin' Rosie*, *"We packed up the baby and grabbed the ol' Lady"* and off we went.

When we got to Arizona we found that body shops were too plentiful so no real money could be made. (Remember we had no medical insurance, yet medical problems were always in the family, and we needed money.) So, Gil wanted to get into construction. He had a taste of it in Oregon and really enjoyed doing drywall finishing. He seemed to have a real knack for it. So he hired on as an apprentice and we shared a house for a couple of years, until he became a journeyman. Of course this meant that I needed to find another area of interest that we could share, and I needed to feel like I was supporting his efforts to make a living. It was also important that he knew I wanted to spend time with him. So, I went to Phoenix College and took some night classes in architecture, construction, cost and estimating. I got a contractor's license and together we went into the construction trade for the next 19 years. I am not saying that all women must go to that extreme, but what I am saying is that because I showed interest in him, in what he was doing and participated in it (as well as letting him know I *KNEW* he was doing the best he could to be our *"bread winner"*) we drew closer. He showed his love and respect in many ways, like respecting my opinion when I spoke, not questioning every dime I spent, baby sitting when I needed a couple of hours of quiet time, etc. We bonded even more, becoming best friends. Am I saying all was hunky dory? *NO*, but we are still together; and the rest of that story would take another chapter.

I guess it is not a surprise when I say roles of men and women in the 21st century have sure changed. No longer is it the "Lucy and Ricky" of yester-year where the man walks through the door yelling, *"Lucy I'm home."*, to find Lucy all dolled up in a dress and high heels, greeting him at the door with a smile, a kiss, dinner on the table and only her daily news of shopping the day away with Ethel, to share with him.

Today's relationships have a much more complex theme. Both couples are usually working, not paid enough, tired and then come home to face whatever else was in the mail, so to speak.

Our generation has also worked very hard to try and help men learn to *"get in touch with their feminine side"*. This creates some other problems or expectations. Now, both men and women are expected to learn how to nurture each other as well as the children *equally,* and still define their roles. Some men might think that to show nurturing qualities like their wives means that they would lose their masculinity or might be in danger of losing the title of *"head of household"*. This is in their own minds of course. A form of insecurity may be the culprit.

One of the most difficult things to do when both are stressed and have had a hard day, is to come home and say thoughtful positive things to and about each other. But I know from experience that it can be done, and with good results. When my husband and I realized that there was a change to be made from all the negativity that we projected at each other at times, we sat down and talked about how it made us feel and what changes needed to be made; this meant being honest, without being defensive. I learned that one must say what needs to be said by using tact, instead of blurting out things, no matter how much you might want to be blunt at times. Here is an example of the _right_ way: *"I really need you to be my friend and my sounding board. I know you think you need to fix whatever is bothering me but sometimes I just need for you to listen to me. I just want to share my feelings with you, and I don't expect a solution."* (This will eliminate the response, *"What do you want ME to do about it?"* from him.) The _wrong_ way would be: *"Every time I want to talk to you, you act like a jerk. You won't let me finish my thoughts, without interrupting me to tell me what I need to do to solve this or that."* See the difference?

Even if I had to cry to relieve some pressure, I always tried to let my husband know it was not him who was making me cry, but circumstances; and that I feel better just having him there, listening, and sharing my distress. One of the most productive decisions my husband and I made with each other was to use the *"golden rule"* in our marriage. To *"Do unto others what you would want done unto you."* We took the rule and applied it to how we would treat each other. So we practice being kind to each other, being courteous, and using manners with each other. We each try to please the other one rather than ourselves, and always work to fill the others wants or needs. When put into practice, it works wonderfully and you're both happy. We learned that when you are busy pleasing the other person, both needs are met; you both feel full and satisfied emotionally. Of course this takes time, *COMMUNICATION,* and humble willingness on both sides. Gil (my husband) is much more humble and patient than I am. I have been fortunate in that regard.

My grandmother used to say that there are two types of men and I suspect this goes for women as well. Type one, shows love and respect to family members more than to outsiders and type two shows more respect and support to those outside the family. She always told me to watch how a man treats his mother and family. She would say that I wanted a man who would care for the family more than for strangers. You can depend more on type one to be there when you need them and to show love and support.

I have also found that men will be much more receptive and willing if they are not made to feel bad; as if they are resented or rejected by women for being different in the way they deal with matters, or, NOT deal with matters in some cases. Yes, there are those men and women who are forever in denial that there is a problem to deal with, either in their relationship or with having a VCFS/22q child (this can extend to grandparents and other family members). Some refuse to take any action because they are either too lazy or prideful to face the problems either in their relationship or their child's unsure future. Whatever the reason, perhaps embassasment, feeling uncomfortable, not willing to spend money or some other reason, the person will someday have to deal with that flaw of character.

One of the sad outcomes I've seen is that couples grow apart because they have not had a meeting of the minds or a game plan made together to deal with life's trials and tribulations. Resentment is inevitable when this is the case and self-worth becomes an issue; usually for the woman (but not always). A friend of mine Cathy Corella once took me to a seminar presented by Barbara Burley, author of the book, **Selfable**. (This is a self-esteem, self-help program.) At first I wondered, *"Why do I need this?"* but I really did gain from this seminar and enjoyed her book. It taught me that it is alright to feel the way I do, but it's not helpful to give in to self-pity and let it _rule_ my life. (By the way I do still think self-pity can sometimes be constructive, if in short duration.) I feel it is a type of grieving and we are entitled to grieve for things lost, things no longer an option or for things that we were not able to do, because of being a full time care-taker. Cathy taught me to set goals, organize my life and to make sure I _allow_ myself some *"me time"*, recognizing that it is not selfish to need and want some *"me time"*. Sometimes we get into a rut by taking on the subconscious martyr role; becoming overly responsible, and almost obsessed to find answers to 22q/VCFS' symptoms, that we forget to live. For the first time I knew what the expression *"get a life"* meant. So I did. One of the goals was to go to nursing school (which I did) and another goal was to write this book!

In relationship expert, John Gray's older book **Men, Women and Relationships**, he teaches that, "There are 4 keys to creating mutually supportive and rewarding relationships; and since I think he is on target, I would like to share them with you:
• **Purposeful communication:** Communicating with the intent to understand and be understood.

- **Right Understanding:** Understanding, appreciating and respecting our differences.
- **Giving up judgments:** Releasing negative judgments about ourselves and others.
- **Accepting responsibility:** Taking equal responsibility for what you get from the relationship and practicing forgiveness.

I honestly believe that if these principles are applied there will be a solid foundation for all relationships upon which to build, even if the relationship has many miles on it. When we are dealing with children, especially children with disorders, it is important that the child feel calm and secure at home. So if the adult relationships are secure the child will absorb it and will thrive, then ALL will have a more enriched life.

Poem on life

Life can seem ungrateful and not always very kind.
Life can pull at heartstrings and play games with your mind.
Life can be so blissful, joyous and free.
Life can put such beauty in the simple things we see.

Life can place great challenges right there at your feet.
Life can make good use of the hardships that we meet.
Life can overwhelm us and put our head in a spin.
Life can reward those who are determined to fight and win.

Life can be so hurtful and not always is it fair.
Life can surround us with people, who will show us that they care.
Life clearly can offer us, its ups and its downs.
Life can bring us both some smiles or unwanted frowns.

Life teaches us to accept, the good with the bad.
Life truly is a mixture of the happy and the sad.

So . . .

Take the life you've been given and give it all your best.
Think positive, be happy and let God guide the rest.
Take the challenges that life offers, which were laid at your feet.
Take pride and be thankful for each one that you meet.

To yourself give forgiveness, if you stumble and you fall.
Take each day that's dealt you, and give it your all.
Take the love that you're given, and return it with great care.
Have faith that when its needed, it will always be right there.

Take time to find the beauty, in the things that you can see.
Take life's simple pleasures, let them set your heartstrings free.
The idea here is simply, to even out the score.
As you are met and faced with challenges through life's great tug of war.

(Author unknown)

Medical Trivia

According to an old Egyptian text, a delicate nerve runs from the fourth finger of a person's left hand to their heart, thus explaining why that finger is the "wedding finger".

"Remember, we all stumble, every one of us. That's why it's a comfort to go hand in hand."
Emily Kimbrough

CHAPTER 17: FAMILY SUPPORT

SUPPORTING FAMILY MEMBERS AND OTHERS

Parents are not the only people who have to adjust to the fact that a child has 22q and may have limitations. The child's sibling(s), relatives (grandparents) and friends have to adjust as well. Many are just as fearful for the child's health prognosis as we are and therefore need to be educated on the disorder.

When we have other children in the home, it is vital that we are sensitive and conscious of the sibling(s). We want to ask ourselves, *"Do I ignore them because I am so consumed with my ill child?"* Caring for all the needs of a child with 22q can be a consuming job. At times it is hard not to ignore the needs of the other children if they don't seem to *need* us, so we assume they are fine. When, in fact, they may be sad and angry at the attention that the 22q child is getting; they may feel that the parent(s) doesn't care as much for them. Sibling(s) may even have a fear or anxiety about what is going on with their VCFS brother or sister.

Being respectful to the other children in the family is important as well. An example might be, not burdening the older sibling with being a babysitter or having them take the VCFS child with them when they go places or see their friends, etc. on a continuous basis.

Is it possible that the sibling(s) might wonder if VCFS is catching like a virus? VCFS must be explained to them, and here is a suggestion on how to do that.

State that VCFS is <u>not</u> a disease that can be caught like a cold or flu. The body is like a giant puzzle with many many pieces, each piece goes to a different part of the body. VCFS happens because a little piece of the puzzle came up missing when the baby's body was being formed. It still wasn't there when it was time for the baby to be born. So our baby came to us with the missing piece. It can't be fixed or put back and no other piece will fit. So our baby will need special help from all of us. We don't know yet in what ways our little one will need help, but we want to be there when our baby needs us.

This may not be the greatest explanation, but I think it depends on the age group. It should at least ease their mind and help them to understand a little better.

One subject that came up in our family was *"How long will my brother live? Is he going to die?"* I told my daughter and anyone who might ask, *"No one knows that answer."*

Is VCFS fatal in itself? No, but there might be a condition or complication that can't be treated or a secondary infection that can cause concerns to arise that were totally unforeseen. So we do the best we can each day.

I was once asked "Does my son know about his VCFS, and should we tell our family?"

This is a personal decision, but it felt natural for us to tell T.G., of course he was older. If I had known earlier in his life, I probably would have done the same. Then he could promptly get answers and understand certain things and would know why he was sick and why things continued to happen with his health. Does he totally understand? No, but who does? The important thing, is for the children and family to know as much as they can about the disorder so they can be prepared. This also helps the child to know that anything that their parents have to do is not to restrict their life, but to enhance it; by keeping them safe and as healthy as possible. Another benefit is that the child can then tell you if there is anything _different_ going on with their health. This might not be the answer for all children with 22q. But, remember, my son was 18 by the time we received diagnosis. So take age into consideration.

As far as other family members being informed, a couple of parents said they told the grandparents, but the grandparents were in denial, refusing to accept that there was anything wrong. Others said they were glad to have everyone know so that they could work as a team, cooperating and looking after the child. All anyone can say is, you know your family and friends and use your own judgment.

How to move past the pain in a family crisis

When a child has been diagnosed with 22q it can put a family into a crisis situation. It is easy to become trapped beneath a mountain of problems that comes with this diagnosis. This can come by way of weighty decisions to make, worry of the immediate future and financial considerations. In order to crawl out from underneath the weight of pain and overwhelming responsibilities there are choices and sacrifices that must be made. On his web site, Dr. Phil offers suggestions about what families can do to survive a crisis and move forward. He wasn't referring to VCFS but as I read I could see where these suggestions are common sense and can be used in any kind of family crisis.

- **Put blame aside, it doesn't matter:**

This is not the time for finger pointing. Your energy needs to be focused on solving problems and coping, not assigning blame. Feelings like anger and resentment need to be put aside so that the family can work together.

- **Prioritize the problems of the moment:**

When dealing with more than one problem, tackle the most pressing issue first, then move onto others. Conquering one problem at a time is vital to moving successfully through a crisis with multiple issues.

- **Stand in the shoes of each person who is involved:**

During a highly emotional and tense time, it's easy to get lost in your own emotions and forget about those of others. You need to imagine how each person involved in the situation is feeling to understand the whole picture and to make decisions that may affect the whole family unit. Then, you can figure out the best solution or make the right decision. This is also true if you are a grandparent or a good friend wanting to support the parents and family.

- **Restructuring the family unit:**

Ask yourself, is your family unit working the way it should be? If not, you will need to change the way you all relate. Does a parent give more attention to one child? It might be the VCFS child or the one(s) without VCFS. Does one sibling communicate more than another, and make you feel more endeared to that one? A shift in the family dynamic could be an important step toward healing hurt feelings and strengthening bonds. The spouses need to have some confidential talk on important matters like this. Make sure the attention and love are shared equally; in harmony of thought and action. This is much like the meeting of the minds with regard to raising children and matters of discipline.

- **Recognize that everyone's affected:**

What happens to one family member happens to the entire family. When you focus only on the person in need, you are only dealing with half of the situation. Be sure that everyone has the chance to get involved and talk about how they feel and let the rest of the family know if they believe something needs to be addressed. Communication is the key.

- **Don't get stuck in the past:**

Don't waste time on things that you can no longer control, such as wishing you had handled things differently or made a different decision. Start today with your new-found knowledge and concentrate on your family's future.

Medical Trivia

Thermometers date back to the 1600's. At the time, however, they were filled with brandy, not mercury

"And in the end, it's not the years in your life that count. It's the life in your years." Abraham Lincoln

CHAPTER 18: GRIEVING

GRIEVING

Forgiving yourself when losing a child

The ultimate loss is that of a child. When this happens many people suffer from feelings of guilt, even though they *KNOW* the death was not their fault. In most cases (if it is not an accident that takes the child), it is a health issue, and no one can control that. Yet, it isn't uncommon for us to ask ourselves, *"What if?"*, *"What if I would have tried this?"*, *"What if we had taken her to another doctor?"* etc. When you have lost a child, do you permit yourself to grieve and then not feel guilty if you find yourself having a good time or merely enjoying life again? Although there is no set timetable for grieving, it should not take over your life forever, bringing it to a standstill. People grieve in many ways and over different lengths of time; one may regroup in a year, where someone else might take much longer. The main thing is to allow yourself to move past the debilitating grieving process in a reasonable amount of time. Should you grieve? *YES* by all means, but don't allow yourself to be crippled with guilt for something that was not your fault and not within your control. If you are crippled with guilt; then you need to find out why you can't seem to come to grips with the loss, and start living life to the fullest again.

According to psychologists, in any situation in life, even one like this, people don't engage in a behavior for a prolonged period of time without getting something out of it in some way. Is the fact that you can't move forward a payoff in itself somehow? This might sound odd, but some people are obsessed with the negatives in their life. You may wonder, "How could anyone say that?" I have known people who have been in a depression for so long, that it doesn't feel normal not to be in that state of mind. This kind of behavior is usually a chemical imbalance that can be dealt with by a doctor. The task is getting yourself, or that person, to go see the doctor.

Are you punishing yourself because you feel you deserve to be punished for being a bad parent? Or, do you think you were a bad sibling/friend/spouse/relative for not supporting the family enough, and now it's too late?

If you won't or can't seem to move past the grieving process, ask yourself, if *YOU* were the one who had died, wouldn't you want your family to move on? Well, you are someone's family and they need you; and no one wants or needs to grieve forever. You keep the child alive in your memory, by remembering pleasant things, through pictures and through the love you both shared. Don't focus on the moment the person died; instead focus on the moments they lived, the joy they brought into your life, and what you learned from them. More than likely you became a better person because of them. Think of ways to use the experience in a positive way in your life; perhaps by helping others. As a hospice nurse I try to practice what I preach.

While I was researching for this book, I spoke with many parents. Some were parents who were overwhelmed with guilt because of the times, when life was so difficult, they had wished that their child had not been born. Could this be a reason of unnecessary guilt? This is a normal feeling at times if you are a caring person. There is nothing about which to feel shameful or guilty. It is our compassion for our children that causes us to feel this way and the overwhelming exhaustion that can come with diligent care. We would never really wish this on our child, we love them so much; but we can only take so much heartache and stress. You need to forgive yourself.

There comes a time when we have to say, *"Enough is enough."* and give up the pain. <u>How long we grieve or how deeply we hurt does not reflect how much we loved.</u> It doesn't mean that we love her any less if the grieving is not eternal. It doesn't mean we've forgotten them. When we are ready, it might help to speak out loud to our child's picture, expressing our feeling and continued love for them. We could tell them that we love them and always will, but we have to go on with life. Then don't let anything or anyone make you feel any different. Grieving is a very personal experience for everyone. No other person can or should tell us how to grieve.

I hope this will help you identify what you are feeling. Also this will let you know we are not alone in these feelings. *We grieve even when there is no death but from a loss of some sort;* if we feel loss, we all must find ways to cope with it.

The 5 stages of grief

According to professionals, the biggest challenge people who are grieving deal with is getting their minds connected to what they're facing. If you are in the grieving process, you can expect to go through these stages, though not necessarily in this order:

1. Denial/shock: This is the immediate and first stage, that of disbelief. It is a feeling of unreality and feeling numb. During those first days there is a feeling of being-out-of-touch. If you were used to going someplace like a hospital, it will feel like you need some place to go and it takes a while to break the habit of feeling this way. Some people experience a sort of panic.

2. Bargaining: This bargaining may be with ourself or it can be a petition to God to "please let our child pull through and we will do . . . " or make some such bargain. Often we will offer something to try to take away the reality of what is happening. It is only human to want things to be normal or back the way they were.

3. Anger: The anger can manifest itself in many ways. We can blame others for our loss and lash out. We can become easily agitated having emotional outbursts, including weeping uncontrollably. We can even become angry with ourselves and become self-destructive. Care must be taken that this does not turn inward. Some individuals feel anger at whatever *"caused"* the loss of the loved one; and some get angry at God for allowing it to happen.

4. Depression: This is a likely outcome for all people who grieve a loss. One can feel in the grip of mental instability. This is what I would consider the most difficult stage to handle. There can be the feeling of listlessness and tiredness. You may burst helplessly into tears, feeling like there is no purpose to life any more. Pleasure and joy can be difficult to achieve even from things and activities which once gave you pleasure. You can find yourself wandering around aimlessly, forgetting things, and not being able to finish what you started. Physical symptoms also can appear. There may be an inability to resume *"business-as-usual."* Please recognize that this is a normal phenomenon. A grieving person's entire being – emotional, physical and spiritual - is focused on the loss that just occurred. It can be all encompassing, but if it lasts too long, you are encouraged to seek a doctors advice.

5. Resolution of grief: This is the final stage of grief. It is when you realize that life has to go on. You can, at this stage, start to accept your loss. You should now be able to regain your energy and have goals for the future. It may take some time to get here, but you will begin to feel like there is a way past the grief, an end to the sadness. Balance in life will start to return little by little, much like healing from a severe physical wound. There are no set time frames for healing this kind of injury. Each individual heals differently. Now there begins to be *Hope*. The ever-present pain of grief will lessen and hope for life emerges. Plans are made for the future and you are able to move forward with good feelings knowing you will always remember and have memories of your loved one.

Getting through the grieving process

Emotions can be overwhelming in the midst of grief—so much so that just getting through each day is difficult. During this time it's important to remember that there are no real guidelines for the recovery process. People heal in their own time and in their own way. Here are some things you can do that have helped others. They would like to share them with you.

• **Know that it is not a betrayal** to the memory of your child to begin the healing process.

• **Take your time.** Don't let anyone rush you into "getting over it". This is your loss and the feelings are yours.

• **Don't make any major decisions.** The time of grief is a time of instability and there is too much room for error in making major decisions that can affect your family.

- **Avoid the temptation to use alcohol or drugs** to numb the painful feelings. This is very important because it can lead to making wrong decisions, cause more grief to the other family members, and add other negatives to your life.

- **CRY!** By all means cry. Tears are the healthiest expression of grief. Don't try to hold back the tears. Others may need to cry also, but think they are being brave for you; and thus, you for them. Then you all need to grieve together; cry together.

- **Know that there will be good days and bad days** for quite some time. Pangs of intense grief can surface and immediately overwhelm you, especially during significant dates, like a birthday or the anniversary of their death.

- **Remember your loved one often.** Honor your child by talking about them, sharing with others what you miss the most. Look at photographs, and retell your memories to friends and other members of the family. This is important.

- **Seek out people who will understand your need to talk about what happened.** Seek out people who will really listen to your remembrances. You will need a whole network of support persons. Support groups are set up for just this purpose. Seek out grief counseling if you feel you cannot cope alone. Grief counseling is available through community resources, churches and licensed therapists. Join a grief support group. Use the Internet and join an electronic bulletin board dedicated to supporting individuals. Remember your grief is individual to you. Not everyone's grief is identical to yours. You will share some similarities with others, but grieving is a very personal and very individual process.

- **Pay attention to your health.** Make sure you are getting enough sleep, eat healthy. Get outside in the sunshine and fresh air, maybe take a walk. See your doctor if you need some help with any concerns.

- **Don't be afraid or too prideful to ask for what you need** from others. It doesn't make you a weak person. Accept a kind offer. Be honest and open about what it is you need. Now is not the time to try to do everything yourself; others may want and even need to help. Perhaps they are grieving also, and *NEED* to do something.

- **Dealing with a severe illness of a child or death is like any significant wound that can leave a scar.** This scar is not visible to the eye, but it is, none the less, there; and so are the memories of the child. Always remember that there is no right or wrong way to grieve AND it is not a betrayal to the memory of your child to begin the healing process.

Giving support to a grieving friend

- **Be available.** If you are always too busy or can't be there every time they call upon you, they will eventually quit calling. Even when it may be a serious call for help.

- **Remember that your friend is in a very different emotional place than you are.** Try to put yourself into their shoes for the moment and be kind.

- **If you don't know what to say, just ask, *"Do you feel like talking about this right now?"*** If they do, be there for them as a listening ear. Then don't be critical about what you might hear. If they don't want to discuss their heartache, don't press the issue. Let them know that you are there for them regardless.

- **Don't treat your friend like they are an invalid.** Encourage your friend to get out and get busy doing normal day-to-day activities. They won't want to at first, but encourage them to just test the waters, so to speak, and go for an outing of coffee or tea for short periods of time. Be supportive, but not smothering.

- **Recognize that you may need your own support system.** Sometimes you can give support, and other times you'll need to receive it. Don't expect yourself to always be the leader and the only one carrying the load.

- **Watch out for a shift into depression.** If you see your friend withdrawing into an emotional cocoon, it's time to intervene.

Blank page for personal data:

Medical Trivia

Blondes have more hair than dark haired people.

Blank page for personal data:

"You grow up the day you have your first real laugh . . . at yourself." Ethel Barrymore

CHAPTER 19: PARENT TO PARENT

PARENT TO PARENT TALK

I don't want you to view this chapter as lecturing because that is not the intent. The purpose of this section is to help you notice if you need more encouragement and reassurance, now that life has changed. We can be the most confident person in the world in our niche (that which we know best) but when it comes to 22q/VCFS and the future of our children it is natural to feel less confident. Perhaps we start doubting ourself, our marriage, our job, and our parenting skills; until our self-esteem starts to suffer. So please read this section and if it can help in any way, it has done its job.

First take the small test below and see how you score on this today. Be honest!

Self-esteem quiz

• Is your self-esteem based on what others think of you or your child?

• Do you do things to make other people happy, even if it makes you feel bad? Do you do things even if it goes against your better judgment?

• Do you have a hard time being happy for others when they brag about how smart and quick-to-learn their child is?

• Do you call yourself names like *"stupid"* or *"dummy"* when you make mistakes—or sometimes even when you don't make a mistake? Perhaps wondering if the VCFS gene could have come from you? (For those who do not know)

• Do you have a hard time taking risks in health treatments and rely on other people to guide you in each direction?

• Do you feel like the worst parent and feel uncertain about whether or not you can handle this disorder?

Scoring:

If you answered yes to any of these questions, your self-concept or self-esteem may be hurting. So now is the time to concentrate on YOU and get built up.

In an earlier chapter, I mentioned a book called **Selfable** by Barbara Burley. She also offers a seminar. I attended one of her seminars and it helped me when I was feeling down. Barbara Burley is a professor at Arizona State University. For more information on her book or seminars, go to: **www.selfable.com** (I am not affiliated with her in any way, it is just suggested reading.) Sometimes all we need is a little lift like this seminar gave me. Some people will need much more.

Another thing that I did to help myself was to attend some free library classes. I had to force myself to go because I knew I was depressed at the time. I also knew that if I didn't do something I would continue to feel the way I did and go downward in self-esteem. I could not afford to let that happen. I had a family that needed me, so I decided to *be kind to myself*.

Be kind to yourself

The more people I meet the more I can see that we are not much different from one another, no matter where we live, or how we grew up, or what situations we are dealing with in our personal lives. We all have similar feelings in given situations, such as *LOSS*. I learned *LATE*, that it is *OK NOT* to be perfect; and that we must allow ourselves to grieve over things like loss of what society calls _normal_, whether it is within us or in our children.

If you are reading this, you're either involved in dealing with a disability directly or are someone who is acquainted with disabilities. In either case I hope this article will help you or someone you know.

First let me say, that we, as humans, somehow have gotten the notion that we have to be almost perfect to be worthy of love and respect. We think that, if we fail or perceive ourselves as failures, others will also. I am primarily speaking about us as parents or care givers, but it can be in any venue of life. I start out with this, because I know first-hand what it is like to get overwhelmed emotionally, feeling as if *"I am dropping the ball"* of my responsibilities. I know what it is like to have circumstances in life change; where you feel you are losing everything financially, perhaps even being on the verge of bankruptcy. I know all too well, what it is like to either be self-employed and not have your invoices paid on time and then having to struggle to make payroll, or to work for a company that you thought was secure, only to find one day you're given a pink slip after many years of service. So I think that on this subject it is important to remind ourselves, *"It's OK to grieve about all kinds of losses"* and that can mean in any way we have to do it. Many people have the wrong idea that when we show emotion it is simply feeling sorry for ourselves. In reality, it is nature's way of finding an outlet for the pressure that trauma can build up in us. It can also be perceived as part of the healing process, allowing us time to sort things out. We are more capable of making rational decisions after the pressure has been released.

Everyone, at one point in their life, will suffer a loss; it could be loss of a loved one in divorce or death, it might be of a job or, yes, loss of health. When this occurs we find we are at a loss as to what to do and many times we try so hard to make things right that it back-fires and makes matters worse. I think that for some reason, we start thinking that we are stupid and asking, *"What's the use?"*, *"Why me?"*, and many who believe in God, start to ask, *"Why is God allowing this to happen to me?"*, *"Am I being punished for something I've done?"*, or *"Is God keeping account of everything I have done wrong?"*

The truth is we are our own worst enemies in times like this; this response is actually guilt. But before taking on any guilt, we need to ask ourselves, *"Is this feeling of guilt really warranted?"* Usually it is not. In the case of a child with a genetic disorder for example; most of us did not know anything about VCFS/22q and would have given anything not to have our child suffer. So, is guilt in this scenario warranted? NO, and neither is it realistic to think that if we get in financial straights through no fault of our own (such as in the case of medical bills) that we are failures. Remember guilt can be useful if we *NEED* to change, but it can be destructive and paralyzing if it is not the appropriate emotion. Guilt can plunge us into depression and despair.

Many of us, as parents, find it hard to accept the fact that our children are not society's *normal* *child* and probably never will be. This form of denial is common at first; when our children are young and there is sometimes little difference between our children and the children of their own age group; but when the differences start to show, it dawns on us that we can't run away from the truth any longer. We feel like we could kick ourselves for not getting the child special education or therapies because we, parents or grandparents, could not accept it, or chose to ignore it.

One of the most powerful emotions is *ANGER*. This can actually be a good thing, if it is properly channeled and understood. It can help in healing our depression and getting us motivated to hang-in-there and keep going in the right direction, to get informed, to not give up, to do all we can no matter what obstacles get in the way. (This is where you should hear the theme song from *Rocky*®, the movie.)

I would encourage anyone who has suffered a loss of any kind to try to get back into a routine as soon as possible; and to find a support group to help you to see and understand that there are many families out in this world who have undergone the same ordeals. It helps to bond with people who have similar situations, people who you know *really* understand. Some other things that can help when you are dealing with a loss or crisis in your life is to read up on the *"problem itself"* and perhaps keep a journal of things that are happening. Ask about these unusual things you have noticed or felt. No, no one will think you are odd or crazy, and especially not the people in the support groups.

There is a wholeness of a person or couple that has come to terms with reality. This wholeness comes when we learn we are strong, strong enough to survive each trauma or crisis in our lives. We come to know that our marriages are strong enough to weather the storms that get thrown in our path. Remember, "*A successful marriage isn't finding the right person--it's being the right person,*" and that is an understatement in marriages that have the added pressures of caring for a disabled child.

It is important to know that life is *NOT* a trap set by God. He is not looking for any failures or weaknesses He can find, so he can condemn us to punishment. I like to remember the scripture of Ecclesiastes 9:11, where it says that *". . . unforeseen occurrences befall us all,"* so it is not fate or some predestined choice by God when something negative happens to us personally, or to our children.

When we accept the fact that imperfections are a part of being human; we can live life and appreciate it. We will have achieved a wholeness that many can only dream of achieving.

So, of all the things I have shared with you, it can be summed up in one sentence: <u>*"Exercise patience and you're allowed to be kind to yourself."*</u>

Improve self-esteem and family happiness

Goal attainers in this life have plans. This is a fact! A person has to know what the plan or goal is, and then how to execute it. Whether it is healing a relationship, getting a new job, learning all you can about a health issue or finding inner peace within yourself; having a plan comes first. Consider these characteristics that are common in people who succeed.

Have an <u>*idea*</u> of what you want and define it on paper (ex: writing a book). Keep the visualization strong if it is to become attainable, so that it keeps you motivated. Determine in your own mind, to what point the idea or goal can be deemed a success, so that you don't feel you are not ever satisfied. Nothing works out perfectly. Be <u>*flexible*</u>. Life is not a success-only journey. Even the best-laid plans sometimes must be altered and changed. Be open to input and consider any potentially practical alternative. Be willing to start over, without becoming discouraged and losing your zeal for your goal. Be willing to take a certain amount of risk. People who attain goals are willing to get out of their comfort zone, talk to strangers (such as other parents), and try alternatives. Be willing to thrust yourself into the unknown and leave behind the safe and familiar (such as your local hospitals and doctors) in order to bring about your goal(s). Of course, for the most part, I am referring to investigating the best way to learn about your child's disorder, what choices you have, how to arrive at the best form of treatment, and by whom. But these steps work for most any goal.

Have a _strategy_. People who achieve their goals in life have a clear and thought-out strategy. They learn what they need to do to accomplish/attain their goal. They write the strategies down in a ledger and _prioritize_ them. They document the results to avoid getting side-tracked, and to be able to alter the approach, should the first plan not work. They might even realize at some point that the goal was not reasonable or attainable.

Have _enthusiasm_ for your goal, a determination. Get excited when you think about your goal and its outcome (like I did when writing this book). You need to live and breathe what it is that you want (without becoming obsessed); be passionately invested in both time and energy to attain the goal. Don't let your past become your future, think positively. Recognize and pursue truth. People who attain their goals have no room in their lives for negative self-talk, fantasy or unrealistic pursuits. They are faithful to themselves, rather than self-deluding; and they hold themselves to high but realistic standards. They deal with the truth.

Surround yourself with a group of people who want you to succeed, people who might want similar goals. These people can share their experiences and move with you toward your goal. Choose and bond with people who have skills, talents and abilities that you do not, so you compliment each other. Support groups and/or classes are good places to start. The main thing is to take action. Don't just sit and think about what you want or need. Plan a meaningful, purposeful, directional, plan of attack and go forward.

All of this is great and dandy, but don't forget the most important thing. Take care of ourself as an individual. You are the most important resource in achieving your goals. Be conscious of managing your mental, physical, emotional and spiritual health above all.

Remember, it's not an accident. _"When you think better, you feel better. But you've got to believe what you're thinking."_

Defining external factors

I thought these *"factors"* below were very interesting when I read them on the Dr. Phil's web site at **www.drphil.com**. The actual numbers might be debatable, but it is the concept that is important. So I thought them fitting in this section because we are a product of all the external factors we encounter throughout life. It is these experiences that can answer many of the questions we want to know about ourselves. I think that realizing some of these factors can help us understand ourselves and could, in fact, help us to deal with whatever life throws at us. Much of what is thrown at us is because of VCFS in our lives; but knowing why we handle circumstances the way we do can be helpful in changing or reinforcing in us what we need to know to carry on.

Dr. Phil says that life has three types of external factors:

- **Ten Defining Moments:** In every person's life, there have been moments, both positive and negative, that have defined and/or redefined who we are. Those events entered our perception with such influence that they changed the very foundation of whom and what we thought we were. A part of you was changed by those events, and caused you to define yourself, to some degree by your experience of that event itself.

- **Seven Critical Choices:** There are a surprisingly small number of choices that rise to the level of life-changing ones. Serious choices are those that have changed your life, positively or negatively, and are a major feature in determining who and what you will become. They are the choices that have affected your life up to today, and have set you on your life's path. It is choices, not fate, that determine the outcome.

- **Five Pivotal People:** These are the people who have left indelible images on your impression of self, and therefore, the life you live. They may be family members, friends or co-workers, and their influences can be either positive or negative. They are people who can establish whether you live consistently with your genuine self, or instead live a phony life controlled by an imagined self-image that has crowded out who you really are. We don't want to compare our reality to everyone else's facade and therefore create a mask that is not genuine.

Defining your internal factors

Even though these ideas happen inside you, it's easiest if we think about them as behaviors, because they are actions that you choose. By choosing how to identify yourself, you can either perform to success or perform to fail. For example, if you believe you are competent and extraordinary, you will try to live up to that fact. If you believe you are incompetent and worthless, you will live down to that belief.

The powerful internal factors that shape your self-concept are:

- **Internal dialogue:** This is the continuous mental conversation that you have with yourself about everything that happens to you and provokes a physiological change. Each thought comes with a physical reaction. This is why it is important to keep more positive thoughts than negative. Positive thoughts are empowering and according to many doctors physically healthier as well. I know that when I was going through my own depression, my internal dialogue was not too flattering. I could sure see a difference when I forced myself to change the channel so to speak; when I began speaking more positively and uplifting to myself. I felt its influence within a few days. It was powerful. *(See Psychophysiology on page 121.)*

- **Labeling:** Humans tend to organize things into categories or stereotypes. We even categorize other humans by labeling them into groups, classes and the functions they perform. But were you aware that we label ourselves, also? For better or worse, these labels have a powerful impact on our perception of ourselves because we tend to "live" up to the categories we've attached to ourselves, such as, *"I'm a loser."* or *"I'm one of the fortunate ones."* Self-defeating behaviors and thoughts are easy labels to give ourselves when we are feeling down or disappointed but it is difficult to change these labels and believe it. This is especially true if these labels were put upon us and we were convinced that what others said was true. Many of us have had someone in our lives who has said or done something that really hurt us to the quick of our very being. Now we know that we can over come it, and so we must start by picking a known positive truth about our self like, *"I'm a good parent."* Believe it and build upon it so that our inner label is something that makes us feel proud.

- **Tape-like recall:** These are beliefs and perceptions that have become so deeply ingrained that they *"play"* automatically in our heads like a tape recorder and can influence our behavior without our being aware of it. These tape-like mental reoccurrences can be dangerous and potentially self-defeating, because they have the power to set you up for an outcome you would not choose. This is why we must be careful what we tell our children.

- **Fixed beliefs/limiting beliefs:** Fixed beliefs are the beliefs we hold about ourselves, others, and life's circumstances that have been repeated for so long they have become ingrained and are difficult to change. Limited beliefs are the beliefs we have about ourselves that limit what we reach out for and achieve because of our confirmation that we can't or shouldn't. They also cause us to block any conflicting (positive) information while confirming any new negative information. These are really an enemy of self-esteem.

This was one area where I personally worked to overcome. I really believed I had no talent; I was the only person I ever knew who got bad grades in art at school. But I did know I loved people, I knew that I related to people, I had fellow feeling, etc.; so I changed my thinking process and started to look for ways that I could change that belief about myself. Now, you are reading some of the ways I changed my beliefs about myself.

We talked about labels before, but "How do you label yourself?"

Are we consciously aware of even a fraction of our labels, whether they come from the outside world or from within ourselves? We must acknowledge the existence of labels, confront the *"fit,"* and deal with the impact these labels have on our concept of self.

Ask yourself the following questions in order to start identifying and evaluating your labels:

- **How do you label yourself?** Are you a career person, a parent, a kind person, a lawyer? Are you a failure or a winner? Are you fat or pretty? Write down all the labels you attach to yourself, going back as far as you can remember.

- **Where did these labels come from?** Did they come from you? Your parents? Another family member? A friend? A teacher? Look at each label you wrote down in the first question, and identify where each one originated.

- **Are you living up to your labels?** How are your labels working for you, in a positive or negative way? What are your payoffs from the labels, both good and bad?

Once you can identify your labels; you can work to change the labels within yourself, if you don't like them. The person who labeled you is not the authority and has no merit. You control your own label.

The cause of stress

Matters of work, money, family and health are the most common sources of stress. Many would say that a life without stress would be wonderful. But it usually is the things that are most dear to us that cause the most amount of stress, or we wouldn't care enough to be stressed. With a certain amount of stress, we do our best. It is the little everyday stresses, not the catastrophic losses in our lives; that cause the most damaging stress, in my opinion.

In the context of this text, it is more than likely, the everyday wear and tear on our minds; being consumed as we are. with family, marriage and our child's health issues. Parenting in itself is inherently stressful with health and safety issues to deal with. But it is a fact, that parents, who have low weight babies or children with disabilities, have shown higher levels of stress; even after the condition is no longer a distressful shock.

Each new stage of life, no matter how old we are, gives us a certain amount of stress. Do you sometimes feel like, *"Life is like licking honey off a thorn?"* Most of us normally learn how to develop the resources to deal with each stressor as we mature, *BUT*, when we have the added challenge of a disabled child piled on top of the other demands of professional, family, and personal responsibilities; the total amount of emotional stress can be overwhelming! With this said, how can we recognize when it is more than just stress?

Stressors like traumas in our lives such as divorce, or dealing with our child's issues on top of divorce, can exacerbate psychological disorders or can occasionally lead to the creation of one; as in the onset of depression or anxiety disorders. Unlike other people who experience stress caused by a specific situation, people with a disorder feel fearful, uneasy or distressed for no obvious reason. Sometimes the trigger of such a condition can be the fact of never being able to feel confident that our life, including our children and marriage, will weather the storm in the face of VCFS. If you feel that your stress level is more than you can bear, or you feel you can't go on much longer with the amount of stress you feel on a daily basis, contact your doctor. He might be able to give you the help you need to live a happier and healthier life. If you need to do this you are *NOT* weak and *NOT* a failure.

Temper that can turn to RAGE . . .

A person may wonder why this paragraph is in this book. Well, let's face it, we are all human and we all have faults and most of us have a temper. People might wonder how anyone could become enraged with any child, especially a _disabled_ _child_, to the point of physical abuse. It happens all the time, because many feel they are out of control. They become impatient and are scared of the future, for their child, and themselves. What should they do? These parents feel they are *"losing it."* They can't stop feeling that they might someday physically harm their child. They also feel guilty about feeling this way and therefore must be a bad person. This creates more guilt and anger. Can you identify with that feeling? If so, you need a plan of action to take, when you experience the very first signs of anger that can turn to rage. Ralph Waldo Emerson made a thought provoking statement when he said, *"We each boil at different degrees."*

It doesn't have to be rage against the child or a known reason, but some rages can surface from frustration, anxiety, lack of coping skills, stresses of work or daily living, or at oneself for not being able to be the kind of person you want to be.

Blank page for personal data:

Medical Trivia:

An average human scalp has 100,000 hairs.

"If we wait for the moment when everything is absolutely ready, we shall never begin." Ivan Turgenev

CHAPTER 20: PARENTS SPEAK OUT

PARENTS SPEAK OUT

The following information and suggestions have been contributed by parents and grandparents of children with VCFS. Names have been changed to protect privacy.

Adoption

- **Adoption:** What would be the odds that I adopted 2 children, not related to each other, and that both would have VCFS? The similarities are that both came from mothers who had no prenatal care and were drug addicts. What kind of drugs? I don't know. We just believe it was the common denominator. I take the children to the genetics clinic at our local hospital for the doctors to keep track of their progress, or lack of it many times. I live in a large city where you would think more than just a handful of the doctors would have heard of VCFS. I am finding that the schools have *ZERO* knowledge of this disorder.

The children are 6 and 7 and they are both in kindergarten. Since I have 2 kids with VCFS, I am aware that our kids have a wide variety of similarities; and yet many differences. The teachers seem to think that since *both* kids have the disorder, they now know all about it and there will be no problems. I keep trying to remind the teachers that, while both kids may have VCFS, they are *different* in many ways. They have many different symptoms, one hyper, one not, yet some similarities. Our biggest issue with both is speech problems.

- **Adoption:** I feel compelled to write this; as it is more of an adoption issue. We have two boys—both adopted at infancy. When we first met our son who is now a teen with VCFS, we were told that he had cardiac issues that had been repaired and that he would lead a relatively normal life. Well, surprise! Surprise! We soon realized there were many other issues and finally, that he had VCFS. But our feelings were no different from many of you who have biological children and received the same diagnosis.

One ironic part of this is that we had discussed adopting a special needs child prior to going through the adoption process but decided that first we would find out about parenthood without the added pressures of special challenges. Anyway, now comes the added irony. We decided, after a couple of years, to try to adopt a second child. We were looking for a healthy child that was not so financially draining. We adopted another little boy at infancy. He had been premature but we were told that he was fine—just small. Well, not long after that, we found out that our second son had health issues as well! When I received the diagnosis I reacted selfishly. I thought, "I don't know if I can handle this—raising two special needs children." And I cried all the way home, with my little boy happily playing in his car seat.

When my husband arrived home that evening, from a business trip, I told him what the doctor had told me and discussed the ramifications for the baby. I ended by saying, "I don't know how we are going to do this!" My dear, pragmatic husband turned to me and said, "Who better to do this?" It put a lot of things in perspective. The bottom line is, we all take chances (whether biologically or through adoption), and sometimes things just don't work out the way we expect; but I would not trade my two boys for any others!

- **Adoption:** We adopted a little guy who has VCFS. We got our son at age 2 and he acted like someone had shot him out of a cannon. He raged and hurt himself and banged his head to go to sleep. (He broke three doors banging his head.) He was diagnosed with ADHD, but stimulants never helped. They made him worse. He seemed to suddenly start talking and normalizing at age 4-5 and his raging stopped. However, he was still hyper. He taught himself to read by sight, but is not good with phonics. He can add, subtract, and multiply; but does not understand word math problems. His IQ tested at 75, but I feel it is higher; however, he can not answer abstract questions well. He was diagnosed with bipolar last year and is on medication. He did not rage, but he did hear voices, which he claims have gone away with the medicine. Sometimes he cycles. Usually he acts like a very hyperactive, immature 10 year old boy; nobody would guess the real problem. He had open heart surgery at 10 days old and has low set folded ears. Since he is African-American it is hard to tell if his nose is too broad or bulbous because of VCFS or is normal for his heritage since he was adopted. I _really_ worry about his future as an adult. I am not a young person and if something were to happen to me, I fear the worst. He would have to take care of himself somehow. We don't have enough money to set up a trust for him. What happens to these kids when we are gone? Who helps them? The other thing I worry about the most is the psychiatric problems. Do the psychiatric problems in VCFS children get better as they get older? There sure isn't much about this disorder, but what there is seems to be the same information over and over again. I love my son very much and want a good life for him. Thank you so much for letting me voice my concerns.

- **Adoption:** I adopted my 2 sons from a sibling group of four, all four children have VCFS/22q. I adopted the 2 boys and their 2 sisters were adopted into another family. (So much for the 50/50 chance of having VCFS children.) We got into adoption as foster parents with infertility problems of our own. The first boy Tod was one of our placements in the foster program who became available for adoption. Then a short time later the other boy Ted was born and the birth-mother asked that he be placed with us, and we wanted him also since he was Tod's natural brother. At the time of the adoptions we had not been given a diagnosis, just a collection of _anomalies_ that no one could figure out. We were given conflicting opinions, of course. One was that Tod was hopelessly retarded and would never be able to attend school with his peers. The oldest, Tod, had his cleft palate repaired and ear surgery.

Since both boys have VCFS and require so much extra help I have decided home schooling is the best for them. I believe public school would have been a real nightmare for them full time. I still have not changed in that opinion. The boys recently had IEPs done at the local charter school where they take a few classes. Both are considered learning disabled. The oldest, Tod, is the age of an eighth grader, but performs in the 4th-5th grade level. The younger, Ted, is a 6th grader. While he has a reading level at nearly 7th grade, the rest of his academics are at 4th grade level. As far as I am concerned, this is the easiest part of raising the boys, because I have teachers and a psychologist helping and offering strategies to make the process of learning easier. The emotional issues and behavioral challenges as they become teenagers and then adults are what I fear.

- **Adoption:** Sonja is 14 years old and was adopted from Southeast Asia at age 2. We didn't get her diagnosis of 22q until she was 11, even though she had Tetrology of Fallot and cleft palate. As parents of two healthy boys by birth, and another healthy daughter from the same country, we felt that we could accept a special needs child. We were aware of the heart defect and cleft palate; both supposedly _correctable_ _handicaps_. At the time, we couldn't put a finger on it, but felt there was more going on with her.

Now, my boys are young men and my older daughter is a senior in high school. I wanted to let you all know that it's kind of nice having Sonja around. She's always game to go shopping or out to lunch with her mom. Oh, we still have our challenges and always will, but she is our special one. Sonja will probably be having surgery again on her heart valve and will soon be starting high school. I have to decide where she will best be served, as the _multi-factored evaluation test_ shows that she could fall into either category, _Developmentally_ _Handicapped_ or _Multi-handicapped_.

Adults: VCFS/22q children over 18

- **Adults-VCFS children over 18:** I thought this was interesting to share: the latest technologies rapidly becoming available. My 20 year old daughter really wants to eventually get married and have a baby. We were discussing adoption which probably wouldn't be approved, BUT, something new is happening. One of my cousins and her husband (who is sterile) will be taking advantage of it. She is looking into adopting _"Frozen embryos."_ It turns out that there are many frozen embryos out there from women who tried this method of having babies and then have extra embryos leftover. Many of these ladies do not want to destroy the embryos and are looking for good homes in which to place them. So my cousin will be able to adopt an embryo, and have it placed in her womb. (They try one at a time to see if it "takes".) Then she has a regular pregnancy, labor and childbirth and has a baby. Anyway, I thought this might be an option for our adult daughters with VCFS and I wanted to put that idea out for future consideration. Of course, all the issues of *"having a mother with a genetic disorder raising a child"* will have to be discussed, but this might be a better option than having a natural child who may have VCFS. *ALSO*, for the men who have VCFS who one day want to father children. There is a procedure where the sperm can be separated into those without the deletion and those with the deletion. Sperm without can then be introduced to the wife's eggs. This procedure is *NOT* paid for by insurance at this time.

- **Adults-VCFS children over 18:** There were many times when I felt that my daughter was stuck way behind her peers and would never accomplish a particular milestone; and then she would have a burst of developmental growth that would amaze me. Now, at 25, even though she can't count change consistently, she has developed wonderfully in her social skills; something I thought would never happen. Our kids struggle to do OK in so many areas of life that they desperately want to be normal, average and for everything not to be so hard, etc. Yes, they want sexual relationships and loving relationships just like we do. I had decided a long time ago that my daughter would never be able to live independently and now she is surprising me at every turn and I am not so sure. Hang in there all of you with adult children. This business of raising an adult child with a disability is a whole new learning experience.

- **Adults-VCFS children over 18:** My daughter is 19 and has a *"low IQ"* around 87, but she is good at computers and art. Our state *Rehabilitation Commission* (every state should have something similar) is paying for her to go to college. She is going to take computer courses and art courses so that she can do Web sites, graphics, etc. She also wants to take video and photography courses and become a video photographer as well as photograph weddings. This would be an excellent side job. All college campuses have *disability services* and they will connect you with the *rehabilitation commission* or similar agency. They also provide modifications so that the young people can take as many college courses as they want. No, maybe they won't be able to get a degree, or maybe they will; depending on their ambition. I personally have a girl in my church who has full blown autism and she received her degree in art. The disability services helped her take every course required for the degree so it is not just a *"give away"* degree. Kids with disabilities might look into the more *"creative"* courses for their degree, such as: art, music, computers, athletics, etc. There are many kids with high intelligence levels and no disabilities, who screw up their lives with drugs and alcohol. My sister is one of them. It is not ability, it is *"hard work"*, strength and courage, and the ability to find solutions, endure and stick it out. That is what we need to be teaching our kids, not trying to limit them. Just work our way around the obstacles, one step at a time!

- **Adults-VCFS children over 18:** Our VCFS daughter, Shell, is now 22. For 30 days, we have been doggedly fighting the good fight to get her more help—just thought I'd share some of our experiences.

She was in transitional housing through community mental health, with 24 hour supervision and medication monitoring. This boarding house attempts, in 24 months, to teach young mentally ill adults how to transition and live independently. They are supposed to provide case management services (medical, employment, life skills, social security application, counseling, and even a payee). I cannot even begin to tell you the moral, legal, ethical mistakes they have made at my daughter's expense. They evicted her and gave her 30 days notice to evacuate. They provided zero help in finding anything else; rather, they just recommended that she go to a mission with the bums, male alcoholics, perverts and who knows what else. I think they hoped they could offload her and be done with it. Part of her *"treatment"* plan was to make appointments, look for work, etc.; *BUT* because she is severely over-medicated, by a doctor (who has never talked to us and has put her on 5 terrible medications) all she can do is eat and sleep. By sleeping, she missed her appointments and therefore didn't keep up with her *"treatment plan"*. Because she is an <u>adult</u>, the case workers and the director of the boarding house will not talk to us, even though Shell has authorized/consented for us to be involved.

I could go on and on. I finally just snapped and called DSHS (<u>Department of Social and Human Services</u>) and because they have a *"No Wrong Door"* policy, they *HAD* to help me at the number I called. I was put in touch with the state mental health network, several layers higher than the community mental health service; and found an *OMBUDSMAN*, who worked magic all over the place. With pressure from him, the mental health people all of a sudden upgraded her to a more serious level, magically found crisis respite housing with 24 hour supervision, and better than where she had been placed previously. She was to be put out on the street that Friday, instead the house director had to take Shell to the headquarters/respite center.

We also met with a Dr. W.G. He was a Godsend. We spent hours with him (most of which Shell slept through) and he is involved with her previous medical doctor. We finally have some hope to change her medication and get her the help she needs and deserves. I finally feel validated by a number of people in the system, and not like a nutty, overbearing, co-dependent mother. What's scary is that up until age 18, things were really going well, but now we are seeing deterioration and total lack of care/help by community mental health. I found nobody, and I mean *NOBODY*, in all of these systems, who had a clue about VCFS. We have learned that a lot of our issues could have been helped by having limited or full guardianship, and finally, we are going to pursue that; not to take anything away from our daughter, but rather to be able to more expeditiously help her. I hope my experience helps someone so that they do not have to go through this the hard way.

- **Adults-VCFS children over 18:** I think my experience is good to know about as your children reach adulthood. Some of you will need to go for the guardianship provision. In California, I obtained a limited conservatorship for my daughter Nancy who is almost 23 yrs now. She agreed to it, but it ended up a bit messy because her estranged father (of all people) decided she didn't need it. (What was he thinking?) He showed up the day we went to court for the *"easy signing of things"* to contest, and told the judge, "blah, blah, blah" which ended up costing me $3000 extra because of all the return court visits. The *Bottom Line*—limited conservatorship is: access to all confidential records; I must approve moves; she can not move out of state and I have to approve any medical changes, surgeries or medications. All seems to be working for now. Her father's objection was that she has a *higher intelligence* than I give her credit for having. It is kind of funny really, we work so hard with our kids, teaching them to fit in and how to communicate (with the intensive speech and language training) and it must work, like in my daughter's case. She compensates well for her limitations, just talking to her a few minutes isn't enough to see my concerns about her reasoning, decision making, and how very vulnerable she is. The good news? It was Nancy's court appointed Public Attorney who informed her father why it was needed.

- **Adults-VCFS children over 18:** My son Steven is now 19 years old. During this year he went to work for a place that taught him to drive a fork lift; he does well but just needs more practice driving. Steven now works full time at the warehouse and is very happy when he is kept busy driving. Now he would love to get his driver's license, but we are stalling.

As far as his music is concerned Steven is very musically inclined. In fact, one of his music teachers encourages him to play his guitar at special events. Steven is earning a reputation for his music and is being recognized in our small town, which is exciting. Our biggest problem is that Steven cannot take disappointment and he gets really devastated when plans fall through, which they do, many times in life. We often *have* to have a *Plan B*, but if that falls through we have to treat him with a mild natural tranquilizer to calm him. The trouble with this disorder is its like living with a human time bomb; and we never know when it is going to explode. Another big problem is that Steven does not have any close friends. The other musicians he meets treat him as a fellow musician and do not see him as someone with disabilities. I just wish his peers would see him that way and support him.

Another issue we must face in the near future is that of Steven eventually wanting to get married and have children. He likes girls and would really love to have a girl friend. How do we go about letting him know that he may marry, but should not have children, when he doesn't accept that he has a problem? There is no way that I am prepared to look after another VCFS baby, a feeling I know many parents with VCFS children must share.

On a positive note, Steven has a marvelous sense of humor, loves to talk on the telephone and remembers numbers. He cannot spell to save his life, but this has not really posed a problem; because if he wants to know something he finds the information or asks someone. Steven is physically fit and does body building and we find this is a good outlet for his frustrations. We started a VCFS support group but in our area there are only small children. Steven was the first to be tested positive for VCFS in our area and we received absolutely *NO* counseling; but it seems that we intuitively did all the right things.

- **Adults-VCFS children over 18:** I have read personal stories on the *Internet Digests* every day for the past four years and much of the information is concerning young children with VCFS. Although it is disheartening to hear stories of adult children, it reconfirms the reality that we are all together in this disorder with our concerns. We love our children so much and feel so helpless at times. We cannot take anything for granted.

My 21 year old son, Monte, has come so far. We have a bond that I don't have with my older three children. As I read different *Digests* my thoughts of the past come to mind thinking *"been there, done that!"* Monte is one of the miracle children who has a very complex medical history. We have been to Oklahoma, Texas, California, Syracuse, and to Philadelphia for evaluations. I'm happy to say that he lives independently in a group home complex. He has his own apartment, goes to school, works part time, rides the city bus by himself, serves dinners at church and has established his own life. There were so many nights that I worried about his future. I now find I can sleep and am content.

- **Adults-VCFS children over 18:** Dionne wasn't diagnosed with VCFS until age 21, two years after she'd become very ill. So our family's journey has been very different from the experiences of those whose children were diagnosed much earlier. In my view, it's not only the child who needs to be on an even keel (even if this means medication), it's the whole family. Dionne's problems surfaced so gradually that we all adjusted inch by inch, and didn't quite understand that the stress we were all under had long reached the point of constant emotional pain for all of us. In a way it's an advantage for parents to know what might lie ahead, and in a way it's not. I think the best approach for parents of younger kids is to try to find a balance between your awareness of the possible problems in VCFS and your need to live an ordinary life without constant fear of the future. It's impossible to predict the course of such a complex genetic condition. It's good to gather a small arsenal of potential resources, and also to have the awareness that they might not be necessary. It's true that we could never have guessed while Dionne was in elementary school, that she might one day need to live in a group home for a while. But we were generally aware that group homes did exist, so when the need came we were able to access a good one by asking trusted professionals for suggestions; and then looking each place over and interviewing the staff, just as if she were about to go off to boarding school.

The two hardest things for us have been: 1) fear of the unknown and 2) giving up cherished dreams. I don't mean to tempt fate, but I do think we're now sufficiently experienced to be able to face new challenges competently, if and when they arise. The "dreams" part has been more painful. We really thought and hoped that Dionne would simply go off to college (in a very structured, supported, part-time sort of way) and gradually put together a normal life for herself with friends, jobs, etc. When her pre-college summer program turned out to include a psychotic break, I thought I'd die of anguish; not only because she was ill, but because someone in my family had a psychotic illness while I was growing up, and all those horrible memories came flooding in. Yet here we are, nearly 5 years later, and Dionne is stable, happy, and reasonably productive; in most of the ways we'd envisioned for her before her episode. Medications, behavioral psychotherapy, and the like, do exist for addressing everything from OCD to depression to psychosis. And yes, it's a good idea to learn what's possible with VCFS and to become aware of available resources. The resources may not need to be used, but they're there. It's not necessary or possible to plan our tactics in advance. Trying to do so would cause a state of hyper-alertness that wears parents down fast. I guess what I'm trying to advise is simply to observe your child's quirks and symptoms, remain aware that they _might_ at some point become problematic, and trust that if that does happen you have the ability to search out remedies and cope, one step at a time.

- **Adults-VFCFS children over 18:** Sean is now well past 18 years old and my best advice to you with younger children is to take each day/week/month as it happens. I think the reason so many doctors cannot answer questions about the future is because they honestly don't know. When you think about it, they couldn't predict the future of our other children accurately, and 22q is a syndrome of such variable symptoms, that no one can tell what is ahead. In some ways, the late diagnosis for us was a blessing and frustrating. It got to be almost scary going to children's hospitals because they always seemed to find one more thing. It was not until he was almost 15 years of age that they did the blood work to see if all his symptoms were connected to one syndrome. Sean has always had obsessions, probably very close to OCD. He writes lists, for himself. He loves movies, so every year for Christmas one of his sisters gives him a huge book listing videos available. At first he listed movies he had, then ones he has seen and now I think he is writing the entire book into scribbles. This is the kid who claims he cannot possibly do homework because his hand hurts! (Nice try!) Another obsession is Elvis. He knows everything about Elvis; his movies, his songs, his family, his birth, his death—oh my, does he know death. Another obsession is death: who died, when they died, where they died and why they died. We finally had to say, "Enough, no more dead people!" It was depressing!

He was respirator dependent for so long because his heart condition was inoperable. He was on a respirator for 3 years and he is now on oxygen 24/7. When they listen to his lungs, they tell me there is very poor air flow everywhere. Fortunately no one has told Sean that he is on *"borrowed time"*, in that instance, it is better not to know.

- **Adults-VCFS children over 18:** Carmel is 18 now, and is out of school; she graduated last year. She is working full time and saving money for a car. She works at a local theater, which is a tremendous amount of help for her social skills. But since she is a very fast typist, I plan on getting an appointment at the rehabilitation commission to see if they can help her find a typing job. She starts part time college in summer. She is of low intelligence, but not so low she can't take a college course. The university says they can give her class modifications. We let her have off a year to work and travel, etc., but we told her that once her *"year is up"* she has to have some kind of college course each semester as long as she lives with us to help her improve her skills. She wants to take some computer courses.

She has several *"higher intelligent"* good guys from church who *"like"* her, but our church follows the courtship theory, where the kids don't date until they are ready to marry after 20 or so; so Carmel has been very blessed to be around a good crowd of kids and has had a very active social life, yet she is still safe. They do group stuff all the time, like go to movies in groups, eat out in groups, etc. She has settled down emotionally enough that we don't fight as much with the mother-daughter thing. So we have every hope that she will be able to stay with us for a little longer, and at least go to college for one or two classes next year. I still think the home-schooling in middle school and most of high school was what helped give her such a satisfying life, because she never got into a *"bad"* crowd. So all parents out there with kids with VCFS should have lots of hope that even though it is sad our kids have this disorder, there are a lot worse things to have and our children do have potential for very good lives.

Advocacy

- **Advocacy:** What is wrong with this picture? Do we have to fight even for reasonable thinking on the part of those who teach our disabled children? I don't understand this. Our VCFS daughter has an IEP in place, but got all D's on her report card. If she has an *"individual"* education plan, that means it is supposed to be geared just for her and her type of disabilities. Right? And if it is *"for her"* how can she be earning D's? Then the teacher wrote this comment, *"Your daughter's willingness to try at activities that she struggles with is an excellent role model for her peers."* Well, that's a real nice comment, but should she be given things that make her *"struggle?"* Challenge her, yes, but struggle and get *D's*? Once again, what is the point of an IEP if she can't do well with it, even if they can see she is trying? There is something wrong with this system. What makes it worse is the fact that this school is for disabled children, and they should know better. This makes me wonder, is this grading because there is a set standard that you have to attain to get better then D's on the IEP, like a *"normal"* school criteria? If so, how can it be individual if it doesn't work to help a child master <u>something</u> and gain any self-esteem? I wonder if this is a problem everywhere. If so it needs to be addressed to *ALL* school boards.

- **Advocacy:** Our daughter, Lee, (14-VCFS) has experienced a fairly significant deterioration in her hearing and learning in the last 6-12 months. She was seen by a local ENT in the last few weeks, and had her hearing checked, which confirmed the deterioration we have observed at home and in her school work Being the helpful person that I am, I took a copy of the report to the school and gave it to the speech therapist. The speech therapist mentioned that she would be contacting the district deaf/hard of hearing specialist, which sounded like a good idea to me. A few days ago I received a written notice that they would be conducting an (IEP) meeting. Then today we got a call from the speech therapist saying that the audiology report was not sufficient; that they needed the doctor's interpretation, and that I should bring it to the meeting. My query is: Why is an audiology report alone insufficient? Insufficient for what? I wasn't asking for anything special that she was not already receiving, only providing them with information. I am now learning to ask questions. Hmmm . . . Are there other services she should have but is not getting? These are questions we need to ask!

Antibiotics

- **Antibiotics:** Repeated courses of FULL strength antibiotics can cause candidasis, also known as a yeast infection or thrush. Systemic candidasis can be due to stress. Candidasis is not uncommon in children with VCFS/DiGeorge. This may lead you to investigate the child's diet because food allergies can stimulate adverse reactions.

Attitudes

- **Attitude:** Many wonder if our children can be productive and give to the community. My daughter is currently a volunteer at the elementary school in our district (small, rural district of less than 2000 students). Her day begins with a bus ride to school. (I question whether she will ever have a driver's license of her own.) She goes to the computer labs, turns on all of the machines and puts in their passwords and remains available to assist kids with their computer assisted reading programs. Next, she helps with food preparation, serving and cleaning-up in the cafeteria; works with kids in the special needs pre-school and ends the day with the office staff doing everything from filing, to running messages, to operating the copy machines. Then she rides home on the bus. She has also done such things as video taping teacher's lessons, reading to non-readers, decorating bulletin boards and *"playing"* with various groups during recess and lunch. This is *"work"* that fits her abilities and need for something more stimulating. Most telling of all, was her comment a few days ago, *"Why wasn't I learning how to do these kinds of things in school that I'm learning now; instead of all that stuff I've forgotten?"* What she means, is that she needed to learn how to apply what skills she had to something useful, instead of struggling to retain certain facts and information that never made sense to her or that she would never use—especially in the last few years of high school. Clearly, for a person with VCFS, IQ test scores do not define current or potential ability. Rather, those dropping scores should be seen as an indication of how far the academic world still has to go before it fully understands the nature of learning and the business of teaching, when dealing with a mind that has been affected by a genetic accident. *(Author comment: I, for one, am awed by both this mother and daughter. We can ALL learn from this experience.)*

- **Attitude:** I recall when my husband and I first realized that our son was not going to be what we had intended in the intellectual department. He was just days old when we met with the doctors and they informed us that our son had VCFS and presented us with a list of symptoms that were possible. I remember them saying that the first hurdle would be to repair his heart; but all I could think about at the time, was that he might be cognitively impaired, wouldn't be good in school and that he might never go to college. I now can't believe how shallow I was, I wasn't worrying about the heart surgery. I figured it was simply mechanics and that it could be fixed. Now I think what a terribly hard thing it must be for doctors to break such difficult news to shocked parents. I no longer have my priorities confused and I am a better and happier parent because of it.

- **Attitude:** Many parents have voiced the thought that they are afraid that they will be guilty of the very thing that they are afraid others in the world will do; and that is to see their own child as the *sum total of all their symptoms,* rather than as a *child* with symptoms. This can be crippling to a parent. We want to be able to look at our child and plan for a good, relatively normal, life. As parents, we can get too involved with the now, and what we will have to worry about next. I was sent this Web site that I thought had a remarkably kind and balanced view. You can see it at **http://www.disabilityisnatural.com**. What a refreshing idea, that disabilities are natural. I also strongly agree with the idea of getting our children involved in the community. (Author's comment: My children are adults now but I used that idea of community and it worked.)

- **Attitude:** I am writing to the young families about fears and concerns. I wanted to write, to encourage. Here is my advice about our kids as they grow older. We need to sit back and enjoy our children. I know that's easier said than done, but you cannot predict and worry about what will happen today or tomorrow. You need to take things as they come and deal with them then. Don't worry about the future too much. Not all children are the same, and this goes for normal children as well. We need to love our children and just be there for them when they need us most. Remember when you worry it makes you feel down; and our kids need to see us happy and healthy for them and for our own sakes. Hang in there all of you. To the moms who are afraid to have more kids. I only have one child and at first I was bummed that I didn't have anymore or didn't want to chance it again; but we are best friends and he is so special to me. I hope this can ease someone's mind . . . one can fulfill you.

- **Attitude:** I am often asked to describe the experience of raising a child with a disability; to try to help people who have not shared that unique experience, understand and to imagine how it must feel. It reminds me of an essay I read called **Welcome to Holland** by Emily Perl Kingsley. To summarize it: When you're going to have a baby, it's like planning a fabulous once-in-a-lifetime vacation trip, to say, Italy. You buy a bunch of guide books and make out your wonderful itinerary of the journey. The Coliseum, Michelangelo's David, the gondolas in Venice. You might have even learned to speak Italian. It's all very exciting! After 9 months of planning and with eager anticipation, the day finally arrives; so you pack your bags and off you go. Then several laboring hours later, the plane lands. The stewardess comes in and announces, *"Welcome to Holland." "Holland?!"*, you say. *"What do you mean Holland? I bought a ticket to Italy! I'm supposed to be in Italy. All my life I've dreamed of going to Italy. I prepared for years for Italy."* But there's been a change in the flight plan and they've landed you in Holland and there you must stay. The important thing, is that they haven't taken you to a horribly filthy place, full of pestilence, famine and disease. It's just a different place and might even be a difficult place at times. So now you must go out and buy all new guide books and learn a whole new language. Now you will meet a whole new group of people who you would never have met had you gone to Italy. It is just a different place. It's slower-paced, less flashy, has a different lifestyle, and different politics than Italy. Then after you've been there for a while and you catch your breath after the shock, you look around. You begin to notice that Holland has windmills, Rembrandts, beautiful tulips and it's actually an interesting place. But everyone you know is busy coming and going from Italy, and they're all bragging about what a wonderful time they had. And for the rest of your life, you will say *"Yes, that's where I was supposed to go also. That's what I had planned."* The pain and suffering of knowing your dream is lost forever will never, ever go away; because the loss of that dream is so significant you may never feel this way about Italy again. But, then, I think that if you spend your life mourning the fact that you didn't get to go to Italy, you might never be free to enjoy the very special wonders of Holland. To some, on the surface, this story may seem like a slightly condescending and simplistic analogy. I first read it shortly after my son Jude was diagnosed with VCFS. I don't know if it was this story that affected my perspective of what having a VCFS kid meant. But when dealing with Jude who had all the usual stuff, heart defects, speech problems and a few others from left field; I really think parenting over the long term is like being dropped in a different country . . . and not necessarily a worse place either. Oh sure, I wish Jude hadn't been born with all these problems. But something tells me that if he hadn't had these problems, he wouldn't be quite the same child who we love and cherish today. It's very hard to explain, but there is something about him that makes you love him; maybe in spite of all he has been through, or possibly *because* of all he has been through.

- **Attitude:** I keep myself from depression by staying focused on what is happening now and plan a short distance ahead. I won't waste any energy worrying about something that isn't there! I tell you, it would be both exhausting and non-productive. I make memories; take care of today and plan next week, but not years ahead. There is no point! Honestly, just enjoy the joys we have and turn those challenges into success stories.

- **Attitude:** I truly believe that what we do with our children can make a difference. You can get a hopeless outlook at what a child can or can't do, but it is up to us as parents and up to the children themselves to discover their true potential. I watched a program on TV a year or so ago, where the development prognosis for a child born very prematurely was bleak, and it was suggested that maybe it would be better for her to be taken off life support. Her parents refused to do this and worked with her. She is now *"normal"* in her development. Had her parents been the kind who were laid back, then she would have been exactly the way the doctors predicted. I believe the same is true of our children and what we do with them.

Bedwetting

- **Bedwetting:** Many parents have stated that they are worried when their child is still wetting the bed at an age where most children are not. First, rule out any physical problem, and then look at emotional problems. It could just be that they are so familiar with the sensation that they ignore it; or if your child is taking medication, they may be in such a deep sleep that they don't feel the urges. If the child is old enough, perhaps set an alarm in the middle of the night and have them go to the toilet at that time each night. Perhaps it will become a habit. VCFS children need routine and it may be in this area also. Lots of VCFS kids take a long time learning bathroom procedures, so with a little help, they should be able to master it.

Books

- **Books:** Two very interesting and inspiring books that I would like to recommend are:
Raising Blaze by Debra Ginsberg and
Maverick Mind by Cheri Florance, Ph.D. There are both very interesting books about boys with serious learning disorders.

- **Books:** The best book, bar none, on the topic of encouragement and hope is ***Expecting Adam*** by Martha Beck. She is a Harvard-educated woman who found out she was carrying a child with Down's syndrome and chose to have him despite pressure from all her *"intellectual"* peers. She says she was scared to death that he would change her life and, *"Thank God, he did!"* It is my favorite book of all time. I have read it several times and have been buoyed by hope with each reading. A few other books are ***From the Heart: On Being the Mother of a Special Needs Child***, edited by Jayne D.B. Marsh. This one is a collection of essays from moms and deals with fear, isolation, all of those things we face. I got a great book for my daughter called **My Brother is a World-Class Pain**, and though it's about a brother with ADHD, most of it applies. It gave my daughter *"permission"* to be able to talk freely to me about the hardships of having a brother who is different. If you go to **http://www.exceptionalparents.com** there are many books about these inspirational topics. There is a book by Amy Baskin that sounds like it covers topics of interest. It was published in 2005. The author's Web site has information at **http://www.amybaskin.com** for those who are interested.

Here are a few of the books I would like to recommend:

Special Children Challenged Parents by Robert A. Naseef. Great book written by the father of an autistic child. Great for dads, too.

After the Tears: Parents Talk about Rasing a Child with a Disability by Robin Simons. Short, easy read, very good.

Changed by a Child by Barbara Gill. Great book. Short stories on everything! A bathroom or tub reader.

From the Heart: On Being the Mother of a Child with Special Needs by Jayne D. B. Marsh. True stories that arose from a support group. This one made me cry a bit.

Nobody's Perfect, Living & Growing with Children Who Have Special Needs by Nancy B. Miller.

It's amazing how these books help us to realize that it is OK to feel the way we feel and that there is no right way to get through the grief/pain of losing the *perfect* child. They also have helped me personally to realize that I have already overcome so many hurdles, and that I am a stronger person because of it. And most importantly, I found that I am not alone. Most of the stories in these books reflect my feelings, thoughts, and actions to a tee.

- **Books:** There is a book called **The Indigo Child**. It is a very interesting book. Whether or not our children have ADHD or ADD, they and we can benefit from this great book.

Cognitive regression

- **Cognitive regression:** When you hear the term _cognitive regression_ it is good to know that there are two kinds of cognitive regression that are mentioned. The term could mean that a child's cognitive abilities are actually declining, for example, the child has a vocabulary of 60 words, then 40, then 20, and so on, over several months or years. This is not usually seen in VCFS kids as far as I know. Or, it can mean that a child's cognitive abilities are increasing, but not as quickly as other kids' abilities. In the second case, which I believe through research is fairly common with kids with attention and/or learning disabilities; you can get IQ scores (for example) that get slightly lower year after year. This is called a _test artifact_, which means it's not an impassive decline (child measured against himself) but a normative decline (child measured against same-age peers). In this case, the VCFS child's vocabulary (for example) may be _INCREASING_ month by month or year by year (say 20, 40, 60 words); but the tests are formed on the performance of non-VCFS children, whose vocabularies are expected to increase even more (say 40, 100, 180 words). The VCFS child in such a situation is _NOT_ losing ground in relation to his _OWN_ skills, but is gaining skills at a slower rate than peers of the same age. We were told at CHOP that, yes, some kids have declines in IQs but that 22q11 is not a regression type condition. That is not to say there are not reasons why regression could happen.

- **Cognitive regression:** Dropping IQ is one thing that alarms us parents who believe there is something physical happening to the brain. My daughter dropped from a 90 at age 3 (she was non-verbal at that time so we were told it was likely higher) to 68 when she was tested 2 years ago at age 17. This lower score qualifies her for SSDI disability payments as it is in the mildly mentally retarded range. We knew from 2 previous test results that it was dropping. The reason is very simple and completely logical in light of what we know and have learned about the intellectual growth of people with VCFS. Our kids tend to be concrete thinkers. Typical-of-VCFS type, learning disabilities tend to be: math disability, language processing disability, and mild ADHD. So, as the tests moved from the concrete world of the very young, to the more abstract world of the older child, her scores dropped. However, her academic test scores improved over the years that the IQ was dropping. Consequently, this dropping IQ is termed a _test artifact_, in that the actual nature of the test fails to give a true picture of our children's abilities. Even at 19, my daughter continues to learn, develop and mature; but not in the same manner one would in order to produce a so-called normal IQ score. On the other hand, it assures her of needed financial/medical benefits.

- **Cognitive regression:** Our daughter's IQ took an abrupt drop when she was 13. We later figured out that was about the time when she started having thyroid failure. Untreated, serious thyroid failure can cause brain damage. Other medical conditions _MIGHT_ show up in an IQ drop if they are causing a lot of physical stress, poor oxygenation, and so on. So please rule out physical organic problems first.

Development

- **Development:** Another area of concern for our children is coordination (fine and gross motor skills). Maybe you have children who have problems learning to button or unbutton, using zippers, tying and untying their shoes. (Use Velcro®, elastic shoe strings that can be tied tight and slipped on, regular slip-on shoes or toggle drawstrings, which are a really good idea that one of the parents mentioned to me.) What if they can't bounce a ball, get a swing pumping, and participate in games at recess like the other kids? It is hard to remember that VCFS children learn these skills at different times. Don't stress too much. I know when I was a child at a preschool, they had a large activity board. This is where we learned to zip, tie, braid strands, etc. If you're handy, you may want to make one of these yourself and mount it on the wall. I can honestly say, that to this day I cannot bounce a basketball, or hit something like a fly, (I miss even when on top of it.) or eat without getting at least one dribble on my clothes (and with my big mouth how can I miss?). So remember, many of us who don't have VCFS can have the same problems. Perhaps set a time for fun exercises to practice in these weak areas. If it is a real worry, ask your doctor for a referral for some therapy.

Ears

- **Ears:** With regard to fluid in the ears, my daughter always has fluid in her ears. I had never before experienced anything like the large amount of fluid that she has in her ears. It's like having a runny nose. When she has a cold or upper respiratory infection the mucus is yellow, green and sometimes brown.

I am familiar with the ENT doctor who is part of the craniofacial team for VCF at Montefiore Children's Hospital located in the Bronx New York. What I do, in addition to drops for her ears, is to take my daughter to a pediatric chiropractor where she uses a candle. This candle gets inserted into the ear canal she lights it with a match. This method is painless. The candle burns half thru and then she removes the candle. Then she cuts the candle open and you would be amazed how much stuff comes out of my daughter's ear. There is always fluid that is in a granular substance, and then there is a clump of wax that is usually behind the granular substance. I have mentioned this procedure to both my pediatrician and my ENT specialist and they both agree that if it's helping my child, I should continue this process. We do this procedure regularly. By no means does this stop the fluid in the ear; but what it does do, is eliminate fluid that would just stay and pool in the inner ear because there is no drainage. My daughter went through 3 surgeries in the past for tubes and they didn't work at all. In fact, she still had fluid in her ears. She also had perforated eardrums, because an infection got so bad that the eardrum burst and the tube came out.

Her hearing, so far with speech, is not bad. While my child is in school she uses a FM unit. (This must be written on a prescription form by an ENT doctor and then it gets put on an IEP so that the school provides the unit.) What this does, is to allow my child to hear just the teacher and not any outside noises that go on in a normal classroom. Both teacher and child wear the unit so my daughter only hears the teacher's voice. It makes it easier for her to do her assignments and to pay closer attention to the lesson. I do hope this information will help someone. Don't try to do the candle trick yourself. Find a professional if you are interested.

- **Ears:** I just learned that the reason my daughter (8 months) is irritable a lot of the time is because she is hypersensitive to noise and outside stimulation. I have 5 other children so it is hard to keep the noise level down. I had not heard that was a problem with 22q children until recently; so I was a little surprised that it was such a simple reason, and one with an almost impossible solution with this many children in the house. But I can see the difference when it is quiet around her.

Education

- **Education:** A parent wrote (the teacher asked) *"One more question, do your children know right from wrong?"* This question came up in our IEP. That's a simple question with a complex answer! My first thought was *"That teacher doesn't have a clue!"* I think I can guess the teacher's thinking. Most kids have a sort of built-in radar that lets them understand basic rules without being taught. If a _normal_ kid breaks a rule, and is caught, he understands what he did and can choose to do it again (or not). He knows he would be breaking a rule. (That might be what the teacher meant by knowing right from wrong.) Kids with VCFS might not have this same radar. Or if they have it, it might not always work well. For example, a VCFS child might get in trouble for spitting in the classroom, and then not realize that he's not supposed to spit in the lunchroom, either. We found this many times with our child when she was growing up (and still now as an adult). She would understand one example of a rule, but she didn't automatically apply that rule to the _next_ example. Most kids sort-of pick up a lot of rules without even being told. Kids who have VCFS might have to be taught the rules one by one, with much patience and practice. This is something a good special education teacher has been trained to do. A regular classroom teacher might not understand it. For sure, it's not something that a parent can *"cure"* at home. It's important that everyone who deals with the child understands this, so parents and teachers are all working together.

- **Education:** My nine year old daughter Sylvia didn't talk except basic jabber, most of which was gibberish, until she was almost five years old; at which time a submucus cleft was discovered and repaired. I did find that sign language enabled her speech tremendously. I was very skeptical at first until I saw the results. The inability to make themselves understood can be very stressful on the little tikes. Sylvia's hair fell out for a time because of it. When teaching sign, we would put posters up with the sign and make sure to say it verbally at the same time. My other kids, all older, really got into this. If it becomes a game, it's great. I was amazed, while driving one day, when she was trying to tell us something that we just couldn't figure out. (I was about to pull my own hair out.) Then she signed the words *red*, *green,* and *lights*. She was commenting on the Christmas lights strung overhead. She was about four years old, and when I finally could listen to her and she knew she could be understood on any level, well . . . she was as bright as the lights!

My experience with math is that, unless the teacher can teach differently than with most kids, pulling the child out into a smaller class is about the only thing that works. Sylvia has to learn in the same sequence for everything. She's predominantly left handed so everything goes from the right to the left. I've been told that if I teach in the same pattern she will learn to process information better and quicker if it comes at her in the same sequence. Math concepts have to be taught over and over every year. Each year she gets a little more of the previous information, and it won't need to be repeated. She's in third grade and still doesn't tell time or count money. From what I understand about VCFS children, the ability to process things is diminished, some times greatly. But like for all of us, if it is repeated over and over it usually sinks in. I've seen progress, since the school has been educated about VCFS, Sylvia is more confident and is starting to work on her own somewhat. Unfortunately it all takes time and persistence. Summer school is beneficial. Sylvia has been in school since she was almost three years old, all year round; but without the consistency and structure that this year round system offers, I believe she would not have prospered as well as she has.

• **Education:** We discovered, through our daughter when she was 13, that she could follow through the steps of an algebra problem sitting at the table with her dad; but she couldn't, absolutely couldn't, do it on her own the next day in class. Math has always been so foreign to her that she never picked up the basics, as many are said to have been able to do. At 23 she truly doesn't have the concept that 2/3 is more than one-half, or even that 1/4 is less than 1/2. She can't make change and can't read a map. Yet she functions well in many other ways. She cooks from scratch, drives a car, etc. She's become *"allergic"* to math from repeated failures to understand concepts that "everybody else" automatically grasps. With <u>*GREAT*</u> tact and delicacy, I managed to demonstrate 1/4 and 1/2 on a pizza recently, but I'm not sure it stuck. I suspect she can't visualize these things as others do. She doesn't seem to have a place in her brain to hang the mental image of 1/4 of a pizza vs. 1/2 of a pizza. She has to learn these things one simple example at a time, and then the learning doesn't generalize to other situations. Her teachers regarded this as lack of motivation, or even obstinacy, but they were wrong. She cooks well because she trusts the cook book, and uses measuring cups and spoons. Just don't ask her to double a recipe!

- **Education:** The best place I have found to get a grip on what school districts must provide in the USA is through: **http://www.wrightslaw.com** You can sign up for a subscription to their newsletter.

- **Education:** In my daughter's school the punishment for unruly students is facing a wall in the principles office for a minimum of 30 minutes. My daughter is eight, and I know she couldn't last even ten minutes in that position with her ADHD. Fortunately, the school understands this. They accommodate her with a one-on-one volunteer aide who gives her the reminders she needs in class; and takes her out of class when she asks, or when the aide can see she is starting to get anxious. This seems to help her to stay on task and complete the work. It also helps to minimize the chances of embarrassing herself in front of her classmates, with the uncontrollable spinning or twitching in her chair. A one-on-one aide has allowed her to stay in a mainstream classroom, have friends, and learn to the best of her abilities at this stage of her education. I am happy that it is the law that whatever each child needs to perform at their best is to be provided. I think all schools should have this aid volunteer program.

Feeding

- **Feeding:** I have a son who is 8 years old and has VCFS. I want to comment about the term *"failure to thrive"* with regard to kids with VCFS. I know it is a common diagnosis when they are infants, but my son is still considered *"failure to thrive"* at 8 years old. My son was tube fed from birth because he had so many feeding problems; but he had his button taken out when he was 6, shortly after he had the pharyngeal flap surgery. I was told at that time that he should not need to be tube fed, because he is totally capable of eating on his own. I felt quite confident that we were doing the right thing for him by discontinuing the tube feedings. However, he has been at about 32.5 pounds for three years now (which is average for a 3 year old, not 8 year old). He did grow about 1-1/2 inches, so that is good; but on the growth chart this last visit to the developmental pediatrician, he actually went down on the growth curve instead of up. I do believe he may have gained about a pound suddenly over this past month or so, which is very encouraging. Maybe he will be going through a growth spurt after all. The developmental pediatrician we see and the pediatric GI doctor we have seen, both believe that our son has a great need to increase calories and that he should be getting anywhere between 1500-1800+ calories a day to maintain his growth. We may have to put the button (for tube feeding) back in place. It is very difficult to know what is best for your child when you have differing opinions from different doctors. The doctor who preformed the surgery has explained to us that VCFS kids tend to go through growth spurts and not to worry, that our son will grow in his own time. While other doctors tell us he needs to catch up *NOW*, or he may never catch up at all. Looking back I wish we had been encouraged to have done something sooner. He is pretty much healthy otherwise and doing well, even in school. I think our next stop might be an endocrinologist to look into hormone therapy.

- **Feeding:** I want to say that I know every child is different when it comes to feeding and eating; so what worked for one may or may not be helpful to you. But I will tell you that I do not think my daughter would be such a great eater today, if it were not for the book I read when she was first being weaned off her tube at 12 months old. The title is **Child of Mine . . . Feeding with Love and Good Sense** by Ellyn Satter. (She also wrote **How to Get Your Kid to Eat . . . But Not Too Much**.) This book was excellent. My daughter's occupational therapist recommended it. I was so happy, I still refer to it today if I worry about her feeding issues. We have always offered her the same foods we eat for a meal. At dinner time, in the beginning, she was only interested in one item out of the entire meal; but we noticed that over time she soon began to try other foods. With her (and the book talks about this) we noticed that she had to watch us eat something several times before she was comfortable trying it herself, and even then she might only try a bite the first time; but when it was offered again, at another meal she might eat several bites. She eventually began to like it and later she would even ask for it. For a while she was asking to eat corn-on-the-cob for breakfast, pretty cute. In the book it stated, *"Studies about children and food acceptance say that a toddler will go through a sampling and removing process 5, 10, 15, or even 20 times before she feels comfortable enough to swallow the food."* It also says *"It may be very discouraging because it looks like food rejection. It is not. It is the toddler's way of gaining comfort with the food."* After reading this section in the book I had an easier time understanding my daughter as I watched her eat. And I would (and still do) continue to offer the same foods on her plate even if the day before she turned her nose up at it or spit it out and said, *"YUCKY"*. I have never forced her to eat anything or try anything, instead I put it on her plate and let her look at it, examine it, or even try it and spit it out, eventually she has liked almost every food. She eats salad with dressing, meatloaf, chicken (she loves almost every type of meat, if cut small enough) potatoes, broccoli and carrots (cooked or raw), almost any type of sandwich, yogurt, bananas, apples, etc. This book also stated *"You are responsible for the what, when, and where, of feeding. Your child is responsible for the how much and whether of eating."* This book has many, many more good points and great advice for feeding.

Feet

- **Feet:** Here are some helpful tips. If you would like to order inexpensive felt inserts for excessive flat feet, I would suggest trying Hapad Inc. We use the Pediatric Comf-orthotic 3/4 length insoles. They cost about $6, and you stick them in their shoes. I am a physical therapist. My son has had foot pain, and his toes stick out on one foot. We saw an orthopedist who wasn't concerned, and ruled out any major problems. The supports are soft, offer mild correction, and could help prevent joint pain. It's worth a shot if your child only needs mild correction. Pronation (flat feet) can overstress joints, and lead to early arthritis. A little prevention may help. **1-800-544-2723** or **http://www.hapad.com**

Group Homes

- **Group Homes:** I have some words of encouragement for parents who feel that they have no choice but to place their child in a group home. It takes a *LOT* of love and courage to accept the help your child needs so that he can grow to be the person he is intended to be. It does *NOT* mean that you are doing anything wrong; it just means you are not equipped to provide for their needs by yourself now. When this happened years ago to me and my daughter I felt so ashamed as a mother; a total failure. I didn't tell but a few people because I was sure they would judge me or think I was nuts or selfish. I kept comparing myself to others who ran the group homes but I didn't have the training, the peers for her to associate with and absolute structure that was needed. Today my daughter *"chooses"* to stay in a group home; she is much higher functioning and all those problematic behaviors went away. Time, therapy, behavior modification, prayer, and consistency all made a difference and it was the best decision for us all. Even my daughter will testify to that today. Stay involved and visit regularly. We cannot do all of this alone, and fortunately, there are kind folks who own and run group homes. Just stay in touch and make unannounced visits; and be sure to view them before placement.

Handwriting

- **Handwriting:** We have struggled tremendously with handwriting. We were told at age 4 by the school occupational therapist (OT) that our daughter would *NEVER* be able to write. One month later, with the help of a private OT, she was writing her name. We now use a program called **Handwriting Without Tears**. It was developed by an OT and information is available at: **http://www.hwtears.com**. We use it at home since we home-school and the OT uses it during her therapy sessions as well. We are currently using a program called **Explode the Code** which is based on the Orton Gillingham method but is much cheaper (I actually called the Orton people and they suggested the **Explode the Code** program!) Anyway if anyone is interested in the **Explode the Code** books their company is Educational Publishing Service and they have many remedial products on their website at: **http://www.epsbooks.com**

Hearing/speech

- **Hearing/speech:** There have been some parents who have asked me if I had heard whether or not tics and/or making noises were common in 22q children. I personally don't know if that is a characteristic of 22q per se, but I do know that there are more than a few children who exhibit this behavior. My own son would make noises at times when he was young. It could be anywhere from a grunt if he wanted something to a hum or a chant-like sound, to a smacking of the lips. I always thought the latter was a way to tell me he was hungry. Now, I realize it could have been a comforting mechanism, an anxiety release. In our case it turned out to be a hearing impairment.

- **Hearing/speech:** Rolando was quite bright and incredibly cheerful, considering that he was in Pedi-ICU without a consistent parent in his life for the first 3 years. He is now 18 years old. He had one word, *"Hi"*, for which he would cover his trachea in the throat to make the sound. When the paperwork was being completed to allow us to take him into foster care, he learned the word *"home."* I assumed it was because on the weekends we were able to take him to our home, and I would go in and say, *"We are going home today."* It would excite my other children; and he got to know that he was about to have a fun day, a day out of the hospital. Near the end of the two months it took, he was being carried by our oldest daughter (a strong teen) and he spied the green van we drove, (to transport all 5 kids) he again said, "Home." I now suspect he might have thought that the van was home. When he finally moved in, he taught us sign language. He had signs for turning on the TV, signs for reading and even a sign to go shopping. We began speech therapy. I refused the offer to teach him proper signing; and requested she just begin with speech. At this time, we had no idea that delayed speech was part of the syndrome. I just felt it was due to having had a trachea in place and not having a parent figure to mimic. We didn't even know there was a syndrome. He was just a confusing list of symptoms for which there was always another reason.

Rolando has trouble with Math. I believe the explanation is a difficulty with abstract concepts. He did like geometry because it was something he could see and relate to such as a triangle or square. Regular math is just a bunch of meaningless numbers, usually involving memory work. At 18, he is still struggles to maintain a grade 3 or 4 math level. I believe that his spells of hypoxia are affecting his mental ability in some ways. The trick is to remember that, when our frustration grows, and also to remind the school.

• **Hearing/speech:** Robert has been getting speech for a year now (he's 2 1/2), without much success. He also had tubes placed in his ears in May. His ENT didn't see any reason not to put them in place. Robert had fluid in his middle ear constantly, with many infections. His ENT specializes in 22q, so he said that the chance of being speech delayed is great, so let's not wait and see, and put them in just in case. I totally agreed. It was so nice having a doctor be proactive about things. He has also been very proactive about Robert's airway issues. Robert has been signing a little more for his therapists, and using some sound cards more often. We are happy with his progress.

• **Hearing/speech:** Our daughter started making "uh uh uh uh" type noises, almost a gulping, shortly after the birth of our last baby. She initially did it when she was at the table eating and the baby was crying. We decided it must be a sensory overload. So we asked the developmental pediatrician and he said it seemed to be some sort of vocal tic. We remind her to stop when she does it. We just say, "Please stop." and she does. That tells us she knows it's a problem but we don't call attention to it in any other way. A lot of the time it completely disappears and then, in times of stress, she begins it again. I think it falls in the OCD type behaviors our kids with 22q can develop.

Infections

• **Infections:** For those who have repeated urinary tract infections, ear infections, etc., they might be caused by a variety of things including thrush, fungus infections or some other culprit that we may not normally consider. If your child has symptoms of severe dandruff, ringworm, athletes' foot, tiredness, catarrh (inflammation of the nose and throat) caused by fungal infection; working or living in a damp atmosphere can increase the chances. See your doctor.

• **Infections:** I know there is a lot of concern about giving too many antibiotics, and for good reasons in terms of the general population; but my experience is that to put your child on an antibiotic regime *"for a specific duration,"* if your doctor suggests it, is the right thing to do. Our son had constant illnesses from birth through early childhood. A single course of antibiotics would appear to cure him of the ear infection, strep-throat, pneumonia, bronchitis or any other infection; but within a week of stopping the antibiotic he would get sick again! One year he missed so many days of school that he never caught up. I know there were days we sent him to school out of guilt when he was ill. His body simply couldn't fight-off infections. Our doctor finally put him on a steady low dose of antibiotics for 3 months and he started doing better all the way around. While it's true, that overuse of antibiotics for people who aren't chronically ill is creating some disease-resistant germs; this is not the case when dealing with a child with a low immune system.

• **Infections:** My daughter was sick all the time until she was 9 months old, when we discovered that she didn't have a thymus and her T-cell count was very low. She had been put on an antibiotic, which was the only time she had been well. Then when she was taken off the antibiotics, she was well for only a month before the upper respiratory infections started again. She is now back on the antibiotic therapy, because she attends a nursery school 2 days a week. Her doctor felt it might be too risky to keep her off antibiotics over the winter with so many *"bugs"* around. I find that her stools are extremely loose, which makes it hard for potty training, but I feed her yogurt culture to help with that.

• **Infections:** Every kid is different but chronic infections in our 22q kids seem common with the low immune count. One of the things that worked for us was to have our daughter on a continuous low dose of antibiotics. The longest we did this was 6 months. If the drainage hasn't been cultured, have it done. There are strains out there that are especially resistant to antibiotics and need a longer course of treatment (or a different antibiotic). Sometimes they will out-grow the illness cycle and start to develop some immunity. At age 16 our daughter has reduced her infections to about one a year, since age 12; but we have problems with scar tissue becoming infected in her ears now, from the numerous tubes she had. In her case, the tubes served as a vent for the drainage to prevent build up; they did not eliminate the infections, but prevented pain from excess pressure and preserved her hearing.

Labels

- **Labels:** I personally do not like the term *"learning disability."* I don't believe it is a very useful term. I think of my son as having learning *difference* rather than a disability. I like to appreciate his true worth, without the stigma.

Medical records

- **Medical records:** Another general tip I've learned is to have (2) 3-ring binders, one for medical records and one for school records. It is simple and it has been a lifesaver for finding IEPs, or goals quickly. The medical record binder has been great so you don't have to repeat histories for the millionth time. When my son started having seizures, it was great to document those also, as well as ear infections, antibiotics etc. I have dividers in the medical record binder for each specialist, as well as home notes (seizures, fevers etc). I slip a current pictures of my son in the cover of the school record keeper, which I have in front of me at all IEP meetings, etc. I believe that it helps the school personnel remember that we're talking about a *little boy* and not a disorder. It keeps it personal. The school record has divisions for IEPs, OT, speech, teacher reports, etc.

Medications

- **Medications:** My daughter, who is 13, was recently put on antidepressants and her teacher says she seems happier at school. But I still see the disturbing sneaky behavior I asked the doctor to help with from the beginning. She smuggles things out of the house after being told numerous times she can't take "whatever" to school. It appears that she just feels like she doesn't have to listen or obey. I can explain to her in simple language, why she can't for example, take her Game Boy® to school: it could get stolen; she is in school to learn; she can play her Game Boy® before and after school; sometimes people have to follow rules, whether it be at home, at school, work, etc., even if they don't agree; and mostly, because I am the mother and *I SAID, "NO."* After our little talk, she will look at me and say,

"But why can't I take my Game Boy® to school!" It's like talking to a wall. If I take it away from her she drives me nuts asking for it back, like an obsession. I have been told it may be a form of <u>oppositional defiant disorder</u>, but I am not so sure. I feel at times as if I'm throwing our money away with counselors as nothing seems to help.

• **Medications:** My son is on a mood stabilizer and Ritalin. He is having a fantastic year mainstreaming in school with the help of a teacher's aide, who monitors him and helps him one-on-one. He is a great kid, and with the help of the medication I am able to enjoy him now. He is not having behavioral issues this year because he is very motivated to please his teacher, and he is not bored to tears, now that he is calmer. The aide has helped tremendously, and the class has really embraced him. He is very bright, and could read by kindergarten (as an encouragement to parents of younger children with 22q).

• **Medications:** Many are reasonably familiar with psychiatric medications and other methods for treating depression and anxiety. But it is always advised that you seek out a psychiatrist, preferably a child psychiatrist. They will know what medication or combination of medications would address anxiety and depression without lowering blood pressure, if that is an issue. It's not necessary to consider only anti-anxiety drugs, as many antidepressants work well to reduce symptoms of anxiety. There are also mood stabilizers and (please don't gasp!) anti-psychotic medications from which one can benefit, even if one is not psychotic. With our kids it's often important to take a mood stabilizer along with an antidepressant, because our kids can have some mood instability which an antidepressant, all by itself, could make worse. Working on symptoms can give <u>HOPE</u> for feeling better, which is extremely important (as I'm sure a parent knows) especially in a teenager who "hates being me". There are also plenty of <u>behavioral</u> methods for treating anxiety, depression, and social issues. These methods are generally used by psychologists, especially psychologists who follow the cognitive-behavioral model. It focuses on the day-to-day issues, including depression and panic attacks, and better ways to deal with them.

Non-verbal learning disorder (NVLD)

- **NVLD:** My son was non-verbal until he was about 2, then he said about 4 words. He is now 7. When we put him in a special education *inclusive* preschool class, he lost all ability to speak because they focused on his hearing loss, and did not feel he needed speech therapy, but needed to learn sign language more. He lost all ability to speak those words after 6 months in this *inclusive* program. Long story here, but after going to CHOP to get the medical backing I needed for him at age 5 (he was non verbal then too), I finally was able to get the county program to agree to speech therapy, although he only received 5 weeks of it before school let out for the summer. It was not until he reached Kindergarten, that a wonderful angel of a speech therapist believed in him as much as I did, and was determined to get my son to talk. By the end of his kindergarten year, he was saying a few words, and by the end of 1st grade he was speaking in sentences. I don't know if your child has a hearing impairment like my son, but don't let anyone tell you that your child does *NOT* need speech therapy when he is 2 or older and not speaking. Find someone good, who believes in your child, and is up to the challenge, to get him to speak. It is amazing what the right person can do for your child! Good luck and God bless you all!

Obsessions

- **Obsessions:** My son 10 with VCFS is fixated with his shoe laces. It has become such a big deal, that we have gone to slip-on shoes only. I didn't realize it was such a distraction until he played basketball this year and spent half the time on the court looking at his shoe laces, rather than looking at the ball during the game. We survived Little League only because he had a tongue flap that covered most of the shoe laces. Yesterday, both pairs of his slip-on shoes where dirty, so I pulled out a rarely worn pair of sneakers for him to wear to church. He got so upset that I was making him wear these shoes with laces, that he started crying. I asked him, "Does it bother you too much to have laces on your shoes?" and he said, "Yes." He also has other ritualistic actions that tend to be distracting and irritating at times. The positive side of this is that he will practice his golf by the hour. He loves to hit golf balls on the range, and seems to have a real gift for the game . . . the next Tiger Woods?

- **Obsessions:** A lot of our kids have obsessions and odd behaviors. My daughter is attracted to strings; she will *"twiddle"* or play with a string all day and night if allowed. We have developed rules for her to follow, such as; she can't have a string at the dinner table or take it into stores with her. They have limited her use of strings at school or she won't pay attention. We can't have anything with tassels in the house or she will strip them out. All of our towels have to be hemmed, no fringe or she will pull and pick at it until there is nothing left of the towel. We can have no rugs with fringes. Warning! All fringes are at risk! Hanging strings anywhere are not safe from her. We have to watch her carefully when we are out visiting somewhere; there might be hanging strings about. We had to take all her pants that had string-ties and replace the strings with elastic. That's just one of her obsessions. We also use slip-on shoes or Velcro® fasteners because the tie shoes were impossible for her to handle. She also likes to have everything the same. Change and any messes are difficult for her. She is the only one of my children who doesn't need to be told to clean her room. (So there is always a positive aspect to these obsessions.) I could go on and on but I think you get the picture. We are planning to move next month so this should be interesting. I'd better hide the twine. All of you who are dealing with obsessions, you're not alone!

- **Obsessions:** My daughter is very obsessive (this seems to be quite common). Her latest obsessions are nail polish and acquiring pets. The nail polish is dangerous, because she gets it all over the house. The obsessing about pets is tiresome. (We already have a dog, a cat, 2 hermit crabs, a snail and fish.) The tantrums are not fun. Next month it will be something different. She is less obsessive when she is sure about her schedule. The neurologist says it's not truly obsessive compulsive behavior since the obsession changes and it is not repetitive actions like washing hands. It seems to be an obsessive behavior peculiar to VCFS. We do not need to medicate and hopefully never will; but I know there are many families who have had to medicate for this issue, and who have found medication to be essential for the well-being of the entire family. Perhaps she's finding her niche, wanting to be a pet groomer, and putting polish on poodles?

- **Obsessions:** One parent wrote that the way she dealt with obsessions was through an OT (occupational therapist) who decided it was a sensory issue. The OT gradually introduced her daughter to different sensors. The parent said *"It worked!"*

Parent stories

• **Parent story:** Speaking as a parent who has had another child after my VCFS child, I can't even put into words how scary that decision was!! We have four children in all, so I did have 2 <u>normal</u> children before having our daughter, Suzy. My older children are better people because of having a sibling like Suzy. They are kinder, more patient, more accepting of people's differences and just more well-rounded people than they would have been otherwise. Yes, they would love it if we could afford trips that didn't involve seeing another specialist; they would love to go to Disneyland®. We would love to have our days not always be so difficult. (We are battling behavior and some emotional/mental health issues right now.) But if you ask them if they could have anything they wanted, even on a bad day, they wouldn't wish for her not to be a part of our family. They would wish for things to be easier for her, for her to be well and happy. Then along came our youngest. He is almost 2 now, and my only wish is that I would have had him when Suzy was younger! She has made so much progress since he came along. He doesn't respect her sensory boundaries like we do, so he has helped her to tolerate touches and other intrusions into her world. As he develops language, her language is going forward as well. Our developmental pediatrician said it's like she's getting another chance to learn these things when her brain is more *"ready"* to learn them, like she's being *"reprogrammed!"* I get a lot of comfort from looking at my healthy little boy and a sense of accomplishment. I think in the back of my mind that I felt that I had done *"something"* to cause Suzy's problems; so now that I have another child, and he is fine, I am not so hard on myself. It's like I have started living again. He is a wonderful developmental model for Suzy in other areas as well. He is also a blessing to us because we had forgotten what <u>normal</u> looked like! He can do things at 2 that Suzy still can't do and it is hilarious how *"gifted"* we think he is, until we look at his age mates, and realize they are all doing these things. I think I stopped looking at other kids because it hurt too much with Suzy. Now I am looking again and seeing she isn't so far off and she is learning so much each day. Everyone has to decide for their families what is right for them, but I can tell you, my only regret in having another child is that I didn't do it sooner. Everyone has to weigh what their family can handle, whether you have the resources to help when one child is hospitalized or whether whatever need they have might take you away from your other children. But realistically you will have that, even if you have completely normal healthy children. You could still have to be away from some of them for whatever reason. Your support systems will be a bigger factor in the case of a child who does require a lot of medical care but it's not an impossible situation! Take care, and know that whatever you do it will be right for *YOUR* family.

- **Parent story:** In the fall, I am planning on sending my beautiful child to a special school. He will only be 3 and he needs something that I can't give him—language. There the teachers are experienced at special education and will take my baby boy, who has VCFS, and teach him in ways I don't feel confident teaching. He has a complex heart defect, a trachea and uses a vent at night. He has a significant developmental delay, who knows if it is from the VCFS or the many weeks of intensive care, or both. But, my child is not learning like other children. It was hard for me to make this decision, but for my child and my family, it was the right decision. This is a respected school for profoundly mentally handicapped children. A large part of the population has Down's syndrome. Others have had strokes, severe CP, Rhett's syndrome, etc. It is a little school, with about 60 children, from birth through high school. My child is not as profoundly disabled as some of these children, and I would love for him to stay home with me for the next 3 years and then go to public kindergarten. But, that's not what's best for him. He needs the attention and the knowledge of professionals, these special education teachers. They are fluent in sign language, have teaching plans for each child, and have 3 teachers in a classroom of 8 children. Every child, every family, deserves the best education possible for their child. Let society label my child; let them say that he goes to a school for the *"retarded."* If this is the place where he will be accepted, where he will be challenged, where he can learn, where he will be valued for what he is and for the accomplishments that he will make; so be it. All of us have to make choices and decisions that are hard, but one day, if he is ready, I will send him to public school. But until then, my baby will go to the special school, because he is special, and for my child I just know in my heart it is the best for him.

- **Parent story:** Our son Omar is 10 years old and he was diagnosed with VCFS at birth, due to an interrupted aortic arch. He had surgery for that, and had a myriad of other heart problems that required subsequent surgery. An immune system problem was discovered when he was a week old, and was severe enough to require reverse isolation. He was diagnosed with DiGeorge/VCFS/22q shortly thereafter. The available literature at that time only discussed physical symptoms. There was some mention of a high risk of mental retardation, but no other discussion of problems. We started him in an early intervention program, and were prepared to look into alternative schooling. But when he was getting ready for kindergarten, his brothers school, (a private school) literally insisted that we permit them to have a week with him in the classroom to determine whether he could go there. His preschool teacher also indicated that she felt he was a *normal* kid. He ended up going there, and has been very, very happy. It was always obvious he wasn't going to be at the top of his class, but he learned to read and write. Years of speech therapy paid off, and he was basically the sweetest, most loving child around. He tried so hard to please. He did have some interesting little quirks—like an obsession with power rangers, followed by pokemon, etc., but I thought no more so than your average little boy. Last year, in third grade, the problems that so many have discussed about education surfaced. It was clear to me something was wrong, but everyone around me, including my husband and teachers, denied it. I began to do more and more research and saw information about NVLD, and it has been an eye opener. Omar appears to be a little quirky, but sweet, and I would expect him to grow up and have a fairly normal life. I see no signs he would need assisted living. He has some minor indications of obsessive compulsive (mostly that he will pick at himself just a tad when under stress) and occasionally has a tantrum. But nothing that seems completely out of line. Now, I wonder, will all of these things worsen as he grows older? Is there anything that I could do prophylactically to keep him on an even keel? Will there ever be an answer to these kinds of questions? Our pediatrician is not really helpful in this area. I adore him, but he just doesn't seem to realize the scope of the potential problems with VCFS.

- **Parent/guardian story:** I was sitting here wondering if there are other older siblings out there living in the same household with people with VCFS, when I decided to answer a request by this author. I was so relieved to know that there are possibly many. It is so wonderful to know there are other understanding people out there in this world, living through the same issues that I deal with in my sister everyday. I became Elaine's care taker after the death of our parents almost ten years ago. Elaine has obsessive/compulsive behaviors and argues with me all the time if I don't give in to her. She also asks the same questions over and over again before she finally stops asking. I wonder if it takes that long for it to sink in. Elaine never has interacted well with others, so school was a nightmare. It was very difficult for me when our parents died and I took over her care, not knowing fully with what I was dealing. She is now an adult and I can't imagine her ever going on her own, but she has surprised me before. I just wish we had known her diagnosis back when she was a child. It is almost impossible to get her any real help now that she is an adult, especially with such a new diagnosis and few professionals knowing all the aspects of VCFS. I am hoping that this is the beginning of a new awakening.

- **Parent story:** My daughter was diagnosed when she was one year old; she is almost 9 yrs old now. She was not born with a heart problem but she did have VPI, and she had surgery to correct that problem when she was around 4 years old. She also had problems with her kidney and had surgery to correct that problem when she was one year old. Today, she has lots of behavioral problems. She has been diagnosed with having *ADHD*, *obsessive compulsive disorder* and *generalized anxiety disorder*. She has been put on different types of medications but nothing really seems to help as of this time. I recently took her for another psychological evaluation. I was advised to get her into some kind of counseling to help her deal with her anxiety and to try medications again. I live in a very rural area and it's hard to find good doctors who are well informed. She has been having a lot of difficulties in school as well. Like any concerned parent, I want all the information I can get to help with the future of my daughter. It's hard dealing with some of the behavioral problems that we are going through. We all need someone to talk to that can understand our situation.

- **Grandparent Story:** I am a grandmother who just finished getting legal guardianship. My 18 month old granddaughter just went into the hospital and had a heart cauterization done. What the cardiologist and the surgeons said was that they will have to go in and close up the VSD. Her artery is so small and soft that it is behind the heart. When the heart becomes smaller, they are hoping that she will stop having so many pneumonias and breathing problems. There are so many risks with these children; I feel like I'm in a catch 22 all the time to do the right thing. If I say no to the surgeries then what? If she has the surgery; she will be on a ventilator, because the left lower lobe of her lung is partially collapsed and compromised. That poses another risk. I never thought I would have to make such decisions, at my age. I guess this is why I now believe that it is not a good idea for kids with disorders to have children. We, the parents, are getting older and may end up raising more children; and the child suffers as well.

- **Parent story:** I have 10 month old twin boys, only one of whom has VCFS/22q11 deletion. He has no cardiac, parathyroid or immune problems. His main problems were swallowing and terrible reflux. He spent his first 6 months in the hospital, had a gastric tube inserted and then a Nissen procedure to curb the reflux. He uses oxygen; mostly at night or when his pulse oxymeter numbers dip down. Our biggest frustrations are the swallowing and tube feedings. He has secretions coming out of his nose because this is where most of what he swallows goes. He has a healthy twin and it is difficult to juggle. Everyone tells me to relax and enjoy them and that the VCFS is not the end of the world; and I am sure it isn't. But it _IS_ hard! Having twins is hard enough, but to have one ill when you are a new parent makes it even tougher. Thank goodness VCFS usually is not in both twins! I can see where it can be a blessing to have a normal twin. As they will do things together and the healthier one can take the lead in teaching. They will be good friends and we will have a built in tutor.

- **Parent story:** My daughter Megan is only a year old and I always feel like I'm in nurse/case manager mode when it comes to her care. The mother in me wonders, if I were to die today, who would know my Megan the way I do? The doctors each know their specialty regarding her condition. My husband knows how to do basic care, but doesn't seem to want to get too involved in *"all that stuff"* because, he says, after all, *"You're the nurse."* Well, what do parents do who have *NO* medical background? They learn don't they, and fast? Don't get me wrong, my husband is willing to help like many husbands, but I think the big picture is just too overwhelming for him. I'm the one who sets all the appointments, talks with the doctors and therapists, schedules the feedings and medication, works the therapy with her, etc. I'm not angry that I do this; I am just scared that, if something happened to me, no one would be able to do it all for her, and things would get missed. There are times I just want to be her mommy and have some other people do the case managing for a while. Before I know it, time will have flown by and our daughter will be more and more dependent on me. If her dad barely knows the diagnosis, he can't stand in for me if I need him to be there for her. After all that is why there are *TWO* parents.

- **Parent story:** I want to address the fact that too many people, including family and friends are so oblivious to the fact that parents with 22q children need breaks from constant child care. People either don't want to see it or can't see it; but we desperately need someone to watch our disabled child for a weekend or maybe even for a week once in a while. Parents need to reconnect and unwind from the constant pressure. I would love a trip or mini-vacation to anywhere! I need to get to know my husband again; and sleep through a night, go to a restaurant with no need of special accommodations, walk hand-in-hand and laugh like we used to do. I would take pictures to remind us that not all is lost in our relationship; and that we can do this kind of *"therapy"* once in a while. The only time we were able to leave home; we had a turbulent flight and landed in a foreign country. Our luggage got lost; it probably ended up in Italy! The next day we were mugged. We couldn't really do much sightseeing because we had our sick child with us. No one back home wanted to take care of her so that my husband and I could have a much needed romantic get-away. Please, if you really want to help someone, give them the gift of personal time. Don't be afraid to take on a day or two of what we do *ALL* the time, *EVERYDAY!*

- **Parent story:** Here are a few recommendations I found useful for developing phonemic awareness and learning to read, which many of our children need. One is **Fast Forward**, a computer based, neurological, *"train the brain"* program that works to develop phonemic awareness, improve auditory processing and working memory. A less expensive alternative is **Earobics**, but it is not as low a level and is more useful to use after **Fast Forward**. When this neurological foundation is laid, these children may then need to have a good phonemic-based reading program, such as **Orton-Gillingham** or **Wilson** so that they can learn to read. After the child can decode, VCFS children often have comprehension problems that can be addressed on a neurological level with the **Lindamood Bell Visualizing and Verbalizing program**. This is a one-on-one intensive tutoring program that lasts about 3 months (a good summer program). It is also helpful to have the child evaluated for vision problems by a behavioral or developmental optometrist, as many VCFS children benefit greatly from vision therapy if they have visual processing and oculomotor problems. If visual problems do exist, children avoid reading, as it is extremely tiring and difficult to struggle through reading materials. Vision therapy is anywhere from 3-9 months and is intensive (another good summer program). The child must do the home exercises every day for about 15 minutes and visit the optometrist once a week. It takes a lot of commitment to do these interventions, and an out-of-pocket outlay of cash. Some school districts are paying for some of these services, so it might be worth approaching the school district. Even if you are home-schooling, the district is still supposed to be involved. The other recourse is borrowing the money. Sallie Mae has a k-12 low cost, government insured loan program that parents can borrow from for such things as private school or tutoring. The terms are very favorable and it is very easy to qualify. All of the above items and more are on the web. There are a lot of exciting research and intervention programs out there. When all is said and done, the money spent is well worth it. It is also a lot cheaper than paying for a private special education program that may or may not provide these cutting edge *"train the brain"* interventions. I Hope this helps.

- **Parent story:** My daughter Lisa fits a common pattern of doing OK up until about 3rd grade, but I knew there was something unusual about her learning style. When the learning began to be a little less concrete and more abstract, the difficulties become evident. Now that she is older I have other concerns. My main concerns with Lisa living independently, are:
  1. She has difficulty with understanding numbers and everyday math. She is at high risk for being unable to handle financial matters.
  2. Many entry level jobs in the service industry require handling money; something she would not be able to do reliably.
  3. Sometimes her common sense evaporates for no apparent

reason. She is quick to make decisions purely on the basis of emotion, especially as her ability to logically analyze a problem and predict the outcome is impaired.

4. She can be very moody

5. I don't know if she will ever be a safe independent driver. My goal for her is to be able to drive a car locally in non-rush hour traffic; to get to a local job or school, etc. But who is going to keep her operating within those constraints?

6. Will she be able to acquire the skills that will enable her to hold a job that pays a comfortable living wage in this generation?

Pets

- **Pets:** Many families think that an animal may be just what is needed in the house so a disabled child can bond. If you are one of these families, there is an incredible organization on-line, called Canine Companions®; and they have all kinds of wonderful dogs who are trained to be companions for different disabled people. **http://www.cci.org**. A friend of mine, who is a speech therapist, uses one at work with wonderful results. A few months ago, our family was blessed with a one year-old German Shepherd, who we got from another organization. To watch her with our son is to behold a small miracle. He talks to her, sneaks treats to her (We're trying to break that habit.) and asks for her at school. We did not go the Canine Companion® route because we only wanted a family pet, and our dog truly belongs to our whole family; but I can only imagine the joy, comfort, confidence and friendship that a companion dog could bring to a child or adolescent. Our dog was fully trained when we got her. Our sweet dog takes a little work, yes, with feeding, grooming and walking; but the rewards are truly astronomical. My son's life is a lonely one and it is so sweet to watch him sit with the dog. She has also been a Godsend to my daughter, whose life can also be isolated. As for me, having struggled for seven years now with knowing I don't have the strength to have another child, even though I desperately want one, the dog has filled that void as well. Nothing is perfect, as we all know, but if a family can deal with a little extra work, a good dog can make it more than worthwhile.

Psychological

- **Psychological:** It is probably worth saying that although my son has psychiatric issues, it is not the end of the world. It is like any of the other significant health issues our kids have; when there is a crisis, it can be overwhelming. At times we fear for their lives; the rest of the time we make adjustments to play whatever hand we've been dealt this round. Those of you whose kids have tracheas; it looks scary to people on the outside, and was most likely overwhelming initially to you; but now it's just part of the fabric of your lives. Ditto for those whose kids have g-tubes, etc., right? I was scared when I first learned about the heightened risk for major psychiatric disorders, back when my son was 2 years old. As things have turned out, he is experiencing some psychiatric manifestations of the deletion; but it has come as no shock. It is just another part of the package.

The training we get while trying to keep our kids alive when they're little, pays off later. We learn how to take things in stride and how to deal with the powers that be, on behalf of our children. We learn to accept our children for who they are and to celebrate things that other parents take for granted. My son (when not in crisis and on the wrong medication) has a good sense of humor, he is affectionate, smart, and I wouldn't trade him for any other child. On effective psychiatric medication, he is reasonably stable. So, please don't be afraid. Even if your child develops mental illness, he or she is still your darling child. My son is the same person I have always loved, and that is what it all comes down to. Doesn't it?

- **Psychological:** Concerning mental/emotional illnesses, we do need to educate our kids, family and friends about emotional illness in general so that it will not be a big *STIGMA* if it does happen. In other words, scientists and doctors are finding that emotional illnesses really are a chemical imbalance. The chemicals are out of whack, so this can be a physical illness just like heart disease, diabetes, etc. However, the prejudice about these conditions are still well-known. Most don't want to admit they have it and so they won't take their medications. However, if we educate our children and others about this being a chemical imbalance, then they might not be so reluctant to take their medication, if needed. For example: If I see a homeless person on the road who obviously has something wrong mentally, I tell my kids, *"That person could have a chemical imbalance which might have caused bipolar or schizophrenia; if they would regularly take their medicine and stay on it, they might be able to live a normal life. But a lot of times, people with these imbalances won't take their medications and so they have a lot of problems."* I have been trying to teach my child about VCFS. We read in some studies that the combination of B-6 and magnesium helps some people with autistic-type tendencies. My daughter can already see she is better in situations when she takes her medication. So I explain to her that it is because of her chemicals being unbalanced and taking her medication helps to balance them so that she feels better. This type of education will help our kids if they ever do need to take psychiatric meds.

- **Psychological:** Based on our own experiences, and on professional literature which says that kids with VCFS are at risk for bipolar disorder, I learned the hard way that this is true. My child was given psycho-stimulant medication. She was diagnosed with ADHD and it was, in fact, a childhood form of bipolar disorder. I found in such cases, taking a stimulant can worsen the behavior (more mood swings, wild irritability) and even trigger psychosis. So finding doctors who are knowledgeable about VCFS is important.

Seizures

- **Seizures:** Our daughter Gloria (7 years old) is now on a medication for her *generalized tonic-clonic seizures* (grand mal). She had 2 febrile seizures when she was 13 and 26 months with very high fevers. She never had any others. Then in July of this year, to our surprise, she had her first unprovoked seizure. The neurologist said her EEG was basically normal and she might never have another one. However, in October, they called from school to say that she was having another one. This time we did a sleep deprived EEG but that showed no abnormalities. However, the neurologist said with her history of febrile seizures and the fact that she had two in 3 months, it would probably be wise to start her on medicine. I know that CHOP has done research on children with the deletion and unprovoked seizures. It hasn't been published yet, but from what I understand, it did show a significant increase in seizures with the 22q11 deletion. Good luck to all who have children with seizures. I know from experience, they are extremely scary to watch. Hang in there and know that you are not alone.

- **Seizures:** I am a parent with a child who has seizures and I have learned that there is an important risk factor for seizures in teens with VCFS; due to hypoparathyroidism causing hypocalcaemia. During the growth spurt and other metabolic stresses of adolescence, it can be hard for a minimal parathyroid gland to keep up. Adolescence is probably the second highest period of life for new-onset seizures in patients with VCFS after the newborn period. One patient has been reported who had first time seizures in pregnancy! While there can certainly be other causes of seizures, my daughter benefits from a serum chemistry to check calcium levels and/or ionized calcium to see what the level is. This is a very easy problem to control by taking a simple supplement, like daily calcium, which might prevent seizures! Other signs of low calcium levels are pains in the hands and feet, diarrhea, fatigue/sleepiness, twitching movements.

Sleep

- **Sleep:** My son is now 23 years old and still requires lots of sleep. As a young child he could sleep 12 plus hours easy. Even if he would go to bed right after dinner, it was still a battle to get him up in the morning! I thought that he would have outgrown it by now, as most kids do; but he still requires lots of sleep. He loves to sleep. Of course I now believe that since our body heals while we sleep that maybe it was his body's way of keeping him alive with all the illnesses he has had most of his life. Now it might be the mix of medication, illnesses and a long time habit. I figure, what can it hurt?

- **Sleep:** When speaking with parents of children with VCFS, many have related that their children sleep a lot and soundly. Our daughter, who is 20, is no exception. She has a weird sleep pattern. She falls asleep as soon as it gets dark. It doesn't matter what time that is, she can *"saw logs."* She can sleep so long that it starts to worry me and I go to check and see if she is breathing. This also makes it difficult to get her up in the morning if she has a job; and it was hard when she was in school. I have noticed that she did not get as ill when we would allow her to sleep as long as she needed. She seemed to get ill if she got overly tired. I noticed she doesn't function as well mentally either when she doesn't get enough sleep.

• **Sleep:** My son had a sleep study recently and after getting the results back we found he has sleep apnea and will need to be on CPAP machine at night. I was a little bit surprised but I shouldn't have been, since I was told there was a possibility of this after his palate surgery. We have suspected it off and on since he had his pharyngeal flap surgery about two years ago. Since his surgery he became a mouth breather and I know that in order for the CPAP to be effective he needs to breath through his nose. However, I understand that there are a lot of different things they can do with masks. CPAP stands for <u>Continuous Positive Airway Pressure.</u> It is a triangular-shaped face mask that covers the nose only, and is strapped around the head with a soft Velcro® head gear. A hose connects to the machine providing the positive airway pressure. A humidifier attachment provides moisture so the airway does not dry out, making it more comfortable to use. The machine itself is the size and shape of a shoe box, with the humidifier being about half that size. Sometimes medical insurance will rent these machines for their clients.

• **Sleep:** My son Paul is 6 (VCFS), has gone to bed at 6:00 p.m. for almost five years. He does wake during the night (we're not sure for how long) but he lies in bed quietly and listens to music, then goes back to sleep for a few hours before I wake him for school. Daylight savings time, as it starts staying lighter later, does a horrible number on him. His circadian rhythm (the biological time clock) gets all out of whack and so this year his developmental pediatrician started him on melatonin. He takes it most nights, but not all, and it helps to keep him in balance. It can help to reset the circadian rhythm. Recently, he's been going to bed about 6:30, and I have to drag him out of bed at 7:00 a.m. He clearly needs the sleep. Just in case I checked with the cardiologist and he says it's not heart-related. My husband and I think that everyday activities just wear the little guy out. He has also begun napping on weekends. It's funny, we started the 6:00 bedtime when he gave up his nap around 18 months. We've been told by several doctors that it's much better for him to get solid nighttime sleep than a nap and choppy sleep. But I have been told that many VCFS kids sleep a lot, so I guess his body needs the extra is gets.

Spine

- **Spine:** I was told there was a new finding in VCFS: a high incidence of cervical fusions in the spine, that have only recently been noticed. Sure enough, our son has an C3-4 abnormality and possibly a C1 to skull fusion as well. These showed up on flexed x-rays but were not confirmed by the MRI done at our home-town lab. I was surprised because I thought MRI's showed everything. 2003 was a very bad year for us. Our son was wheelchair/crutches bound and sleeping 20 out of 24 hours a day. He dropped over 20 lbs last fall and nearly quit eating. His NG tube wouldn't go in any more and we finally went with the G-tube as was advised. Unfortunately (or maybe not) the first tube did not stay put and started leaking into the muscles outside his stomach and had to be replaced with an old fashioned type. Anyway, all the pain woke him up and forced him to try to walk without crutches. That was early summer of 2003. Now he is walking normally, and eating like regular people, almost. He is only sleeping like an average teenager, and is now off all medicines! He is still underweight, but not having the nausea episodes since the second tube operation. Although his activity level is fairly low he has been able to resume mowing the yard this summer (as exercise) and spends lots of time on the Internet with his buddies who never get to see or hear him in person. Where would we be without a computer?

Teens

- **Teens:** I am knocking on wood when I say this: My son is 16 and has not turned into Mr. Hyde with emotional problems yet! He has the normal teenage grumps sometimes but is easily coaxed out of them. He is a genuinely caring son/ brother/ friend and only voices self-pity occasionally. The only other VCFS kids I know of are through the Internet. I gather that, even though there are some common threads, the VCFS children vary enormously in personality, health and abilities. I agree that we must take each day as it comes. We don't know what is around the corner but rushing to meet problems is unlikely to help us to deal with them, when and if they come. We can be pleased that so far, many of us have not seen any of the signs of mental illness, that are common to others. Of course, it is early in the game for some of us, but I wanted to give some hope to those who have young children and are frightened of this aspect. I also know of adults with VCFS who have led happy lives without psychiatric problems.

- **Teens:** Because my daughter is pretty, she has had several boyfriends. Once she realizes they like each other more seriously, she always tells the boyfriends about the VCFS, for which I am proud of her. She doesn't tell them right away, only when it looks like it is more serious. All of the young men have handled it very well and a couple even did research on the Internet, asked more questions and kept dating her. It has been about even, the number of times that she broke-up with them or they broke-up with her; and usually it was over something else (such as *"liking someone else"*). Anyway, this shows that honesty early in the dating relationship doesn't have to be traumatic. I always tell people *"there are a lot of high intelligence kids who screw up their lives with alcohol and drugs"*, so having a disability is relative, it's how you handle life in general that counts. I hope this gives courage to the young adults who, when the right person comes along, will be able to handle the news of VCFS. If they don't; it means they might not be able to handle any problems very well. Then, when it blossoms into an engagement, together the couple will be willing to explore the options of a permanent relationship and parenting; and it will all be fine.

- **Teens:** My heart aches for all you parents who are going through what I did when my daughter was a teenager. I've never felt so frustrated, despairing, and sometimes out of control (or having a lack of control) at any other time in my life. It doesn't happen now, thank God, but it was so complex that it took me years to understand what was happening. I will give a try at explaining it to you, in hopes it will help someone else. This is my understanding. What looks like, sounds like, and feels like willful disobedient behavior, and *would* be in a child without VCFS, is often *NOT* willful behavior in a VCFS child. It results from structural and neuro-chemical differences in the brain, or a chemical imbalance. Kids with VCFS have brain differences in the frontal lobes where the execution of systems is generally located. This can have a strong effect on their ability to plan and organize, to make inferences, to think things through, to act on important information, or to carry out a task as promised. Also, they often have expressive and receptive language difficulties. They may be able to understand what you say *at that moment* and say it right back to you; but this does *not* mean the information has really been received intellectually, or gone to where it needs to go, in their brain; so that it will be available to them when they need the information. If this were happening in a child who'd gone through a windshield head first in a traffic accident, we parents, teachers, etc. would readily see it as the result of a brain injury. But in VCFS kids it's the result of invisible and developmental processes in the brain. Much of the time the child seems relatively normal. So it's natural for us to chalk it up as willful disobedience if they don't do as we say; or we think the child does not care enough about what the parent or teacher is telling them to cooperate. Unfortunately, because of the way their brains develop, they sometimes can find it difficult to do as they should. Every now and then, they succeed in an area, which backfires by raising our general expectations for them. The good news is that they can learn to behave better. The bad news is that they can learn *BUT* it proceeds very slowly. Hours of counseling won't help; years might. A trial of one medication at one specific dose might help, but trials of several medications, at different doses and in different combinations, taking months and years to work out, are more likely. (It is not an exact science.) Parental efforts can do part of the job, but coordinated effort on multiple fronts (parents, teachers, counselor, case manager, pediatrician, child psychiatrist, and pediatric neuropsychologist, etc!) can be even more effective.

- **Teens:** I have a son with VCFS who will be graduating this year from high school. I am very proud of him for getting to this point and thought it might be interesting for some to see his educational history:
- Infant stimulation: PT
- Head Start Preschool: OT, PT, Speech (age 3 to 5)
- IEP from kindergarten to 2nd grade in the classroom, some OT PT Speech
- No IEP from 3rd grade until 5th grade, repeated 3rd third grade (The school system said he was performing up to his IQ and didn't qualify for the 504 plan since end of 5th grade.)
- He missed a lot of school due to being sick. He missed most of fourth grade, missed 126 days of 5th grade, and missed 89 days of 6th grade. (He did well during this time because he had a lot

of one-on-one tutoring and teaching, with no distractions and therefore no pressure.)
- 6th Grade testing was done after discovering VCFS. His IQ went up 30 points. Then he started slipping. The healthier he got the less school he missed, yet the more he slipped.
- There was no consistent pattern as he progressed through school because one manifestation or another of the syndrome would surface. As classes became more difficult, and homework more demanding, anxiety caused whole-system shut downs. So a lot of communication with the school was necessary to ensure we knew exactly what was being assigned, how much of it, and when it was due. Short term memory problems and very prominent cognitive skill problems kept him from really understanding his class work. ADHD medication helped with that. OCD (obsessive compulsive disorder) started in a big way, accompanying the increased difficulty in school, and his anxiety increased. Severe picking at everything started. He couldn't have any clothes with rubber iron-on transfers. He'd pick the transfers off to little bits; he couldn't keep his hands off of stuff like that, and he would pick at small sores, like bug bites, and scabs until they bled and festered into quarter sized wounds. He tugged and scratched at hair until there were thin spots. He chewed on his own clothes; a lose button on a shirt or a coat would be like a magnet. His coat has no buttons, neither do his sweaters, collared or no collar shirts; in other words, there are *no* buttoned garments in his wardrobe. He still has a dexterity problem (such as problems tying his shoes, zipping zippers, snapping the snaps, etc.)

His behavior in high school started declining rapidly. By his sophomore year he was back on an IEP and the school pulled him out of classes that were beyond his abilities, added a study hall, and we've had the school test and re-test him. His IQ has definitely dropped; cognitive skills are on the decline, severe compulsive behavior is a constant problem. This year, at our meeting before school started, the school officials said they had never seen a kid slip cognitively so much and so fast. His doctors did some non-verbal testing on him and in some areas of the testing he tested in less then 1% for his chronological age.
- Socially, making and keeping friends is very difficult. His social skills is nowhere near his chronological age, impulsive behaviors are a real inhibitor there. Age-inappropriate responses to social situations make normal social interaction impossible. When we get one OCD behavior under control another seems to surface.

Such as hoarding food or hiding things; not wanting to leave the house (I guess that would be a phobia); or obsessing over school work but not being able to do it. This all sounds terrible, but at times, it really has even been worse. When he was younger he loved school, he loved going to school. Right now he even hates missing school. It's just that it keeps getting harder and the new materials, which are difficult to learn are coming too quickly. His obsessing over school work could no longer continue; and with his inability to learn and comprehend, his anxieties have been increasing. It has been a very fast downward spiral since beginning high school. Why did we send him? We thought about pulling him out, but then felt it would just make him give up and he already had issues about leaving the house, which we did not want to magnify. I hope adulthood is a brand new beginning.

• **Teens:** I worry about adulthood for my 22q.11 deleted child. Chuck is nearly 16 now and, except for music, clearly not going to do very well at testing for college. Although with a bit of work from time to time, he does make astonishing improvements, Chuck's academic learning doesn't seem to stick for very long. Yet in every other way he is an intelligent, fairly well organized, and perceptive person. I don't think it is just mother love talking. For example: Chuck went by train 500 miles to meet his older sister last week. He arrived at the station on time with a pre-purchased ticket. It was his _normal_ sister who kept him waiting for almost an hour. Thank goodness for cell/mobile phones! Sometimes things don't work out very well because he doesn't think them through, but follows his heart rather than his head. He gets impatient to get to where he wants to be; but I was exactly the same at his age and I do not have the deletion. I would like to know if a sense of humor is a sign of intelligence. Chuck was slow in the humor department, although he did have his own sense of fun when he was little. Between 6 and 12, he learned to laugh at jokes he didn't really understand by picking up on other clues (clever, I think) but now he is genuinely witty himself and I hear him spontaneously laughing aloud at TV programs that he is watching alone. The range of difference with 22q11 (VCFS) seems very wide and I suspect that we still have a lot to learn about it. I have far more questions than answers at this point in Chuck's life.

Tests

- **Tests:** For those of you who want a FISH test, it should be covered by private insurance. Usually, what is covered comes down to the term *medically necessary*. And this term is generally broadly construed. If you have been denied a FISH test by your plan, there are other options. There is also a big price discrepancy between what a private individual would get charged by a lab, and what price the provider has *"negotiated"*. If you aren't insured, and still want the FISH test done, you should be able to negotiate a price close to the $350. Even if they won't agree to that price up front, their accounts receivable should take it in a heartbeat. It's worth a try.

- **Tests:** My daughter is almost nine and for the past 7 years she has picked at herself until she creates huge wounds. On top of this she seems to always have a sinus infection. I never thought they may be related until now. The doctors have put her on a low dose of a medication to help with the anxiety; but it wasn't till recently, that I was told they both may be related to allergies. Off we go for tests.

- **Tests:** My son Matt (who is 13) has what they are describing as inflammatory bowel disease. He has not been diagnosed yet because he has yet to have all the necessary tests done. He has been on a drug they use in Crohn's disease which does great when he uses the high dose, but when we reduce it the diarrhea starts back up again. My own mother has Crohn's disease and is on an immunosuppressant and steroids. She is, however, much better on the treatment. There are so many different combinations of drugs out there for this. Matt has had bouts of these symptoms on and off for years and we were always scared to press for treatment because Matt has autism and there is quite a bit of controversy over autism/Crohn's. I know that Matt always has a bad bout after he eats popcorn (which he loves!). I sometimes wonder if lack of growth has anything do with it; although I have not heard of anything like that contributing to this condition. He is small in stature. All I know is that far more boys are diagnosed with this bowel problem than girls! We were also told that if someone else in the family has it, it is possible for others to get it as well. We have the same doctor as my mother and he has always been wonderful about not rushing us into going for all the *"things"* out there, like experiments or investigations. So much goes on with children with VCFS that my poor Matt could be spending most of his life going through exhausting tests. Because of this, we are taking it slow while erring on the side of caution.

- **Time for a break?:**

Looking for something fun for your kids to do while you catch a break from life for a few hours? Check out Easter Seals Web pages for exciting camps and adventures.
www.easterseals.com

Medical Trivia

Alexander the Great had epileptic seizures

Blank page for personal data:

CHAPTER 21: MEDICATIONS

MEDICATIONS
Here are the drug classifications used in VCFS

If you still have questions ask your doctor or pharmacist.

Analgesic	Relieves pain
Anesthetic	Partial or completely numbs or eliminates sensitivity, with or without consciousness
Antiarrhythmic	Corrects cardiac irregular heart beats
Antibiotic	Stops or controls the growth of infection-causing microorganisms
Anticonvulsant	Prevents or relieves convulsions or seizures
Antidepressant	Prevents, cures, or alleviates mental depression
Antidiabetic	Helps to control the blood sugar level
Antidiarrheal	Prevents or treats diarrhea
Antiemetic	Prevents or relieves nausea and vomiting
Antihistamine	Opposes the action of histamine, which is released in an allergic reaction.
Antihypertensive	Prevents or controls high blood pressure.

Anti-inflammatory	Counteracts inflammation in the body
Antitussive	Controls cough due to various cause.
Antiulcer agent	Treats and prevents peptic ulcer and gastric hyper secretion (too much)
Antiviral agent	Treats various viral conditions such as serious herpes virus infections, chicken pox, and influenza A
Bronchodilator	Expands the bronchial tubes by relaxing the bronchial muscles
Calcium channel	Treats hypertension, angina, and various abnormal heart rhythms.
Diuretic	Increases urine secretion
Hormone	Treats deficiency states where specific hormone level is abnormally low.
Hypnotic	Induces sleep or dulls senses.
Immunosuppressant	Suppresses the body's natural immune response to antigen, as in treatment of transplant patients
Laxative	Prevents constipation or promotes the emptying of the bowel contents with ease
Sedative	Exerts a soothing or tranquilizing effect on the body, skeletal muscle, relieves muscle tension, relaxant
Vitamin	Prevents and treats vitamin deficiencies and used as a dietary supplement, usually with feeding problems

Blank page for personal data:

Blank page for personal data:

CHAPTER 22: DICTIONARY

DICTIONARY AND GLOSSARY

A

Aberrant: Pertaining to a wandering from the usual or expected course
Adenoids: Pads of immune tissue located behind the nose and nasal cavity.
ADHD: Attention-Deficit Hyperactivity Disorder (ADHD or AD/HD) is a neurobehavioral disorder. It is primarily characterized by "the co-existence of attention problems and hyperactivity". Symptoms starting before the age of seven. ADHD is the most commonly studied and diagnosed psychiatric disorder in children globally. It is a chronic disorder with 30% to 50% of those individuals diagnosed in childhood continuing to have symptoms into adulthood.
Alar: Pertaining to wing-like structures, such as a shoulder
Alleles: Alternative forms of a genetic locus; a single allele for each locus is inherited separately from each parent
Alveolus: Gum of the mouth
Amniocentesis: A method of screening a pregnancy for some birth problems, including chromosome disorders, biochemical disorders and gene abnormalities; a needle is inserted into the mother's abdomen and a small amount of amniotic fluid (containing cells) is removed for testing. This test is preformed at 16 weeks of pregnancy.
Amniotic fluid: The liquid that surrounds the fetus in the mother's womb
Aplastic: Not exhibiting growth or change in structure (aplasia - incomplete or faulty development of an organ or part)
Apnea: An absence of spontaneous respiration; not to breathe
Articulation: The process of the structures of speech; such as the mouth, tongue, teeth, etc., coming in contact or proximity to alter or change sound by the vocal cords, thus producing speech. An articulation disorder occurs when the wrong contact or omission is made.
ASD: <u>A</u>trial <u>S</u>eptal <u>D</u>efect; this is a hole or defect in the wall

between the two upper chambers of the heart
Asthma: A condition often of allergic origin, that is marked by continuous or paroxysmal (whooping like cough) labored breathing accompanied by wheezing, by a sense of constriction in the chest, and often by attacks of coughing or gasping
Aspiration: Liquids going into the lungs by mistake
Asymmetric: Unequal in size and shape; different in placement or arrangement
Ataxia: An abnormal condition characterized by impaired ability to coordinate movement, such as in a staggering gait (walk), means without order
Atria: The two upper chambers of the heart
Attention Deficit Disorder (ADD): A syndrome affecting any age, where the person has a short attention span and poor concentration; the symptoms may be mild to severe in association with functional deviations of the central nervous system without signs of any neurological or psychiatric disturbances; usually normal to above average in intelligence; other symptoms can include: language; impairment in perception, memory and mo-tor skills; decreased attention span; increased impulsiveness and emotional liability. This disorder is 10 times more prevalent in boys than girls and usually results from genetic, biochemical factors, or from an injury or disease.
Attention Deficit Hyperactivity Disorder (ADHD): A mental childhood disorder characterized by symptoms of inattentiveness, impulsiveness, and hyperactivity.
Autism: A disorder of neural development characterized by impaired social interaction and communication, and by restricted and repetitive behavior. Signs begin before a child is three years of age. Autism affects information processed in the brain. There are three recognized disorders in the autism spectrum.

B

B-Cells: Cells in the body that make antibodies (IgM, IgA); their job is to kill germs such as viruses, bacteria and fungi (They work with T-Cells)
Base pairs: These are chemicals that make up the DNA molecule. The letters on a helix stand for the base pairs: <u>A</u>denime & <u>T</u>hymine or <u>G</u>uanine & <u>C</u>ytosine
Bifurcation (fork): A splitting into two branches; such as in the trachea, which branches into the two bronchi
Bilateral: Affecting both sides of the body
Bipolar mood disorder: Varying degrees of grandiosity of self,

decreased need for sleep, pressured speech, flight of ideas, distractibility, increased inactivity, marked impairment of occupational functioning; Many of the increases in activity can have disastrous consequences, such as: spending huge amounts of money and making decisions that are not rational. In extreme cases there are psychoses; delusions or hallucinations during these abnormal mood states.

C

Cardiac: Having to do with the heart
Cardiologist: A doctor specializing in diagnosis and treatment of heart problems
Cataract: An abnormal progressive condition of the lens of the eye, characterized by the loss of transparency, a gray white spot within the lens of the eye
Catch 22: A label sometimes used in the UK for VCFS; this is not a proper name but a slang term that many find distasteful
Cell: The smallest thing you can see under a microscope; contains chromosomes in the nucleus
Cerebellar: Relating to, or affecting the cerebellum
Cerebellum: The part of the brain that is located in the posterior (back) of the head, behind the brain stem; it consists of two lateral hemispheres
Chromosomes: The genetic structure of cells containing the DNA that bears in its nucleotide (chemical sequencing) *(See base pairs.)* the linear array of genes for each individual which determine everything about us (eye color, height etc.); they are rod-like bodies which are only seen under the microscope
Chronic: A disease or disorder that slowly and persistently continues for a long time, sometimes for a lifetime. Compared with **Acute** which describes a disease or disorder that begins abruptly and peaks with intensity. and then subsides after a short period of time
Chronionic villus sampling (CVS): A method to screen a pregnancy for some birth problems including chromosomal abnormalities
Circle of Willis: An interconnecting network of vessels at the base of the brain
Cleft: opening
Cleft lip: Failure of the lip to close during development; open lip
Cleft palate (velum or velo): The failure of the embryonic (first 8 weeks) fusion of the hard and/or soft part of the roof of

the mouth and floor of the nose to close *(See submucous and occult clefts.)*
Clinical geneticist: A doctor who specializes in hereditary (genetic) diseases and disorders
Coarction of the aorta: A congenital cardiac (heart) anomaly characterized by the localized narrowing of the aorta
Cognition: The process of perceiving, thinking, reasoning and analyzing information
Coloboma (defect): A congenital or pathological defect in the ocular tissue of the body; usually affecting the iris of the eye
Commissure: A band of nerve fiber or other tissue that crosses over from one side to another, usually connecting two structures or masses of tissue, such as in the corners of the eyes and lips.
Communication skills: The ability to use language to receive and express information.
Conductive hearing loss: hearing disorder that is caused by disruption in the sound-conducting mechanism of the outer or middle ear so that a reduced level of sound reaches the inner ear and nerve conduction; the sound that goes to the brain.
(See also sensori-neural hearing loss.)
Congenital: A problem which exists at birth.
Conotruncal heart anomaly: A congenital heart defect particularly involving the ventricular (lower chambers); outflow of the heart includes sub-arterial ventricular septal defect, pulmonic valve astresia and stenosis, Tetralogy of Fallot and truncus arteriosus.
Contractures: An abnormal and usually permanent condition of a joint. This prevents the easy movement of the joint and is in a fixed position; a permanent shortening (as of muscle, tendon, or scar tissue) producing deformity or distortion.
Craniosynostosis: Premature closure of the fontanels (soft spots) on the skull.
Cryptorchidism (hidden testes): A developmental defect characterized by failure of one or both testes to descend or drop into the scrotum. They are retained in the abdomen or in the inguinal canal.
Cyanosis: A bluish skin coloring caused by low oxygen levels in the blood.
Cyclothymic: Relating to, having, or being in an affective disorder characterized by the alternation of depressed moods with elevated, expansive, or irritable moods without psychotic features such as hallucinations or delusions
Cystic: Relating to, composed of, or containing cysts (*cystic* tissue) (a *cystic* tumor)

D

Deletion: Missing base pairs of information in a DNA molecule.
Desaturation (to fill): Formation of an unsaturated chemical compound from a saturated one
Diaphragmatic: Of, involving, or resembling a diaphragm
Diastastis recti abdominis: The separation of the two rectus muscles along the middle line of the abdomen wall; in newborns it is caused by incomplete development
Depression: A long term or chronic mood disorder
Development: The process of growth and learning
Developmental delay: When learning and motor skills are falling short and are slower than expected
DNA: The molecule that carries and encodes genetic information.
Dysgenesis: A defective or abnormal formation of an organ or part, primarily during embryonic development; the impaired or loss of the ability to procreate
Dysphagia: Feeding or swallowing difficulties
Dysphasia (speaking): An impairment in speech
Dyspraxia: A partial loss of the ability to perform skilled, coordinated movements in the absence of any associated defect in motor or sensory functions
Dysthymia: A mood disorder characterized by chronic mildly depressed or irritable mood often accompanied by other symptoms (as eating and sleeping disturbances, fatigue, and poor self-esteem)
Dystonia: Referring to impairments of muscle tone, often excessive increase in tone, when the muscle is in action; it is hypotonia when it is at rest, often resulting in postural (orthopedic) abnormalities

E

Ear tubes: Myringotomy tubes; small tubes placed in the eardrum to allow fluid to drain from the middle ear
Echocardiogram: An ultrasound of the heart
Electrocardiogram (ECG/EKG): A cardiac test which measures the electrical impulses of the heart
Embryotoxon: A congenital defect of the eye characterized by an opaque ring around the margin of the cornea
Endocrinologist: A physician who cares for patients with problems of the glands; (ex: the parathyroid gland which controls the hormone that regulates calcium levels) specializing in growth and managing diabetes

ENT: An <u>e</u>ar, <u>n</u>ose and <u>t</u>hroat doctor
Epiglottis: Structure that typically folds over the opening of the trachea to direct the traffic of the food down the esophagus; and preventing it from going into the lungs.
Esophagus: The "food pipe" that connects the mouth with the stomach
Esophageal reflux: An abnormal backward flow or return flow of a liquid, such as when the food or drink comes back up into the esophagus
Expressive language: The ability to use gestures, words and written symbols to communicate

F

Fine motor skills: Involves the movements of the small muscles of the body such as hands, feet and fingers and toes
FISH test: <u>F</u>luorescence <u>I</u>n <u>S</u>itu <u>H</u>ybridization; this laboratory test uses radioactive tags to identify extra or missing pieces of chromosomal material which are often too small to detect by just looking under a microscope
Fissure: A cleft or a groove on the surface of an organ
Fontanels: The soft spots of the skull; front and back
Fossa: A hollow or depression; usually on the surface of a bone at the ends
Fused: To cause to undergo fusion, to fuse or join together, permanently

G

Gastroesophageal reflux: Backward flow of the gastric contents into the esophagus resulting from improper functioning of a sphincter at the lower end of the esophagus
Gene: The unit of heredity
Gene expression: The process by which a gene's coded information is converted into structures that operate within the cell
Genetic: A problem which may "run in the family"; be present from one generation to another, occur in siblings or other close relatives or it may be a mutation which is the start of a new change in the cell of an offspring
Genetic code: There are thousands of base pairs linked together to form a DNA molecule; this is the triplet coded (by 3's) sequences of the sub-units that make up the individual DNA

Genome: ALL the genetic material in the chromosome of a particular person
Genotype: This is the order of the sequencing of the genes (base pairs - their specific combination and location) of an individual
GI System: Gastrointestinal system; including the esophagus, stomach and intestines, which facilitate food digestion
Glottal stop: A type of articulation error; when saying a word it is sometimes confused for an omission of a letter and it will come out as a grunt. In VCFS this glottal stop is used to substitute the sound needed to say the word.
Glottis (glottal): The space between one of the true vocal cords and the arytenoid cartilage on one side of the larynx and those of the other side; *also*: the structures that surround this space.
Gross motor: Involves movement of the large muscles of the body.

H

Hemi: Meaning "half"
Hepatoblastoma: A cancer of the liver that usually is found in childhood
Hereditary: A problem which "runs in the family" or is considered genetic
Hernia: A protrusion of an organ or part through connective tissue or through a wall of the cavity in which it is normally enclosed; called also *rupture*
Hirschsprung's Megacolon: Megacolon is caused by a congenital absence of ganglion cells in the muscular wall of the distal part of the colon with resulting loss of peristaltic function in this part and dilatation of the colon proximal to the aganglionic part
Human gene therapy: Insertion of normal DNA directly into cells to correct a genetic defect
Hyperextensible: Having the capacity to be hyper (over) extended or stretched to a greater than normal degree; such as in joints
Hypernasal: Too much air passing through the nose; which makes the voice sound like they have a stuffy nose.
Hypospadias: A congenital condition in which the urethra meatus is located on the underside or undersurface of the penis
Hypertelorism: A developmental defect characterized by an abnormally wide space between two organs or parts
Hypocalcaemia: A deficiency or low calcium in the serum that may be caused by hypoparathyroidism.
Hypomania: A mild mania especially when part of bipolar disorder

Hypoparathyroidism: A condition of insufficient secretion of the parathyroid glands; may be caused by a dysfunction of the parathyroid or a by an elevated serum calcium level
Hypoplasia: Incomplete or underdeveloped organ or tissue
Hypothyroidism: Decreased production of thyroid hormone by the thyroid gland
Hypotonia: A condition of diminished or abnormally low intra-ocular fluid or muscle tone that can affect any body structure.

I

IEP: Individualized Education Plan
Immunization: The use of administering a medication by injection to prevent someone from contracting a disease (ex: polio)
Immunologist: A physician who cares for children having problems involving the bodies immune "defense" system
Infant stimulation: Early interventions for young children
Inguinal hernia: A hernia or tear that allows a loop of intestine to enter the inguinal canal; in males this can sometimes fill the scrotum sac
Intervention: Providing therapies and specialized programs to help children to reach their fullest potential as quickly as possible
IQ: Tests that measures a child's intelligence or cognitive ability as determined by a standardized scale

J

JRA: Juvenile Rheumatoid Arthritis

K

Karyotype: A picture of an individual's chromosomes in a standard format showing the number, size and shape of each chromosome type

L

Language: The expression and understanding of human communication
Larynx: The modified upper part of the respiratory passage of air-breathing vertebrates that is bounded above by the glottis, is

continuous below with the trachea, has a complex cartilaginous or bony skeleton capable of limited motion through the action of associated muscles, and in humans, most other mammals, and a few lower forms have a set of elastic vocal cords that play a major role in sound production and speech; called also *voice box*
Laryngeal: Anything pertaining to the larynx.
Locus: A specific place or position of a certain gene on a chromosome

M

Mainstream: Involving children with disabilities in a regular school environment
Malacia: Abnormal softening of a tissue
Malrotation: Improper rotation of a bodily part and especially of the intestines
Maxillary: Pertaining to the upper jawbone
Microcephaly: Small head size
Morphea: A skin disease consisting of patches of yellow or ivory colored hard, dry, smooth skin; more common in females
Multi-view videofluoroscopy: A diagnostic procedure using x-rays to obtain a three dimensional view of the velopharyngeal function and recording it on a video tape, barium contrast is used to help see the movement of the speech musculature; at least 2 different views must be used for an adequate diagnosis
Mutation: Any hereditary change in DNA sequencing

N

Nasopharyngoscopy: A diagnostic procedure used to assess velopharyngeal function during speech
Nasopharyngoscope: A fiber-optic instrument (endoscope) which can be inserted through the nose to see the pharynx and larynx (voice box); it is video-taped through the endoscope

O

Occult submucous cleft palate: Many submucous clefts have a bifid (split) uvula as a clue to diagnosis; some findings may be normal, yet the palate may still be missing muscle tissue which results in abnormal function; this is referred to as "occult" meaning "mysterious"; the palate looks normal on oral exam, but speech is nasal

Oral motor: Use of the muscles in and around the mouth and face
Oral transport: Phase in which food is pushed backwards by the tongue to swallow
Ophthalmologist: Medical doctor who specializes in the eye
Osteopenia: A condition of subnormal mineralized bone
OT: Occupational therapist; a medical professional who specializes in improving the development in fine motor skills
Otolaryngologist: A physician who specializes in ENT (ears, nose and throat)
Otitis Media: Fluid in the middle ear behind the eardrum (Ear infection)

P

"P" arm: The short arm of the chromosome.
Palpabral: eyelid
Paresis: Motor weakness or partial paralysis related to some case of neuritis
Patent ductus arteriosus (PDA): An abnormal opening between the pulmonary artery and the aorta; caused by the failure of the fetal ductus to close after birth; usually seen in premature infants
Peri: prefix meaning "around"
Peri-ventricular: Situated or occurring around a ventricle especially of the brain (*periventricular* white matter) or (*periventricular* calcifications)
Phalangeal: Of or relating to a phalanx or the phalanges (such as fingers and toes)
Pharynx: The throat
Pharyngeal flap: Soft tissue layer of mucous membrane and its underlying muscle; elevated from the back wall of the throat and then inserted into the top surface of the palate
Phenotype: The *visible* or *observable* expression of the gene; what we see when we see a patient who has a genetic problem
Phobia: An exaggerated and often disabling fear to something often inexplicable to the person; may be a logical cause but usually illogical reason; may be a symbolic object, class of objects, or situation
PKU (phenylketonuria) is a blood test taken in many states at birth to determine if the baby has a toxic metabolic disorder that may lead to retardation if not treated. Affects 1-16,000 births in the US.
Platelets: Blood cells that assist in clotting; thrombocyte

Pneumonia: Inflammation of the lungs resulting from infection.
Polydactyly: The condition of having more than the normal number of toes or fingers
Posterior: In the back part of a structure
PPF: Posterior Pharyngeal Flap; this is a surgical procedure to correct VPI
PT: Physical Therapist; a medical professional who specializes in improving the developmental gross motor skills and overcoming physical problems such as weak muscle tone
Polymorphism: Difference in DNA sequence among individuals at different stages
Post-axial: "post" meaning after or behind; "axial" meaning pertaining to or situated on the axis of a body structure
Pre-axial: "pre" meaning before or in front of; "axial" meaning- pertaining to or situated on the axis of a body structure
Proband: First family member coming to medical attention with a specific known genetic condition
Psychosis: A severe form of a psychiatric disorder characterized by a disorganization of normal mental processes, thinking, personality and temperament
Pulmonary Artresia: A congenital heart defect of the right ventricular outflow tract; an extreme form is "Tetralogy of Fallot" with 4 defects

Q

"Q" arm: The long arm of the chromosome

R

Raynaud's sign: Intermediate attacks of ischemia (starved of oxygen) of the extremities of the body; usually the fingers, toes, ears and nose
Receptive language: The ability to understand the spoken and written gestures and words in communication
Renal: Pertaining to the kidneys

S

Scapular deformity: Affecting part or all of the shoulder blade
Schizo-affective: Relating to, characterized by, or exhibiting symptoms of both schizophrenia and bipolar disorder

Schizophrenia: A psychotic disorder characterized by loss of contact with the environment, by noticeable deterioration in the level of functioning in everyday life, and by disintegration of personality expressed as disorder of feeling, thought (as in hallucinations and delusions), and conduct
Scoliosis: Curvature of the spine
Seizure: A sudden lack of consciousness resulting from abnormal electrical activity of the brain
Sensori-neural hearing loss: Hearing disorder caused by an abnormality in the sound sensing mechanism (the inner ear) or nerve which conducts sound to the brain
Septum: The wall (divider) of cardiac tissue between the chambers of the heart; also a septum divides the nostrils in the nose
Social skills: The ability to interact with other people and to function in groups
Special education: The education of children with disabilities
Spina bifida: Congenital defect of the spine; opening in the spine; spinal cord often protrudes out of the vertebrae resulting in spinal cord injury
Sprengel's deformity: a congenital elevation of the scapula (also called the shoulder blade)
Standardized test: A test in which the child's performance is compared to the performance of other children of the same age; this is done for placement in school
Strabismus: An abnormal visual condition in which the visual axis of the eyes are not directed in the same direction, there is more than one type of strabismus and different causes
Stroke: Sudden diminution or loss of consciousness, sensation, and voluntary motion caused by rupture or obstruction (as by a clot) of a blood vessel of the brain; called also *apoplexy, brain attack, cerebral accident, cerebrovascular accident*
Sub: is a prefix that means "under"
Submucous cleft palate: In this cleft there is an observable opening between two separate halves of what should be one solid piece (roof of the mouth); a submucous cleft is easily detectable on oral examination; because of the presence of a split in the uvula (bifid uvula), a notch in the bone of the hard palate (which can be felt with a finger)
Sylvian fissure: A deep fissure of the lateral aspect of each cerebral hemisphere that divides the temporal from the parietal and frontal lobes in the brain; called also *fissure of Sylvius, lateral fissure, lateral sulcus*
Syndactyly: A webbing of skin; a union of two or more digits;

this is normal in some animals and can occur in humans. It often occurs as a hereditary disorder marked by the joining or webbing of two or more fingers or toes
Syrinx: A pathological cavity in the brain or spinal cord

T

Talipes: A congenital deformity of the foot; in which the foot twists and is relatively fixed in an abnormal position (club foot)
T-Cells: Cells in the body that are protectors against infection; they include killer, helper, and suppressor cells
Tethered: A short radius in which to move, a limited range of movement or ability
Tetralogy of Fallot: Congenital cardiac anomaly consisting of four defects of the heart
Thrombocytopenia: Persistent decrease in blood platelets
Thymus gland: A small gland in the chest that is involved with immune function
Tortuous: Having or making twists and turns
Trachea: "Wind pipe"; connects the mouth and nose to the lungs
Transposition: An abnormality where parts are reversed; things that are typically on the right are on the left and things usually on the left are on the right
Tri: Prefix meaning three
Tricuspid artresia: A congenital cardiac anomaly characterized by the absence of a tricuspid valve so that there is no opening between the right atrium (right upper chamber) and the right ventricle (right lower chamber)
Truncus arteriosus: The embryonic arterial trunk that initially opens from both ventricles of the heart and later divides into the aorta and the pulmonary trunk; the two parts separated by a septum or wall

U

Ultra sound: The use of sound waves to produce an image of an organ inside of the body (ex: detecting gallstones)
Umbilical hernia: A soft skin covered protrusion of an intestine through a weakened wall of the abdomen at the umbilical area
Unilateral: Involving only one side
Urologist: Physician who cares for patients with problems of the urinary tract; such as kidneys and bladder; also performs vasectomies

Uvula: The tissue or "punching bag" that hangs down in the back of the throat

V

Vascular ring: A vascular ring is a type of vascular compression, which represents a mixed bag of anomalies that share the common feature of compromise of the esophagus or airway by adjacent arterial structures
Velum: Soft palate; rear portion of the "roof of the mouth"
Velopharyngeal valve: Area behind the palate; which opens and closes during speech to modulate the flow of air through the mouth and nose during speech
Velopharyngeal insufficiency (VPI): *(Also known as Velopharyngeal inconsistency)* VPI is the failure of the muscular portion of the soft palate (velum) and the throat to close completely during normal non-nasal speech; if air leaks into the nose during speech, this is what is called VPI; this is what causes the hypernasal sound in speech
Ventricles: The lower chambers of the heart
Vermis: A structure resembling a worm; such as the median lobe of the cerebellum
Vertebrae: The bones of the spinal column; hemi-vertebrae (failure of the vertebrae to develop completely)
Vesico-ureteral reflux: Reflux of urine from the bladder into the urethra
Voice: Sound usually produced by the vocal cords which then resonates in the throat above the larynx; vocalization (sounds of the voice)
VPI: *(See velopharyngeal insufficiency.)* This is the failure of the palate to meet the throat during crying, swallowing, and speech and can allow nasal regurgitation (fluid coming through the nose)
VSD: <u>V</u>entricular <u>S</u>eptal <u>D</u>efect; a hole in the septum between the two lower chambers of the heart.

W

Wheezing: a sound that occurs when a child or adult tries to breathe through air passages that are narrowed or compressed. Wheezing is most common when exhaling. It is sometimes accompanied by a mild sensation of tightness in the chest.

Z

Z-plasty: Originally used to repair skin defects by the transposition of two triangular flaps of adjacent skin, for relaxation of scar contractures. This procedure is commonly seen in Burn wards. Also a technique used in the repair of Cleft palates.

Zygote: the initial cell formed when a new organism is produced by means of sexual reproduction.

Blank page for personal data

CHAPTER 23: BIBLIOGRAPHY

ACKNOWLEDGEMENTS

We want to give credit where credit is due. Thank you to those organizations whose contributions in research and diligent pursuit of knowledge have made undertaking this writing possible. A special acknowledgement goes to those who gave their personal and professional contributions to this book:

• To those families who so willingly and unselfishly shared their lives with us to benefit others. We give you special thanks!

• To the VCFS Education Foundation in Syracuse N.Y. **www.vcfsef.org**

• To the 22q.11 Group in the UK **www.maxappeal.org.uk**

• To the Children's Hospital of Philadelphia (CHOP)

• To Children's Rehabilitation Services (CRS) of Phoenix, AZ a division of St. Joseph's Hospital

• To the following individuals for their contributions: Kyrieckos Aleck, M.D., Genetics and Dysmorphology; Monica Alvarado, M.S., C.G., Assistant Professor USC; Maureen Anderson; Stacia Cain, L.P.N.; Cathy Yoshida Corella, B.A.; Dianne and Alan Detrick; C. Garrison, Ph.D. and staff of St. Joseph's Hospital CRS; Gail Klein (CHOA); Jeff Gaydish; Dr. Karen J. Golding-Kushner, Ph.D.; Gilbert Gomez; Kathleen Hamel, Librarian; Donna Landsman, M.S.; Debra Leach, H.M.A.; Debra Lightfoot, CCC-A; Paul Mesa; Merry Nessinger; Sylvia Nobel; Dr. Carol Ordynsky, O.D. Optometrist; Greg Peoples; Marty Peoples, herbalist; Lynda and David Purdin; Lori Rehder, R.N., Department of Public Health, Arizona; Kelvin P. Ringold; Patricia Schiffman, D.M.D.; Dr. Robert J. Shprintzen; Tamara and Jerry Smith; Tishri and Jeff Solmon; Carolyn Tanquary, R.N.; Margi Stevenson, M.S.W.; Nadine O. Vogel, MetLife; Peter White, P.N.P.; Gloria Warren; and Kas Winters

BIBLIOGRAPHY

Books and articles:

Anderson, Kenneth N. *Mosby's Medical, Nursing & Allied Health Dictionary, 6th Edition*. Mosby.

Berkow, Robert, Editor. *Enteral Nutrition. The Merck Manual of Diagnosis and Therapy, 16th edition*. New Jersey. Merck. 1992.

Children's Hospital of Philadelphia. *Faces of Sunshine: The 22q11.2 Deletion, A Handbooks for Parents & Professionals*. New Jersey. Cardinal Business Forms & Systems. 2000.

Fast, Julius. *Body Language in the Workplace*. New York: Penguin Books. 1994.

Gray, John, Ph.D. *Men, Women and Relationships*. New York: Harper Paperbacks, A division of HarperCollins Publishers. 1996.

Howard, Lyn. *Enteral and Parenteral Nutrition Therapy. In Harrison's Principles of Internal Medicine, 14th editiion*. New York: McGraw-Hill. 1998.

Lagua, R.T., and V.S. Claudio. *Nutrition and Diet Therapy Reference Dictionary. 4th ed.* New York: Chapman & Hall, 1996.

Merriam-Webster Medical Dictionary. Massachusetts. C & G Merriam Co. 2002.

Morava, E., Lacassie, Y., King, A., Illes, T., Marble, M. *Scoliosis in Velo-Cardio-Facial Syndrome*. Children's Hospital, New Orleans, Louisiana 70118, USA. mmarble@bellsouth.net

Morris, Desmond. *Body Talk*. New York: Crown Trade. 1995.

Orman, Suze. *Nine Steps to Financial Freedom*. New York: Crown Publishers, Inc. 1997.

Parents' Guide, Everything you need to know about Velo-Cardio-Facial-Syndrome – DiGeorge Syndrome. UK. 22q.11 Group in the UK. www.vcfs.org. 2000.

Pena, Alberto, MD, VACS, FAAP, Chief of Pediatric Surgery. *Management of Anorectal Malfunctions*. New York: Schneider Children's Hospital.

Vatter, Glenn. *Group Homes & Other Alternatives*.

Worthington S., Colley, A., Fagan, K., Dai, K. and Lipson, AH. *Anal anomalies: An uncommon feature of velocardiofacial (Shprintzen) syndrome*. Australia: Department of Clinical Genetics, Sydney Children's Hospital.
http://www.sch.edu.au

Organizations:

ABIL, Arizona Bridge to Independent Living
Independent Living for People with Disabilities
1229 E. Washington Street
Phoenix, AZ 85034
Telephone: 602-256-2245
www.abil.org

Arc of the United States (Advocacy)
The Arc of the United States works to include all children and adults with cognitive, intellectual and developmental disabilities in every community. **http://thearc.org**

Central New York Center for Cleft & Craniofacial Disorders
Eileen Marrinan, MS, MPH
Clinical & Research Coordinator
SUNY Upstate Medical University Hospital
750 East Adams Street
Syracuse, NY 13210
Telephone: 315-464-6580

Easter Seals
National site provides links to local organizations
www.easterseals.org

MUMS: National Parent to Parent Network
Julie J. Gordon
150 Custer Court
Green Bay, Wisconsin 54301-1243
1-877-336-5333 (Parents only please)
920-336-5333
1-920-339-0995 (fax)
E-mail: mums@netnet.net
Web Site: **www.netnet.net/mums**

The National Chronic Pain Outreach Association
National Chronic Pain Outreach Association (NCPOA),
Laura S. Hitchcock, PhD., Director.
7979 Old Georgetown Road, Suite 100,
Bethesda, MD 20814-2429,
phone 301-652-4948,
Fax 301-907-0745.
http://neurosurgery.mgh.harvard.edu
International Association for the Study of Pain®
Public Information:
www.iasp-pain.org/pinfopen.html

Web sites:

Bipolar disorder:
www.bpkids.org
wwwcmell.com/bipolar/topics/html
www.nimh.nih.gov/publicat/bipolar.cfm
www.bipolarhome.org
www.nlmh.nih.gov

Body language:
www.bodylanguagetraining.com/
www.positive-way.com
www.mediamagiconline.com/4.htm

Bullying:
www.lfcc.on.ca/bully.htm
www.winmarkcom.com/gorp.htm (child's book)
www.drphil.com

Grieving:
http://depts.washington.edu/counsels
www.helpguide.org/topics/grief.htm
http://ub-counseling.buffalo.edu
www.MedicineNet.com

Palate and feeding issues:
www.emedicine.com/ent/topic136.htm
www.craniofacial.net
www.cleftadvocate.org

Psychology-coping:
www.psychlinks.ca

Psychophysiology & Biofeedback:
www.psychophys.com
www.aapb.org
www.medwebplus.com

Raising kids with special needs:
www.disabilitytraining.com

Raising special kids of AZ
http://www.raisingspecialkids.org

Self-help
www.selfable.com

VCFS:
VCFS Education Foundation **www.vcfsef.org**
22q11 UK Group **www.maxappeal.org.uk**
22q11 Ireland Group **www.22q11ireland.org**

Vision therapy: a type of physical therapy for the eyes and brain at the Vision Therapy Network:
www.visiontherapy.org

Visual spatial skills:
www.newhorizons.org/spneeds/inclusion/teaching/-
    stockdale.html
www.visionandlearning.org/visualskills2.htm
www.visualspatial.org

Blank page for personal data:

CHAPTER 24: INDEX

INDEX

A
Abandonment, by doctor, 295
Abdominal anomalies, 5
Aberrant subclavian arteries, 6
Abnormal behavior. See Behavioral problems; Psychiatric/psychological problems
Abstract thinking, 6, 377 378
Abuse by other children, 210 211
Acidophilus drink, 79
Action, listening and, 334
Acute pain, 61
Adaptive functioning, 167 168
ADD/ADHD. See Attention deficit hyperactivity disorder
Additional time for tests, 147, 151
Adenoids
    absent or small, 9
    definition of, 501
    removing, 181
Adolescents. See Teens
Adoption, personal stories, 443 446
Adults with VCFS
    caregivers, 305
    group homes, 301 305, 306, 357, 469
    jobs, 355 356
    personal stories, 355 358, 446 454
    special needs trust for, 141, 304, 310, 311 317
    transition plan for, 260
Advanced directives, 296
Advocacy, 257 306
FAPE Solutions, 294
    group homes, 301 305, 306
Individualized Education Plan (IEP), 150, 259, 260, 290
    medical care, 294 296, 306
    negotiating skills for parents, 249 255
    parent centers, 262 263
    personal stories, 257 259, 454 455
    records to keep on file, 261
    schools, 145 146, 150, 259 260, 293, 299 300
Technical Assistance Alliance for Parent Centers, 261 293
    See also Support groups
Affect, 9, 133
Air conduction, 194
Air transport, medical, 306
Airway anomalies, 9
Akil, Dr. Mayada, 244

Alabama, parent centers in, 265
Alaska, parent centers in, 266
Alleles, 18, 501
"Allergic shiners", 6, 50, 375
The Alliance Project. See Technical Assistance Alliance for Parent Centers
Alternate living arrangements, group homes, 301 305, 306, 357, 469
Amniocentesis, 16 17, 501
Anal anomalies, 5, 38 39
Analgesics, 497
Anderson, Maureen, 321 333
Anesthetics, 497
Anger
   parents' anger, 201, 434, 441
   stage of grief, 425
   of VCFS child, 235
Anne Ford Scholarship, 173
Anomalies in DiGeorge syndrome, 3
Anomalies in VCFS, 4 11, 15, 19 20, 21
   anal anomalies, 5, 38 39
   ears, 7, 41 46
   eyes, 6 7, 21, 47 48, 50, 507
   face, 6 7, 49 50
   fingers, 8, 50
   growth hormone deficiency, 7, 21, 54, 55
   heart defects, 5 6, 21, 25 29, 360 361, 365, 381, 395
   hernias, 5, 21, 38 39
   hormone difficulties, 7, 53
   hypocalcaemia, 7, 53 54, 65, 378, 487, 507
   immunologic anomalies, 8, 21, 58 59
   kidneys, 5, 21, 37 38
   leg and foot anomalies, 8, 62 63
   leg pain, 10, 62, 248
   pulmonary and respiratory system, 8, 21, 33 36
   teeth and dental problems, 6, 65 66, 349, 368
Anti-anxiety drugs, 474
Anti-inflammatory drugs, 498
Antiarrhythmic medications, 497
Antibiotics
   function of, 497
   prior to dental work, 26
   thrush (candidiasis) after taking, 79, 455
Anticonvulsants, 497
Antidepressants, 497
Antidiabetics, 497
Antidiarrheals, 497
Antiemetics, 497
Antihistamines, 497
Antihypertensives, 497
Antitussives, 498
Antiulcer agents, 498
Antiviral agents, 498
Anxiety, 9, 123, 167, 405, 474
Anxiety disorder, biofeedback for, 123

Aortic anomalies
   coarctation, 5, 25, 503
   interrupted aortic arch, 5, 25, 28
   right-sided aorta, 5, 25, 376
Aortic valve anomalies, 5, 404
Aplastic kidney, 7
Apnea, 403 404, 488, 501
Argumentativeness, 229
Arizona, parent centers in, 266
Arkansas, parent centers in, 267
Arteries, function of, 30
Arterioles, function of, 30
Arthritis, 10, 21
Articulation impairment, 10, 178, 501
Arytenoid hypoplasia, 9
ASD. See Atrial septal defect
Asperger's syndrome, 202
Aspiration, 35, 76, 80, 501
Assault, 210, 294
Assets, special needs trust, 317
Assignment notebooks, 147
Assignments in school, 146, 147 148, 151, 218
Assistive devices, in school, 147, 151, 195
Asthma, 35
Asymmetrical pharyngeal movement, 9
Atrial septal defect (ASD), 5, 501
Attention, 152 153, 168 169
ADHD symptoms, 216 217
   focused attention, 165
   initial auditory attention, 165
Attention deficit hyperactivity disorder (ADD/ADHD), 6, 129, 213 220, 222, 361
   biofeedback for, 122 123
   coexisting disorders, 217, 239
   confidence building tips, 218 219
   definition of, 502
   diagnosis of, 215
   support groups, 219
   symptoms of, 214 217
   treatment of, 217 219
Attentiveness, during negotiation, 252, 253
Attitude, personal stories, 455 459
Auditory attention, 189, 195
Auditory closure, 195
Auditory cohesion, 189
Auditory discrimination, 189, 195
Auditory figure-ground, 189, 195
Auditory memory, 189
Auditory overload, 195
Auditory processing. See Central auditory processing; Central auditory processing disorders
Auditory synthesis, 195
Autism, 167, 486

B

B-cells, 58, 502
Babies. See Infants
Bacterial infections, 58
Bacterial pneumonia, 35
Bargaining, stage of grief, 425
Bathroom, use of in school, 171
Battery (legal term), 210, 294, 295
Beck, Martha, 459
Bedwetting, 459
Behavior, 168
   assessing your VCFS child, 124
   disapproving of, 200 201, 207
   discipline, 169 170, 200 202, 208
   parental, during negotiation, 254 255
   promoting good behavior, 206
   at school, 169 170, 208 212
Behavioral problems, 199 212
   depressive episodes, 224 225
   discipline, 169 170, 200 202, 208
   guidelines for handling, 200 202
   hypersexuality, 234, 353
   manic episodes, 224
   parents' watchfulness, 209
   personal stories, 480, 490 491
   problem behaviors, list, 202 203
   punishment, 208
   See also Psychiatric/psychological problems
"Being different", 138 139
Belly button hernia (umbilical hernia), 5, 38, 513
Bernard-Soulier disease, 11
Beverages, for babies and children, 76, 79
Bilateral cleft lip and palate, 70, 72
Binaural integration, 195
Biofeedback, 121 123, 237
Bipolar disorder, 9, 221 225, 228 239
   definition of, 502
   parents' stories, 54, 220, 352, 405, 486
Blood pressure, 30-31
Blood vessels, 30
Bloody noses, 34 35, 36
Blowing air through the mouth, 180, 196
Board games, 139
Body language, in negotiation, 251 254
Body temperature regulation, 7, 60
Bone conduction, 195
Books, recommended
   activities and entertainment, 148
   bipolar disorder, 223
   building self-esteem and increasing potential, 150
   bullying, 212
   child development, 112
   encouragement and hope, 459 460
   exceptional children, 459 461

    feeding problems, 467 468
    non-verbal learning, 152, 171
    potty training, 119
    self-esteem self-help program, 416, 432
    tube feeding, 467
    visual processing remediation program, 47
Books on tape, for VCFS child, 147
Boundaries for conduct, 208
Brain. See Neurological anomalies
Breastfeeding. See Feeding problems
Breathing, 34
Bronchodilators, 498
Buffalo model of CAPD, 193 194
Bullying, 208 212
Burley, Barbara, 416, 432
Butterfly vertebrae, 10

C

Calcium, functions of, 53
Calcium channel drugs, 498
Calcium deficiency, leg pain and, 62
Calcium level, 7, 53 54, 65, 487
Calcium metabolism, dental problems and, 65
Calculators, in school, 147
Calendar, using, 159
California, parent centers in, 267 270
Camp, for VCFS youngsters, 89
Candidiasis, after taking antibiotics, 79
CAP. See Central auditory processing
CAPD. See Central auditory processing disorders
Capillaries, function of, 30
Cardiac problems. See Heart anomalies
Cardiovascular problems. See Vascular anomalies
Caregivers, 305
Carotid artery, 6, 10, 11, 68
Cataracts, 7, 503
"Catch 22", 503
Central auditory nervous system, 195
Central auditory processing (CAP), 189 198
Central auditory processing disorders (CAPD), 84, 190 198
    assessment for, 190 192
    Buffalo model, 193 194
    management of in school, 194
    symptoms of, 190 191
    therapy for, 151 152
Cerebellar ataxia, 8
Cerebellar hypoplasia/dysgenesis, 8
Cerebellar vermis, 8
Cervical fusion in spine, 489
CHARGE association, 11
Chart of daily activities, 158
Child abuse, 295
Child development, 91 107
    1 to 3 months, 91 92, 109

4 to 7 months, 94, 109
8 to 12 months, 96 97, 109
12 to 18 months, 98, 109 110, 179
18 to 24 months, 100 101, 109 110, 179
2 and 3 years old, 103 104, 110, 179 180
3 to 5 years old, 106 107, 110 112, 180

Children
  beverages for, 65, 79
  common injuries of, 114
  common problems of, 114 115
  dentist visits, starting, 65
  developmental stages of, 91 107
  foods for, 76 79
  immunizations for, 59
  potty training method for, 117 119
  speech, 175 187
  toys, age-appropriate, 109 113
  See also Children with VCFS; Infants; Teens

Children with Special Health Care Needs (CSHCN), 319 320

Children with VCFS
  adoption of, 443 446
  assessing current status of, 123 125
  behavioral problems. See Behavioral problems; Psychiatric/psychological problems
  "being different", 138 139
  central auditory processing, 189 198
  developmental issues, 81 89
  ear infections, 41, 42 44, 346 347, 363, 376, 389, 395
  education for. See Education
  educational assessment of, 143 144
  educational assets of, 164 165
  educational deficits of, 165 167
  emotional development of, 127 141
  failure to thrive. See Failure to thrive
  feeding problems. See Feeding problems
  friends, 133, 138, 140 141
  hard to understand speech, 180 182
  illness in, 75, 113, 351
  infections in. See Infections
  initial auditory attention, 165
  IQ test scores, 6, 153
  laughter and joy for, 139
  loss of, 423 428
  non-verbal learning disorder (NVLD), 152, 166, 170
  parents' personal accounts. See Personal stories
  preparing for dangerous situations, 125 126
  psychological problems. See Psychiatric/psychological problems
  puberty, 163, 205
  rote learning and memory, 165
  routines for, 129, 130, 153
  self-esteem issues, 131
  sexuality. See Sexuality
  sign language for, 128, 182 184, 464 465
  social skills, 130, 133 141, 167 169

special needs trust for, 141, 304, 310, 311 317
speech evaluation schedule, 185 186
tube feeding, 76, 79, 403, 481, 489, 505
velopharyngeal insufficiency (VPI). See Velopharyngeal insufficiency
verbal IQ and comprehension, 164 165
visual-spatial skills, 47, 166 167
what is your child like now?, 123 125
See also Infants; Toddlers; Teens
Children's Health Insurance Program (CHIP), 319
Children's Special Health Care Services, 319
Chorionic villi sampling (CVS), 16, 503
Chromosome 22 deletion, 13
See also Velo-cardio-facial syndrome (VCFS)
Chromosome 22q11, 13
Chromosome abnormalities, 13, 18
Chromosomes, 14 15, 503
Chronic pain, 61, 62
Cilia, function of, 33
Circle of Willis anomalies, 11, 503
Circles under eyes, 6, 50, 375
Circulatory system, elements of, 30
Civil action, 295
Civil law, 294
Classroom management, for special needs kids, 147, 150 151
Cleft lip and/or palate, 6, 7, 21, 67 72
    definition of, 503
    in DiGeorge syndrome, 3
    feeding problems and, 75 76
    team approach to, 68 69
    testing for, 67 68
Cleft of primary palate, 72
Cleft of secondary palate, 71
Cleft Palate Foundation, 69
Closeness to others, 135 136
Club foot, 10, 114, 393, 513
Co-op group home, 303 304
Coarctation of the aorta, 5, 25, 503
Cognition, definition of, 503
Cognitive problems, 6, 83, 385
Cognitive regression, 461 462
Cognitive skill development
    2 and 3 year olds, 103 104
    3 to 5 year olds, 106
    IQ testing, 6, 153
Coloboma, 503
Colorado, parent centers in, 270 271
Colostrum, 113
Commissure, 503 504
Common sense, 163
Communication
    central auditory processing, 189 198
    definition of, 504
    non-verbal learning disorder (NVLD), 152, 166, 170
    parents with other VCFS parents, 333 335

presenting educational material, 335 336
See also Communication skills; Language; Social skills; Speech
Communication skill development
   1 to 3 months, 92
   4 to 7 months, 94
   8 to 12 months, 96
   12 to 18 months, 98, 179
   18 to 24 months, 100, 179
   2 and 3 year olds, 103 104, 179 180
   3 to 5 year olds, 106, 180
Communication skills
   assessing your VCFS child, 124
   evaluation schedule for children, 185 186
   eye contact, 136
   naming objects, 137
   negotiating skills for parents, 249 255
   recognition, 137
   spontaneous communication, 137
   taking turns and sharing, 138
Communication style, of teacher and parent, 154 156
Community parent resource centers (CPRCs), 262
Companion animals, 484
Complex verbal memory, 166
Computer, uses of, 148 149, 162 163
Computer software, 82, 149, 151 152
COMT gene, 241 248
   obsessive-compulsive disorder and, 246
   pain tolerance and, 242 243
   schizophrenia and, 243 245
Conceptual thinking, 83, 87
Concrete thinking, 6, 377 378
Conductive hearing loss, 7, 195, 504
Congenital indifference to pain, 246
   See also Pain tolerance
Congenitally missing teeth, 6
Connecticut, parent centers in, 271
Conotruncal heart anomaly, 504
Consequences of their actions, 168
Conservatorship, 449
Constipation, 9, 74 75, 346, 385, 386
Contract, with doctor, 295
Contractures, 8
Coordination, 462
Corella, Cathy, 416
Corneal nerves, 7
Courtesy, 136
CPAP machine, 488
CPRCs. See Community parent resource centers
Cranial fossa, 6
Craniofacial anomalies, 6 7, 384
Craniostenosis, 21
Criminal law, 294
Croup, 35
Cryptorchidism, 7, 21, 504

CSHCN. See Children with Special Health Care Needs
Curfew, 130
CVS. See Chorionic villi sampling
Cyclomania, 9
Cyclothymic, 504
Cystic kidneys, 5

D

Daily activities chart, 158
Dark circles under eyes, 6, 50, 375
Decibel, 195
Decision making, by parents of VCFS children, 336 337
Decoding, 165
Dehydration, 50, 74 75
Delaware, parent centers in, 264, 271
Denial/shock, stage of grief, 425
Dental problems, 6, 65 66, 349, 368, 384
Dental visits, when to start, 65
Dental work, antibiotics prior to, 26
Depression, 9, 167, 221, 222
    definition of, 504
    stage of grief, 425 426
Developmental delay, 8, 376, 504 505
Developmental stages of children, 91 107
    1 to 3 months, 91 92
    4 to 7 months, 94
    8 to 12 months, 96 97
    12 to 18 months, 98
    18 to 24 months, 100 101
    2 and 3 years old, 103 104
    3 to 5 year olds, 106 107
Diagnosis
    anomalies, list of, 4 11, 15, 19 20, 21
    dealing with, 421 422
    FISH test, 13, 17, 19, 350, 394, 494, 506
    tests, 494 495
Diaphragm, 34
Diaphragmatic hernia, 5
Diastasis recti abdominis, 5, 505
Diastolic blood pressure, 30 31
Dichotic, 195
Difficulty swallowing, 364, 374, 403
DiGeorge, Dr. Angelo, 3
DiGeorge sequence, 2, 11
DiGeorge syndrome, anomalies in, 3
Digestive system
    constipation, 9, 74 75, 346
    functioning of, 73
    See also Gastrointestinal system
Diotic, 196
Disabilities, viewing people with, 297 298
Discipline, 208
    parental handling of, 200 202
    on school matters, 169 170

Discussion groups, 327
  See also Support groups
Displaced anus, 5
Dispute resolution, IEP, 150
District of Columbia, parent centers in, 272
Diuretics, 498
DNA, 13, 14, 17, 505
DNA testing, for VCFS, 17
Doctors
  changing doctors, 295
  malpractice, 295
  medical records, 296, 473
  specialists and their fields, 22
  what primary care physician should know, 20
Down's syndrome, 3
Dr. Phil, 117, 421, 437
Drinks, for babies and children, 76, 79
Due care, 294
Durable Power of Attorney for Health Care, 316
Dysphagia, 505
Dyspraxia, 10, 505
Dysthymia, 9, 505
Dystonia, 505

E

Ear anomalies, 7, 41 46, 463
Ear canals, 7
Ear infections, 41, 42 44, 346 347, 363, 376, 389, 395
Ear tags or pits, 7
Ear tubes, 41, 42, 43 44, 505
Ear wax, 41, 463
Ears
  anatomy of, 44
  anomalies in VCFS, 7, 41 46
  infections, 41, 42 44, 346 347, 363, 376, 389, 395
  position and size of, 45, 49
  shape of, 7
  tubes in, 41, 42, 43 44, 505
  wax in, 41, 463
  See also Hearing loss
Eating problems. See Feeding problems
ECAC. See Exceptional Children's Assistance Center
Education, 82, 129, 143 173
  accommodations by teachers and schools, 146 149, 466
  ADHD and, 6, 129, 213-220, 222, 361, 502
  advocacy by parents, 145-146, 150, 259 260, 293, 299 300
  assessing your child, 125
  attention and, 152 153
  central auditory processing disorders (CAPD) and, 194
  child's age and, 159 160
  communication style of teacher and parent, 154 156
  computers for, 148 149, 162 163
  discipline on school matters, 169 170
  dispute resolution, 150

    evaluation of child, 125, 143 144
    grade school ages, 157 173
    high school, 205, 209 210
    IDEA Act, 143, 150, 263
    IQ test scores, 6, 153
    legal rights of child, 143
    listening skills, teaching, 194
    mainstreaming, 509
    math, 6, 85, 159, 167, 464
    negotiating skills for parents, 150, 249 255
    non-verbal learning disorder (NVLD) and, 152, 166, 170
    parent advocacy, 145 146
    parental participation in, 82, 113, 149 152
    personal stories, 171 172, 299 300, 354, 357, 363, 369, 382, 388,
       399 400, 464 466, 478, 479, 492
    pre-school, 112
    psychological issues, 152 153
    reading comprehension, 6, 167
    regular classroom, 144
    scholarships for higher learning, 173
    sequencing and, 152 153
    tests, 146, 147, 148, 151, 169
    transition plan, 260
    visual learning style, 86, 157 159
    See also Individualized Education Plan (IEP); Learning issues;
School; School issues; Special education; Teachers
Education strategy, 82
Educational history, 261
Emotional development, 127 141
    of infants, 127 128
    of preschoolers, 129
    of school age children, 129 130
    of teens, 130 131
    of toddlers, 128 129
Emotional functioning, 167 168
    See also Behavior; Behavioral problems; Psychiatric/psychological
       problems
Enamel (teeth), 6, 65
Endocrine problems in VCSF, 7
Enjoying life, 139
Enlarged print, 147
Epstein's pearls, 113
Esophageal reflux, 506
Estate planning, 141, 307 320
    beneficiary, 316
    financial power of attorney, 318
    guardianship, 313 314, 315
    letter of intent of care, 314
    not jeopardizing government benefits, 309 310
    probate, 310 311
    revocable living trust, 308 309
    special needs trust, 141, 304, 310, 311 317
    wills, 307, 308, 310 311
Evaluation, listening and, 334

Event sequencing, 85
Exaggerated startle response, 9
Exceptional Children's Assistance Center (ECAC), 265
Exceptional Parent magazine, 339
Executive function, 168 169
Expressive language, 167
Extended time for assignments, 147 148
Extended time for tests, 147, 151
External ear canals, 7
Extra ribs, 10
Eye contact, 136
Eye exams, 48
Eyelids, appearance of, 7, 50
Eyes
    "allergic shiners", 6, 50, 375
    anomalies in VCFS, 6 7, 21, 47 48, 384, 505
    eye exams, 48
    eyelids, 7, 50

F

Face
    shape of, 6, 49
    anomalies in VCFS, 6 7, 49 50
    facial expression, 9, 49, 50, 133, 183
Facial expression of others, understanding by VCFS child, 167
Facial expression of VCFS child, 9, 49, 50, 133, 182
Failure to thrive, 9, 69, 77, 114, 345, 367, 393, 466 467
Familiar structured environment, 88
Families
    dealing with VCFS diagnosis, 421 422
    VCFS, effect of on, 409 422, 482
Family happiness, 435 437
Family support, 390, 419 422
Family trust, 312
FAPE Solutions, 294
Feeding problems, 9, 21, 67 69
    in infants, 75 76, 114, 176
    personal stories, 80, 344, 373, 376, 466 468
    toddlers, 76 80
    tube feeding, 76, 79, 80, 403, 481, 489, 505
    See also Nasal regurgitation; Reflux
Feeding tube, 76, 79, 80, 403, 481, 489, 505
Feet
    anomalies in VCFS, 8, 62 63, 468 469
    club foot, 10, 114, 393, 513
    flat footedness, 10, 63, 365, 468 469
Fever, 59 60
Finances
    financial power of attorney, 318
    out-of-pocket expenses, 312 313, 316
    Social Security, 309, 316
    See also Estate planning
Financial power of attorney, 318
Fine motor skills, 84, 506

Finger anomalies, 8, 50, 385
Finger foods, 77 79
Fingers
    anomalies in VCFS, 8, 50, 385
    length of, 50, 385
FISH test, 13, 17, 19, 350, 394, 494, 506
Fisher's auditory problem check list, 191 192
Fixed beliefs, 439
Flap surgery, 69
Flat expression, 9, 49, 50, 133, 182
Flat footedness, 10, 63, 365, 468 469
Florance, Cheri, 459
Florida, parent centers in, 272
Fluoride, for teeth, 65
FM system, 196, 463
Focused attention, 165
Following directions, 84
Food
    digestion of, 73
    See also Feeding problems
Food coming out the nose. See Nasal regurgitation
Food preferences, 76 77
Foods
    drinks, 79
    finger foods, 77 79
    preferences for certain foods, 76 77, 176
Foot anomalies. See Feet
Foremilk, 113
Free air transport, 306
Friends, 133, 138, 140 141
Fruits, encouraging eating of, 77 78
Fused vertebrae, 10

G

G-tube, 80
G-U reflux, 7
Gastro-esophageal reflux, 9, 176, 367, 377, 506
Gastrointestinal system
    anomalies in VCSF, 5
    constipation, 9, 74 75, 346, 385, 386
    definition of, 506 507
    workings of normal GI tract, 73 74
Gene expression, 506
Gene therapy, 507
Generalized anxiety disorder, 9
Genes, 14, 17, 506
Genetic code, 506
Genetic counselors, 16
Genetic disorders, 14 15, 18
Genetic testing for VCFS, 16, 396
Genetics
    alleles, 18, 501
    chromosomes, 14 15, 503
    COMT gene, 241 248

DNA, 13, 14, 17, 505
gene expression, 506
genes, 14, 17, 506
genetic code, 506
genetic disorders, 14 15, 18
genome, 506
genotype, 506
karyotyping, 17, 508
mutations, 17 18
obsessive-compulsive disorder and, 246
pain tolerance and, 242 243
phenotype, 510
schizophrenia and, 243 245
sequence, 2
syndrome, 1 2, 3
transcription and translation, 17
of VCFS, 13 14, 15
Genito-urinary anomalies, 7, 21
Genome, 506
Genotype, 506
Georgia, parent centers in, 272 273
Gestures, 155 156, 166
Gill, Barbara, 460
Ginsberg, Debra, 459
Glasses, 47 48
Goals and objectives, for IEP, 161
Golding-Kushner, Dr. Karen, 4
Good behavior, reward system, 168
Government group homes, 302
Grade school, 157 173
    See also Education
Gray, John, 416 417
Grief, stages of, 425 426
Grieving, 423 428
    getting through, 426 428
    guilt while grieving, 424
    stages of grief, 425 426
    supporting grieving friends, 428
Gross motor skills, 84, 507
Group care co-op, 305
Group homes, 301 305, 306, 357, 469
Growth hormone deficiency, 7, 21, 54, 55
Guardianship, 313 314, 315
Guilt
    of parents of VCFS children, 433 434
    while grieving, 424

H

Hairiness, 10, 50, 378, 386
Hand anomalies, 8
Handmade foods, 78
Handwriting, 469
Hawaii, parent centers in, 273
Health, assessing your VCFS child, 123

Health care services, 319 320
Health insurance, 319
Hearing
    hyperacusis, 45 46
    sensitivity to loud noises, 45 46, 464
    See also Hearing loss
Hearing loss, 41 42, 470
    conductive, 7, 195, 504
    ruling out, 181, 183
    sensorineural, 7, 512
Hearing voices, 221, 237
Heart
    anatomy of, 26, 29
    blood flow in, 29 30
Heart anomalies
    in DiGeorge syndrome, 3
    personal stories, 360 361, 365, 381, 395, 405
    in VCSF, 5 6, 21, 25 29
Height, 7, 54
Hemivertebrae, 10
Hepatoblastoma, 5, 507
Heredity. See Genetics
Hernias, 5, 21, 38 39, 507
High school, 205, 209 210
Hindmilk, 113
Hirschsprung's disease, 507
Hoarseness, 10
Holoprosencephaly, 11
Home economics, 85
Homework, 218
Hormone difficulties, 7, 53 54
Hormones, 498
Housing, group homes, 301 305, 306, 357, 469
Hyperactivity
    symptoms of, 216
    See also Attention deficit hyperactivity disorder
Hyperacusis, 45 46, 464
Hyperextensible/lax joints, 10, 21, 507
Hypernasal voice, 10, 68, 177, 507
Hypersexuality, 234, 353
Hypertelorism, 507
Hypertension, 31
Hyperthermia, 60
Hypnotics, 498
Hypocalcaemia
    definition of, 507
    in DiGeorge syndrome, 3
    in VCFS, 7, 53 54, 65, 378, 487
Hypomania, 9, 507
Hypoparathyroidism
    definition of, 507 508
    in DiGeorge syndrome, 3
    in VCFS, 7, 53
Hypoplasia, definition of, 508

Hypoplastic immune system, in DiGeorge syndrome, 3
Hypoplastic kidney, 5
Hypoplastic pituitary gland, 7
Hypoplastic thymus, 7
Hypospadia, 7, 21, 507
Hypotension, 31
Hypothermia, 60
Hypothyroidism, 7, 54, 508
Hypotonia, 8, 81, 85
    articulation disorders, 10, 178
    definition of, 508

I

ICF. See Intermediate care facility
Idaho, parent centers in, 273 274
IDEA Act, 143, 150, 263
IEP. See Individualized Education Plan
Illinois, parent centers in, 274
Illness, 351
    daycare and, 113
    effect on VCFS child, 75
    See also Infections
Imitation, 83 84
Immune system, 57 59
    in DiGeorge syndrome, 3
    in VCFS, 8, 21, 58-59
Immunizations, 58 59, 508
Immunosuppressants, 498
Imperforate anus, 5, 39
Impulsiveness, 9, 130, 168, 236
    ADHD and, 215, 216
    symptoms of, 216
Inattention, symptoms of, 216 217
Indiana, parent centers in, 275
Indifference to pain, 246
    See also Pain tolerance
Individualized Education Plan (IEP), 145 146
    accommodations by teachers and schools, 146 149
    advocacy by parent, 150, 259, 260, 290
    bathroom, use of in school, 171
    bipolar disorders, IEP for, 240
    dispute resolution, 150
    goals and objectives in, 161
    personal stories, 388
    social skills and, 161
    transition plan, 260
Individualized living arrangement (IRA), 303
Individualized Technical Assistance Agreements (ITAGs), 263
Infants
    asymmetric crying facies, 6
    common problems of, 114
    developmental stages of, 91 100
    ear infections, 41, 42 44, 346 347, 363, 376, 389, 395
    emotional development of, 127 128

   failure to thrive, 9, 69, 77, 114, 345, 367, 393, 466 467
   feeding problems, 67 69, 75 76, 114, 176, 344, 466 468
   nasal regurgitation, 9, 68, 69, 80, 176, 180, 197, 345, 385, 386
   new mothers' FAQ, 113 114
   problems with VCFS infants, 9
   sign language for, 128, 182 184, 187, 464 465
   taking bottle to bed, 65
   tooth brushing in, 65
Infections, 57, 471 473
   ear infections, 41, 42 44, 346 347, 363, 376, 389, 395
   personal stories, 346 348, 359, 471 473
   pneumonia, 35 36, 348, 511
   upper respiratory infections, 8, 378
Inguinal hernia, 5, 21, 38, 349, 508
Initial auditory attention, 165
Intelligence, IQ, 6, 153, 508
Intentional tort, 295
Intermediate care facility (ICF), 303
Internet resources. See Online resources
Interpretation, listening and, 334
Interrupted aortic arch
   in DiGeorge syndrome, 3
   in VCFS, 5, 25, 28
Intervention, definition of, 508
Iowa, parent centers in, 275
IQ, 6, 153, 508
IRA. See Individualized living arrangement
IRDS, 35
Iris coloboma, 7
Iris nodules, 7
Irlen lenses, 47 48
Irritability, 9
ITAGs. See Individualized Technical Assistance Agreements

J

J-tube, 80
Jobs, adults with VCFS, 355 356
Joint anomalies, 10, 21
Joint dislocations, 10
Juice, effect of on teeth, 65
Junk foods, 77
Juvenile rheumatoid arthritis (JRA), 10, 21

K

Kansas, parent centers in, 275
Karyotyping, 17, 508
Kentucky, parent centers in, 275 276
Kidneys
   anomalies in VCFS, 5, 21, 37-38
   in normal body, 37
Killer T cells, 58
Kingsley, Emily Perl, 457
Kleinman, Dr. Joel, 244

L

Labeling, of self, 438 440
Labels, for disabled children, 473
Lactate milk, 113
Landsman, Donna, 143
Language
  components of, 184
  definition of, 508
  processing, 167
Language development
  in 2 and 3 year olds, 103
  in 3 to 5 year olds, 106
  assessing your VCFS child, 124
  speech pathologist and, 176
  See also Communication skill development; Communication skills
Language problems, 10, 184 185
  personal stories, 377, 385
  as target area for special education team, 145
  wording sentences simply for better comprehension by child, 154 156
  See also Communication; Communication skills; Speech; Speech problems
Language processing, 167
Lanugo, 113
Laryngeal anomalies, 9
Laryngomalacia, 9
Larynx, 34, 508 509
Laughter, 139
Laxatives, 498
Leach, Debra, 175 182
Learning by memorizing, 85
Learning by repetition, 85, 86
Learning disabilities
  personal stories, 377, 385
  as target area for special education team, 145
Learning issues, 6, 83 88
  conceptual thinking, 83, 87
  familiar structured environment, 88
  following directions, 84
  imitation, 83 84
  language problems, 84
  learning styles, 85
  math, 6, 85, 159, 167, 464
  memory problems, 84
  motor skills, 84
  non-verbal learning disorder (NVLD), 152, 166, 170
  problem solving skills, 163
  researching school paper, 86
  taking notes, 86
  talking in front of class, 85 86
  visual-spatial skills, 47, 166 167
  writing sentences, 84
  See also Education
Learning skills, assessing your VCFS child, 124
Learning styles, 85-86

Leg pain, 10, 62, 248, 349, 361, 386
Legal issues
   conservatorship, 449
   Durable Power of Attorney for Health Care, 316
   financial power of attorney, 318
   living will, 318
   medical records, 296, 473
   special needs trust, 141, 304, 310, 311 317
   See also Estate planning
Lenses (for glasses), 47 48
Letter of intent of care, 314
Lightfoot, Debra A., 189 198
Lightheartedness, 139
Limb anomalies, 8, 385
Limb contractures, 8
Limited conservatorship, 449
Limiting beliefs, 439
Lip, normal anatomy, 70
Listening
   during negotiation, 252 253
   stages of, 334
Literal thinking, 83
Live vaccine, 58
Living arrangements, group homes, 301 305, 306, 357, 469
Living trust, 312
Living will, 318
Long fingers, 50
Loud sounds, sensitivity to, 45 46, 464
Louisiana, parent centers in, 276
Low calcium level, 7, 53 54, 65, 487
Low self-esteem, 131
Lower airway disease, 8, 378
Lower jaw, position of, 6
Lung problems, 5
Lungs, 34
Lying
   perception of during negotiation, 253 254
   as symptom of psychological problem, 235

M

Maine, parent centers in, 276 277
Makaton, 187
Making change, 160
Malpractice, 295
Malrotation, definition of, 508
Malrotation of the bowel, 5
Manic depressive illness, 9, 55, 221, 222, 352
Marriage, parents' marriage, effect of VCFS on, 409 418
Marsh, Jayne D.B., 460
Maryland, parent centers in, 277
Massachusetts, parent centers in, 277 278
Masturbation, 55
Math, 6, 85, 159, 167, 464
Matrix Parent Network and Resource Center, 265

Mattay, Dr. Venkata, 245
Mealtimes, 77-79
   See also Beverages; Feeding problems; Food preferences
Meconium, 113
Medicaid Waiver Program, 319
Medical care
   advocacy, 294 296, 306
   Durable Power of Attorney for Health Care, 316
   free air transport, 306
   insurance for, 319
   living will, 318
   medical history, 261
   medical records, 296, 473
Medical history, 261
Medical insurance, 319
Medical records, 296, 473
Medical tests, 494 495
Medications
   classifications of, 497 498
   psychiatric, 473 474, 485 486
Memorizing, 85
Memory problems, 84, 165
Memory techniques, 86
Meningomyelocele, 8
Mental Health Mental Retardation offices, 306
Mental illness. See Psychiatric/psychological problems
Mental retardation, VCFS and, 6
MetDESK, 309
Michigan, parent centers in, 278 279
Microcephaly, 6, 509
Middle ear infections, 42 44
Milia, 113
Milk coming out the nose. See Nasal regurgitation
Miller, Nancy B., 460
Minnesota, parent centers in, 279
Missing teeth, 6
Mississippi, parent centers in, 279 280
Missouri, parent centers in, 280
Modified tests and assignments, 146, 151
Money
   handling and using, 160
   spending habits, 87
   See also Finances
Mongolian spots, 113
Montana, parent centers in, 280
Mood disorders, 9
Morphea, 8, 509
Motor skills, 81, 84, 123, 167
   See also Movement skill development
Mouth
   normal anatomy, 70
   size of, 49
Movement skill development
   1 to 3 months, 91 92

4 to 7 months, 94
8 to 12 months, 96
12 to 18 months, 98
18 to 24 months, 100
2 and 3 year olds, 103 104
3 to 5 year olds, 106
Multi-step directions, 84
Multi-view videofluoroscopy, 509
Muscle anomalies, 10, 21, 81
Music, 135, 139 140, 399
Mutations, 17 18

N

Nails, 8
NAMI. See National Alliance for Mental Illness
Naming objects, 137
Napping, 128
Nasal anomalies, 8
Nasal bridge, 8, 385
Nasal passages, narrow, 8
Nasal regurgitation, 9, 68, 69, 80, 176, 180, 197, 345, 385, 386
Naseef, Robert A., 460
Nasoendoscopy, 67 68, 177
Nasopharyngoscopy, 509
National Alliance for Mental Illness (NAMI), 306
National Father's Network, 338
National Information Center for Children and Youth with Disabilities (NICHCY), 338
National Organization on Disability, 339
National Parent to Parent Support and Information System, 338
National Patient Air Transport, 306
Nebraska, parent centers in, 280
Negligence, 295
Negotiating skills for parents, 249 255
    body language, 251 254
Neurological anomalies, 8, 21, 385
Neuropsychology, 164
Nevada, parent centers in, 281
Nevus flammeus, 113
Nevus vascularis, 114
New Hampshire, parent centers in, 281
New Jersey, parent centers in, 281
New Mexico, parent centers in, 281 282
New York, parent centers in, 282 283
NICHCY. See National Information Center for Children and Youth with Disabilities
Non-verbal communication, 155 156, 166
Non-verbal learning disorder (NVLD), 152, 166, 170, 475
North Carolina, parent centers in, 283 284
North Dakota, parent centers in, 284
Nose
    anomalies, 8, 49 50, 69
    function of, 33
    shape of, 8, 49

Nose, See also Nasal regurgitation
Nose regurgitation. See Nasal regurgitation
Nosebleeds, 350
Nostrils, shape of, 8
Note cards, 147
Note taking, in school, 147
Nursing problems. See Feeding problems
NVLD. See Non-verbal learning disorder

O

"O" expression, 49, 50
Objects of reference, 158
Obsessions, personal stories, 475 477
Obsessive compulsive disorder (OCD), 9, 129, 153, 167, 237, 246
Obsessive thinking, managing, 152 153
Occult submucous cleft palate, 363, 509 510
Ocular-motor processing, 48
Ohio, parent centers in, 264, 284
Oklahoma, parent centers in, 284
Online resources
    ADHD, 219
    advocacy, 258, 259
    Asperger's syndrome, 202
    author's web site, vii
    bipolar disorder, 223
    camps for disabled children, 89
    classroom setting, 147
    companion animals, 484
    CSHCN program, 320
    Dr. Phil web site, 117, 421, 437
    FAPE Solutions, 294
    flat feet, 468
    handwriting, 469
    IEP assistance, 294
    learning disabilities, 152
    medical air transport, 306
    MetLife's Division of Estate Planning for Special Kids, 309
    non-verbal learning disorder (NVLD), 152
    oral-motor therapies, 196
    parent centers, 263 273
    parent self-esteem, 432
    parent-training and information centers by state, 259
    patient air transport, 306
    schizophrenia, 238
    scholarship funds, 173
    sign language for infants, 184
    support groups, 333, 338 340
    Suze Orman, 307
    Talk Tools, 196
    toys and utensils, 115
    VCFSEF, viii, 4, 324
    visual impairment and testing, 47 48
Oppositional defiant disorder, 474

Oral commissures, downturned, 6
Oral hygiene, 65
Oral-motor therapies, 196, 510
Oral transport, 510
See also Nasal regurgitation; Reflux
Orbital dystopia, 7
Orbital hypertelorism, 7
Oregon, parent centers in, 285
Organization, scheduling, 160
Organizations
   by state, 259
   See also Online resources; Support groups
Organized thought process, 85
Organizing, 160
   executive function, 168 169
   routines, 129, 130, 153, 168
   sequencing, 152 153, 159, 166
   timetable of daily activities, 158
   yearly calendar, 159
Orman, Suze, 307
Orthopedic anomalies, 10
   See also Club foot
Osteopaenia, 10, 510
Otitis media, 376, 510
Out-of-pocket expenses, 312 313, 316
Oxygen desaturation without apnea, 9, 378

P

PACER Center, 263, 264
Pain, 61 63
Pain tolerance, 62, 246 248
   COMT gene and, 242 243
Palatal lift, 198
Palate anomalies
   in DiGeorge syndrome, 3
   in VCFS, 6, 7, 21, 67 72, 384
   See also Cleft lip and/or palate
Palatine tonsils, function of, 33
Palpebral fissures, 6
Panic disorder, biofeedback for, 123
Papolos, Dr. Demitri, 238 239
Paranoia, 352
Parathyroid gland
   DiGeorge syndrome and, 2, 3
   in VCFS, 53, 487
Parent centers, 262 263
Parent training and information centers (PTIs), 262
Parental advocacy. See Advocacy
Parents of Blind Children, 338
Parents of VCFS children
   adoption of VCFS children, 443 446
   advocacy by, 145 146, 150, 257 306
   behavior problems, handling, 200 202

being kind to yourself, 432 435
coaching by teachers, 161 163
coming to terms with reality, 434 435
communication with other VCFS parents, 333 335
critical thinking by, 336 337
dealing with VCFS diagnosis, 421 422
decision making, 336 337
demands of infants, 127 128
discipline by, 200 203
discipline on school matters, 169 170
emotional issues facing, 431 441, 482
estate planning by, 141, 307 320
family happiness, 435 437, 482
family support, 390, 419 422
goals of, 435
grieving, 423 428
guilt, 433 434
helping kids enjoy life, 139
helping with schooling, 82, 113, 149 152, 218
internal dialogue of, 438
labeling of self, 438 439
limiting beliefs, 439
marriage and relationships of, 409 418
negotiating skills, 249 255
out-of-pocket expenses for, 312 313, 316
parent's anger, 201, 434, 441
personal accounts by. See Personal stories
resources for. See Resources for parents
respite for, 482
responsibilities of, 482
self-esteem of, 431 432, 435
sharing with others, 135
special needs trust, creating, 141, 304, 310, 311 317
speech development, stimulating, 178 180
stress in, 440 441, 482
support groups for. See Support groups
taking a break, 482
watchfulness and alertness by, 202
watchfulness for inappropriate behaviors, 206
wording sentences simply for better comprehension, 154 156
Parents' stories. See Personal stories
Patent ductus arteriosus (PDA), 5, 25, 28, 510
Patience, 136
PDA. See Patent ductus arteriosus
Peer groups, 130
Pennsylvania, parent centers in, 285 286
Perception, in listening, 334
Perceptual dominance, 61
Peripheral hearing, 196
Peristalsis, 73, 74
Periventricular cysts, 8, 510
Personal stories, 343 407, 477 484
    ADD/ADHD, 361
    adoption, 443 446

adults with VCFS, 355 358, 446 454
advocacy, 257 259, 454 455
attitude, 455 459
bedwetting, 459
bloody noses, 36
child development, 462
cognitive regression, 461 462
constipation, 346, 385, 386
dental problems, 66, 349, 384
difficulty swallowing, 364, 374, 403
discipline, 88
ear infections, 346, 363, 376, 384, 389, 395
ear problems, 384, 463
education, 171 172, 299 300, 354, 357, 363, 369, 382, 388, 399 400, 464 466, 478, 479, 492
eye anomalies, 384
failure to thrive, 345, 367, 377, 393
feeding problems, 80, 344, 373, 376, 466 468
flat feet, 63, 468 469
friends for VCFS kids, 140 141
group homes, 306, 357, 469
handwriting, 469
health care services, 319 320
hearing loss, 347
heart defects, 360 361, 365, 381, 395, 405
hormonal problems, 54 55
hypersexuality, 353
infections, 346 348, 359, 471 473
inguinal hernia, 349
language problems, 377
leg pain, 248, 349, 386
marriage, VCFS child and parents' marriage, 409 418
masturbation, 55
medical insurance, 319
music in kids' lives, 139 140, 399
nasal regurgitation, 345, 385, 386
non-verbal learning disorder (NVLD), 170 171, 475
nosebleeds, 350
obsessions, 475 477
pain tolerance, 246 248
pneumonia, 348
psoriasis, 51
psychiatric/psychological problems, 228 237, 352 353, 370 372, 405, 480, 485 486
recreational activities, 115
response to loud sounds/noise, 45 46, 464
scholarship funds, 173
school, 171 172, 299 300, 354, 357, 363, 369, 382, 388, 399 400, 464 466, 478, 479, 492
school issues, 171 172, 299 300, 354, 357, 363, 369, 382, 478, 479, 492
seizures, 376, 403 404, 486
sexual obsession, 371 372
sign language, 187, 464 465

    sleep habits, 487 489
    social skills, 139 140, 493 494
    speech problems, 186 187, 380, 385, 470 471, 483
    summer camp, 89
    support groups, 297 298, 340 341
    surgical palate repair, 80
    taking ourselves too seriously, 139
    teens, 489 494
    tics and grunts, 470
    velopharyngeal insufficiency (VPI), 350
    video games, 115
    viewing people with disabilities, 297 298
Pets, 484 485
Pharyngeal anomalies, 9, 385
Pharyngeal flap, 69
Pharyngeal hypotonia, 9
Pharynx, function of, 33
Phenotype, 510
Phimosis, 114
Phobias, 9, 510
Phoneme, 196
Phonemes, 195
Phonemic synthesis (PS) test, 192
Phonological processing, 167
Physical skills, assessing your VCFS child, 123
Physicians. See Doctors
Piano fingers, 50
Pierre Robin sequence, 11
Pituitary gland, 7
Planning for the future. See Estate planning
Platybasia, 6
Play skill development
    4 to 7 months, 94
    8 to 12 months, 96
    12 to 18 months, 98
    18 to 24 months, 100
Play skills, assessing your VCFS child, 124
Playing store, 87
Playing with food, 77
Pneumonia, 35 36, 348, 511
Polio vaccine, 58
Polydactyly, 8, 511
Postaxial polydactyly, 8
Posterior cranial fossa, 7
Posterior embryotoxin, 6
Potter sequence, 11
Potty training, 117 119
Pre-school, 112
Preaxial polydactyly, 8
Preferential seating, in school, 147, 150
Pregnancy, prenatal testing for VCFS, 16
Prenatal testing for VCFS, 16
Preschoolers, emotional development of, 129
Prevident® toothpaste, 66

Primary care physician, what he/she should know, 20
Primary palate, cleft of, 72
Private home, as group home, 302 303, 304 305
Privately-run group homes, 302
Probate, 310 311
Problem solving skills, 163, 218
Procrastination, 137
PS test, 192
Pseudo-hypoparathyroidism, 7
Psoriasis, 51
Psychiatric/psychological problems, 9, 21, 167 168, 221 240
    ADD/ADHD, 6, 129, 213 220, 222, 239
    anxiety, 9, 123, 167, 405, 474
    autism, 167, 486
    bipolar disorder, 9, 54, 220, 221, 222, 223 225, 228 237, 238 239, 352, 405, 486, 502
    cycling bipolar disorder, 220
    depression, 9, 167, 221, 222, 504
    depressive symptoms, 224 225
    dysthymia, 9, 505
    education and, 152 153
    hypersexuality, 234, 353
    hypomania, 9, 507
    manic depressive illness, 9, 55, 221, 222, 352
    manic symptoms, 224
    medications, 473 474, 485 486
    obsessive compulsive disorder (OCD), 9, 129, 153, 167, 237, 246
    oppositional defiant disorder, 474
    paranoia, 352
    parents' stories, 228 237, 352 353, 405, 480, 485 486
    phobias, 9, 510
    schizoaffective disorder, 9, 352, 511 512
    schizophrenia, 222, 237 238, 243 245, 512
    separation anxiety, 232, 234
    social-anxiety disorder, 85 86, 167
    See also Behavioral problems
Psychogenic pain, 61
Psychophysiology, 121
Psychosis
    definition of, 511
    symptoms of, 221 222
    See also Psychiatric/psychological problems
PTIs. See Parent training and information centers
Puberty, 163, 205
    See also Teens
Public schools, 143
    See also Education; School; Special education
Puerto Rico, parent centers in, 286
Puffy eyelids, 7, 50
Pulmonary atresia, 5, 511
Pulmonary stenosis, 5
Punishment, 208

Q
"q" arm, 13

R
Rage. See Anger
Raynaud's phenomenon, 11
Raynaud's sign, 511
RDS. See Respiratory distress syndrome
Reactive airway disease, 9, 376
Reading comprehension, 6, 167
Recall of paragraphs, 167
Receptive language, 167, 511
Recreational activities, 115
Rectum, crooked, 385, 386 387
Referred pain, 61
Reflux, 6, 9, 176, 373, 506
Relationship skills, assessing your VCFS child, 124
Renal anomalies. See Kidney anomalies
Repetition, 85, 86, 165
Researching school paper, 86
Resolution, of grief, 425 426
Resonance, 176 177
Resources for parents
    Alliance Project, 261 263
    FAPE Solutions, 293
    parent centers, 259, 266 293
    S.N.A.P., 258
    by state, 259, 265 293
    See also Advocacy; Books, recommended; Online resources
Respect, 136
Respiratory distress syndrome (RDS), 35
Respiratory system, VCFS and, 8, 21, 33 36
Retrognathia, 6, 385
Revocable living trust, 308 309
Rhode Island, parent centers in, 286
Rib fusion, 10
Ribs, extra, 10
Right-sided aorta, 5, 25, 376
Rote learning and memory, 165
Routines, 129, 130, 153, 168
Rules, 208, 464

S
Sadness, 223
See also Bipolar disorder; Depression
Safety, preparing child for dangerous situations, 125, 136
Satter, Evelyn, 467
Scalp hair, 10
SCAN-C tests, 193
Scapular deformation, 10
Scheduling, 160
Schiffman, Patricia, 65 66
Schizoaffective disorder, 9, 352, 511 512

Schizophrenia, 222, 237 238
    COMT gene and, 243 245
    definition of, 512
Scholarships for higher learning, 173
School
    advocacy, 259 260, 293, 299 300
    assistive devices used in, 147, 151, 195
    behavior at, 169 170, 208 212
    bullying at, 208 212
    classroom management for special needs kids, 147, 150 151
    high school, 205, 209 210
    medical needs cared for in, 171 172
    negotiating skills for parents, 249 255
    obligations under IDEA Act, 144, 263
    personal stories, 171 172, 299 300, 354, 357, 363, 369, 382, 388, 399 400, 464 466, 478, 479, 492
    pre-school, 112
    public schools, 143
    scholarships for higher learning, 173
    school diaries, 130
    See also Education; Individualized Education Plan (IEP); Learning issues; School issues; Special education; Teachers
School issues
    assignments, 146, 147 148, 151, 218
    bullying, 208-212
    homework, 218
    tests, 146, 147, 148, 151, 169
School districts, advocacy, 259 260, 293, 466
Scoliosis, 10, 386, 512
Scotopic vision sensitivity syndrome, 47
Scrapbook of achievement, 206
Seating, in classroom, 147, 150
Secondary palate, cleft of, 71
Sedatives, 498
Seeing double, 48
Seizures, 8, 54, 365, 376, 403 404, 486 487, 512
Self-awareness development
    1 to 3 months, 92
    4 to 7 months, 94
    8 to 12 months, 96 97
    12 to 18 months, 98
    18 to 24 months, 100 101
    2 and 3 year olds, 103 104
    3 to 5 year olds, 106 107
Self-control, 130 131
Self-discipline, 153, 218
Self-esteem, 131
    of ADHD child, 218
    of parents, 431 432, 435
    scrapbook of achievement, 206
Self-help, assessing your VCFS child, 124
Self-image, 131
Self-pity, 416
Sensation, 196

Sensitivity to loud noises, 45 46, 464
Sensorineural hearing loss, 7, 512
Sensory integration dysfunction, 46
Sentence recall, 167
Separation anxiety, 232, 234
Sequence (genetics), 2
Sequencing, 152 153, 159, 166
Sequencing games, 166
Setting boundaries, 130
Setting rules, 208, 464
Sexuality
   hypersexuality, 234, 353
   masturbation, 55
   sexual obsession, 371 372
Shaking hands, 136
Sharing, 138
Shopping sprees, 231
Short nails, 8
Shprintzen, Dr. R.J., vi, ix, 3
Siblings, effect of VCFS sibling on, 419 422
Sign language, 128, 182 184, 187, 464 465
Signal-to-noise, 196
Simons, Robin, 460
Skeletal anomalies, 10
Skeletal muscles, 10
Skin, 8, 10, 378
Sleep apnea, 488
Sleep habits, 487 489
Small head, 6
Small stature, 7, 54
S.N.A.P. See Special Needs Advocate for Parents
Social activities, 133 135
Social-anxiety disorder, 85 86, 167
Social immaturity, 9
Social Security, 309, 316, 355
Social skills, 130, 133 141, 167 169
   definition of, 512
   eye contact, 136
   friends, 133, 138, 140 141
   patience, 136
   personal stories, 139 140, 493 494
   recognition, 137
   respect, 136
   spontaneous communication, 137
   taking turns and sharing, 138
   teachers and, 161
   teens, 134, 163
   tolerating closeness to others, 135 136
Socializing skill development
   1 to 3 months, 92
   4 to 7 months, 94
   8 to 12 months, 96
   12 to 18 months, 98
   18 to 24 months, 100

2 and 3 year olds, 103 104
　　3 to 5 year olds, 106
Software, 82, 149, 151-152
Somatogenic pain, 61
South Carolina, parent centers in, 286 287
South Dakota, parent centers in, 287
Special education
　　accommodations by teachers and schools, 146 149, 466
　　advocacy by parent, 259 260, 293, 299 300
　　classroom management, 150 151
　　Individualized Education Plan (IEP), 145 146, 150, 161, 171, 240, 290
　　non-verbal learning disorder (NVLD), 152, 166, 170
　　obligation of schools, 144
　　parental participation in, 149 152
　　referring child for, 143 144
　　target areas, 145
　　transition plan, 260
Special Needs Advocate for Parents (S.N.A.P.), 258
Special needs trust, 141, 304, 310, 311 317
Specialized behavioral plans, in school, 147
Specialized instruction, 146
Speech, 175 187
　　central auditory processing, 189 198
　　dyspraxia, 10, 505
　　evaluation schedule, 185 186
　　hard to understand speech, 180 182
　　resonance, 176 177
　　sign language, 128, 182 184, 187, 464 465
　　See also Communication; Communication skills; Language problems; Speech problems
Speech and language needs, as target area for special education team, 145
Speech bulb, 198
Speech-in-noise test, 192
Speech pathologist, role of, 175 176
Speech patterns, of parent or teacher, 154 156
Speech problems, 10, 21, 68, 81, 180 181
　　personal stories, 186 187, 380, 385, 470 471, 483
　　ruling out hearing loss, 181, 185
Speech recognition testing, 196
Speech therapy, 361, 475, 483
Spelling, 85
Spending habits, 87
Sphincter-plasty, 69
Spina bifida, 8, 512
Spina bifida oculta, 10
Spinal anomalies, 10, 21, 386, 489
Spontaneous communication, 137
Sports, 135
Sprengel's anomaly, 10, 512
Staggered spondaic word test (SSW), 192 193
Standardized test, 512
Startle response, 9

State Children's Health Insurance Program (CHIP), 319
States, parent-training and information centers by state, 259
Stature, 7, 54, 359
Stealing, 230, 235
Stenson, Margi, 182 187
Story problems, 85
Strabismus, 6, 512
Strangers, preparing child to deal with, 125, 136
Stress, for parents, 440 441
Stroke, 8, 512
Structured environment, 88
Subclavian arteries, 6
Submucous cleft palate, 69, 176, 363, 509 510, 512
Successor trust, special needs trust, 312
Suicidal thoughts, 231
Summer camp, for VCFS youngsters, 89
Supplemental needs trust, 310
Support groups, 321 333, 338 341
    ADD/ADHD, 219
    assessment of, 332
    definition of, 324 325
    fundraising by, 331
    group format, 327 330
    group homes, 306
    leadership in, 326 327
    meetings, 328 330
    membership, 330 331
    mission statement of, 325 326
    non-profit status of, 331 332
    personal view of, 297 298
    promoting, 330
    referral list, 330
    resources, 330
    starting a group, 325
Surgery
    nose, shape of, 49
    temperature regulation and, 60
    for VPI, 69, 177 178
Swallowing, 364, 374, 403
Swimming, 364
Sylvian fissures
    definition of, 512
    enlarged, 8
Symptoms of VCFS. See Anomalies in VCFS
Syndactyly, 8, 114, 513
Syndrome, use of term, 1 2, 3
Syrinx, 10, 513
Systolic blood pressure, 30

T
T-cells, 8, 58, 378, 472, 513
Taking notes, 86
Taking turns and sharing, 138
Taking yourself too seriously, 139

Talipes, 114, 513
Talipes equinovarus, extra, 10
Talk Tools workshop, 196, 197
Talking through the nose, 177
Tantrums, 128, 182, 233, 235, 441
Teachers
    accommodations by teachers and schools, 146 149
    coaching parents, 161 163
    helping with social and communication skills, 161 163
    wording sentences simply for better comprehension, 154 156
Technical Assistance Alliance for Parent Centers, 261 293
    coordinating offices, 264 265
    description of, 261 264
    offices by state, 265 293
Teens
    boyfriends, 490
    bullying, 209 210
    common problems of, 114 115
    emotional development of, 130 131, 163
    friends, 141
    high school, 205, 209 210
    impulsiveness, talking to about, 130 131
    personal stories, 489 494
    puberty, 163, 205
    social skills of, 134, 163
Teeth
    antibiotics prior to dental work, 26
    enamel, 6
    missing teeth, 6
    multiple sets of, 66
    personal stories, 349, 368, 384
    size of, 6
Temper tantrums, 128, 182, 233, 235, 441
Temperature regulation, 7, 60
Tennessee, parent centers in, 287
Test artifact, 461
Test retakes, 148
Testicles, cryptorchidism, 7, 21, 504
Tests, in school, 146, 147, 148, 151, 169
Tethered spinal cord, 10
Tetralogy of Fallot
    definition of, 513
    in DiGeorge syndrome, 3
    in VCFS, 5, 25, 27, 381
Texas, parent centers in, 288 289
Thermometers, 60
Thinking skills. See Cognitive problems; Cognitive skill development
Thrombocytopenia, 11, 21, 513
Thrush, after taking antibiotics, 79
Thymic hormone, immunity and, 8
Thymus gland
    definition of, 513
    in DiGeorge syndrome, 2
    in VCFS, 7, 57

Thyroid problems, 54 55
Tics and grunts, 470
Time sequences, 166
Timetable of daily activities, 158
Title XXI, 319
Toddlers
    dentist visits, starting, 65
    emotional development of, 128 129
    feeding problems, 76 80
    sign language, 128, 182 184, 187, 464 465
Tolerating closeness to others, 135 136
Tone of voice, parent or teacher, 154 155
Tonsils
    function of, 33
    removing, 181
Tooth brushes, 66
Tooth brushing, in infants, 65
Tooth enamel, 6, 65
Toothpaste, 66
Tortuous retinal vessel, 6, 375
Toys, age-appropriate, 109 113
Trachea, 34, 513
Transcription (genetics), 17
Transition plan, 260
Translation (genetics), 17
Transposition of great vessels, 6
Trauma, temperature regulation and, 60
Tricuspid atresia, 6, 25, 29, 513
Triphalangeal thumbs, 8
Truncus arteriosus, 5, 25, 27, 513
Trust, revocable living trust, 308 309
Trustee, special needs trust, 312
Trusts, 307
Tube feeding, 76, 79, 80, 403, 481, 489, 505
Tubes in the ears, 41, 42, 43 44, 505

U

Umbilical (belly button) hernia, 5, 38, 513
Uncertified group home, 302 303
Undescended testicles (cryptorchidism), 7, 21, 504
Unilateral cleft lip and palate, 71, 72
Unilateral cleft of primary palate, 72
Upper respiratory infections, 8, 378
Urinary tract anomalies, 7, 21
Utah, parent centers in, 289

V

Vaccines, 58
Vascular anomalies, 6, 10 11, 21, 385
Vascular rings
    definition of, 514
    in DiGeorge syndrome, 3
    in VCFS, 6, 376

VCFS. See Velo-cardio-facial syndrome
VCFS children. See Children with VCFS
VCFS parents. See Parents of VCFS children; Personal stories
VCFSEF. See Velo-cardio-facial Syndrome Educational Foundation
Vegetables, encouraging eating of, 77 78
Veins, 10 11
 function of, 30
Velo-cardio-facial syndrome (VCFS)
 advocacy. See Advocacy
 anomalies in, 4 11, 15, 19 20, 21
 developmental issues, 81 89
 family support, 390, 419 422
 frequency of occurrence, 19
 genetic description of, 13 14, 15
 history of, ix-x
 learning issues, 6, 83 88
 prenatal testing for, 16 17
 support groups. See Support groups
 symptoms of. See Anomalies in VCFS
 See also Adults with VCFS; Children with VCFS; Parents of VCFS children; Personal stories
Velo-cardio-facial Syndrome Educational Foundation (VCFSEF), 4, 324
Velopharyngealoscopy, 67
Velopharyngeal insufficiency (VPI), 10
 adenoid and tonsil removal, 181
 definition of, 514
 personal story, 350
 speech and, 68, 177 178, 181, 197 198
 speech bulb, 198
 surgery for, 69, 177 178
Velopharyngeal valve, 514
Velum, 514
Ventricular septal defect (VSD)
 definition of, 514
 in DiGeorge syndrome, 3
 in VCFS, 5, 25, 26
Verbal fluency, 167
Verbal IQ and comprehension, 164 165
Verbal learners, 86
Vermont, parent centers in, 289
Vernix caseosa, 114
Vertebrae, 10, 21, 514
Vertebral artery, 11
Vesico-ureteral reflux, 514
Video games, advantages of, 115
Videofluoroscopy, 67, 177, 509
Violent outbursts, 229, 235
Viral infections, 58
Viral pneumonia, 35
Virgin Islands, parent centers in, 289 290
Virginia, parent centers in, 290
Vision therapy, 47 48
Visual analysis, 166
Visual learners, 86, 157 159

Visual processing remediation program, 47
Visual sequences, 166
Visual skill development, 1 to 3 months, 92
Visual-spatial skills, 47, 166 167
Vitamins, 75, 498
Vocal cords, 9, 34
Voice
    definition of, 514
    high-pitched, 377
    hypernasal, 10, 68, 177, 507
VSD. See Ventricular septal defect

W

Washington (state), parent centers in, 290 291
Washington, D.C., parent centers in, 272
Web resources. See Online resources
Weight gain, infants, 69
Weinberger, Dr. Daniel, 245
West Virginia, parent centers in, 291 292
Wheezing, 35
White, Peter, 223
White matter UBO's, 8
Wills, 307, 308, 310 311
Wisconsin, parent centers in, 292
Witch's milk, 114
Word decoding, 165
Word reading, 165
Working, adults with VCFS, 355 356
Working memory, 168 169
Wyoming, parent centers in, 293

Y

Yearly calendar, 159

Z

Z-plasty, 69

About the author

Sherry Baker-Gomez is the parent of a child with VCFS. Her son, now 32, was finally diagnosed with VCFS at 18 years of age after a long medical history and searching for answers. Sherry, herself, had been so desperate for answers after many years of struggling with her son's undiagnosed disorder; that she became a nurse in an effort to understand the symptoms she saw in him and what they meant. Then she realized that many other parents needed answers and needed to know where to turn, so she began writing.

Committed to VCFS education, Sherry started gathering information on resources, and stories that offered support. Working along with other parents and professionals, Sherry has organized this collection of information into a comprehensive handbook that brings information and resources to parents, professionals, and others under one cover.

Originally from Salem, Oregon, Sherry moved to Arizona for her son's health in 1982. Here, she and her husband both became licensed contractors to support themselves and their two children while Sherry returned to school. A cosmetologist at the start, Sherry pursued classes in psychiatry, counseling, architecture and nursing. Along the way, she was also heavily involved in volunteer work at Oregon state mental hospital, Oregon state prison (Bible education and counseling), and St. Joseph's hospital, Arizona as well as financial, marriage and teen counseling, working at an Alzheimer care center, doula duties and physical therapy assistant, to name a few.

Sherry is determined to pursue the campaign for VCFS recognition, and make VCFS a recognized name in the community. This book is produced with her own funds, at her own risk, and is not sponsored by any institution or outside source. The information presented is the result of her own search for answers and resources. A portion of the proceeds from the sale of each book will be donated to organizations that benefit VCFS disabled children and adults.

Photo by Bill Dooley

ORDER FORM

2011 REVISED EDITION
MISSING GENETIC PIECES
STRATEGIES FOR LIVING WITH VCFS
The Chromosome 22q11 Deletion

by Sherry Baker-Gomez

$35.00
Shipping $10.00
e-mail: s4918@ymail.com
for international shipping rates

For credit card payment:
Order **MISSING GENETIC PIECES** online at:
http://www.22qcentral.org

Or send a check or money order
Make payable to: Sherry Gomez
4918 West Phelps Road, Glendale, AZ 85306
Phone/Fax: 602-789-6416

SHIP TO:

Name: _____ Phone: _____

Address: _____

City: _____ State: _____

Zip code: _____ Country: _____

e-mail: _____

☐ Please send information about quantity discounts.
☐ Please send information about speaking engagements by author.
☐ Please put me on your mailing list.
☐ Please put me on your e-mail list for updated information

Care to be Aware

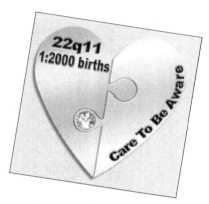

Join 22q Central on-line forum! Become part of the on-line community NOW! The *Missing Genetic Pieces* book is available on the site.

- **The lapel pin** represents the 1:2000 who are born with 22q11 deletion each year.

- **The heart shape** represents one of the main symptoms of 22q11, as well as the way 22q breaks our heart and tugs at the heartstrings of those who live with it.

- **The puzzle piece** that puts the two parts together stands for the genetic puzzlement of the disorder.

- **The different colored stones** stand for all ethnic groups affected. There are NO boundaries!

- **The inscription raises awareness** of how prevalent 22q is and how we would like everyone to "CARE to be aware" of it.

Order your pin today to raise awareness!
Choose the color of the stone you want.
Pins are included *FREE* with the purchase of every copy of the book
Missing Genetic Pieces.
Order the pin only for $5.00 plus $2.50 shipping USD
e-mail: s4918@ymail.com for international shipping rates

www.22qCentral.org